SLEEP – PHYSIOLOGY, FUNCTIONS, DREAMING AND DISORDERS

SLEEP, COGNITION AND EMOTION

FROM MOLECULES TO SOCIAL ECOLOGY

SLEEP – PHYSIOLOGY, FUNCTIONS, DREAMING AND DISORDERS

Additional books and e-books in this series can be found on Nova's website under the Series tab.

SLEEP – PHYSIOLOGY, FUNCTIONS, DREAMING AND DISORDERS

SLEEP, COGNITION AND EMOTION

FROM MOLECULES TO SOCIAL ECOLOGY

IGNACIO RAMÍREZ SALADO
ANA PAULA RIVERA GARCÍA
AND
MANUEL ALEJANDRO CRUZ AGUILAR
EDITORS

nova
Medicine & Health
New York

Library of Congress Cataloging-in-Publication Data

Names: Ramírez Salado, Ignacio, author. | Rivera García, Ana Paula,
author. | Cruz Aguilar, Manuel Alejandro, author.
Title: Sleep, cognition and emotion: from molecules to social ecology /
Ignacio Ramírez Salado, Ana Paula Rivera García, Manuel Alejandro Cruz Aguilar, editors.
Identifiers: LCCN 2020006782 (print) | LCCN 2020006783 (ebook) | ISBN
9781536175615 (hardcover) | ISBN 9781536175622 (adobe pdf)
Subjects: LCSH: Sleep--Physiological aspects. | Sleep disorders. | Cognition. | Emotions.
Classification: LCC QP425 .S6744 2020 (print) | LCC QP425 (ebook) | DDC 612.8/21--dc23
LC record available at https://lccn.loc.gov/2020006782
LC ebook record available at https://lccn.loc.gov/2020006783

Published by Nova Science Publishers, Inc. † New York

In loving memory of our beloved teacher
and mentor, José María Calvo,
whose academic and life teachings
transcend in time.
Thank you for your patience, guidance and everlasting
love to the science of sleep.

From his students, Nacho, Ana and Manuel

CONTENTS

Preface **ix**

Chapter 1 Basic Mechanisms in Sleep Regulation **1**
Anabel Jiménez-Anguiano

Chapter 2 Neuroendocrinology of Sleep **17**
Anabel Jiménez-Anguiano

Chapter 3 Effects of Menopause on Sleep-Waking Cycle **33**
Mónica Dafne García-Granados, Nadia Alejandra Torres-Marcial and Dulce Alondra Guadalupe Islas-Jiménez

Chapter 4 Amygdala Oscillations and Plasticity during Sleep **69**
Zeidy Muñoz-Torres and Corinne J. Montes-Rodríguez

Chapter 5 Altered Behavior and Cortical Activity during Sleep in Neurodegenerative Disorders and Non-Invasive Neuromodulatory Techniques as a Possible Treatment **99**
Ruth Alcalá-Lozano

Chapter 6 Effects of Environmental Pollution on Sleep **125**
Eduardo Sánchez-Islas and José Samuel Mucio-Ramírez

Chapter 7 Mental and Cognitive Phenomena during Sleep and Dreams **163**
Ignacio Ramírez-Salado, Nadia Alejandra Torres-Marcial and José Carlos Aguirre-Barrera

Chapter 8 Affect, REM Sleep and Dreams: Towards Emotional Insight **203**
Ana Paula Rivera-García, Manuel Alejandro Cruz-Aguilar and Ignacio Ramírez-Salado

Chapter 9	Analysis of Sexual Oneiric Experiences by Gender and Sexual Orientation *Ignacio Ramírez-Salado, Manuel Alejandro Cruz-Aguilar and Ana Paula Rivera-García*	**229**
Chapter 10	Sleep Disturbances in Autistic Spectrum Disorder (ASD) Patients *Fructuoso Ayala-Guerrero, Graciela Mexicano and Erik Leonardo Mateos*	**245**
Chapter 11	Effect of Transcranial Magnetic Stimulation on Sleep and Behavioral Disorders in Patients with Autism *Fructuoso Ayala-Guerrero and Horacio Balam Alvarez-García*	**275**
Chapter 12	Developmental Mode and REM Sleep in Mammals Show Correlated Evolution *Pilar Chiappa*	**297**
Index		**315**

Preface

Sleep, Cognition and Emotion: From Molecules to Social Ecology, provides a brand new source of theory, practice and developing topics on sleep science. This book offers a useful guide to history of sleep and emotion science, and outlines future directions for the development of new discoveries on sleep, cognition and affective science.

A plethora of work has shed light on the relationship between sleep and emotion. The first section of the book provides an overview of neurobiological, historic and physiological outcomes regarding sleep and emotion. The importance of several hormones over sleep such as growth and adrenocortocotropic hormones, gonadotropines and melatonin, among others, is described, as well as the changes that female sex hormones have over sleep patterns, particularly during such a critical stage as menopause is in women. The compelling association that sleep and limbic system withholds is excellently addressed as well, describing the oscillating patterns that the amygdala present during sleep and its relation with brain plasticity, mood disorders and memory. The effect that neurodegenerative diseases have over sleep structure and EEG components is thoroughly described, focusing on the relevance that therapeutic methods such as Repetitive Transcranial Magnetic Stimulation (rTMS) and Transcranial Direct Current Stimulation (tDCS) can have over sleep disorders. A topic of very high relevance today, such as the impact of pollutants on sleep, is addressed in a thorough manner, highlighting the disrupting effects that this contaminants have over several components of sleep.

In the second part of the book, a comprehensive review on complex topics such as dream content, affective mentation and evolution are addressed. State of the art research regarding Theory of Mind, Social cognition and Default Mode Network during sleep, and particularly during dreams is described, focusing on several components of the cognitive activity in sleep. The role that dreams may have over Affective Regulation is also reviewed, standing from a historical background of both emotion and dream research, up to revealing one possible function of dreaming as a feedback of emotional states and concerns. Everyday dream content is also explored in function of sexual and erotic experiences in both a heterosexual and homosexual population, revealing that this specific content is much more frequent regardless of sexual orientation.

The relevance that sleep may have over specific developmental disorders such as Autism is also extensively described, indicating that attention should be paid to sleep patterns among this population. Importantly, the beneficial effect of non-invasive treatments such as TMS over sleep in autistic children is excellently described, focusing in the functional relationship of the brain mechanisms involved in circadian rhythms, sleep, behavior and social cognition. Finally, a brand new study focused on a hypothesis regarding the amount of REM sleep among species in function of the mode of development is presented and briefly discussed in order to shed a light on the intricate evolutionary network that led to a complex state such as REM sleep.

This book aims to provide a new source of information -other than the mere clinical approach- regarding the interrelation that sleep can have with different substances, structures, processes, disorders and environmental factors, offering new directions of research in our understanding of sleep, cognition and emotions.

In: Sleep, Cognition and Emotion
Editors: I. Ramírez Salado et al.
ISBN: 978-1-53617-561-5

Chapter 1

BASIC MECHANISMS IN SLEEP REGULATION

Anabel Jiménez-Anguiano*
Área de Neurociencias,
Departamento de Biología de la Reproducción,
Universidad Autónoma Metropolitana-Iztapalapa,
Mexico City, Mexico

In memory of my Mentor *Dr. René Drucker-Colín* (1937-2017), for all his teachings and for introducing me in the study of neurochemical control of the sleep-wake cycle.

ABSTRACT

Sleep in mammals implies an enormous temporal space because a third of life is occupied sleeping. Why do we need to sleep? Does that help us to dream? These are very ancient questions and their study approaches have been diverse. One of them -the neuroanatomical and neurochemical approach- has been the study of several cerebral nuclei and chemical substances in the modulation of the sleep-wake cycle, one of the most fructiferous, since it has given us a great amount of information. The knowledge about brain phenomena during sleep has had a considerable advance, but we still do not know much of them.

In this chapter, a brief historical review of how sleep has been conceptualized from the first studies to the present time is shown. We will review the main scientific findings that have contributed to the knowledge and understanding of the basic mechanisms that the brain regulates, in order for sleep to occur.

Keywords: sleep, REM sleep, wakefulness, neural control

---

* Corresponding Author's Email: aja@xanum.uam.mx.

Introduction

The scientific study of sleep began with the recording of brain activity. Richard Caton (1842-1926) in 1875 at the University of Liverpool in England and Adolf Beck (1863-1942) in 1891 in the University of Krakow in Poland made the first records of brain electrical activity at the end of the 19th century (Siegel 2002). They recorded the cortical surface activity of various animal species using galvanometers, recognizing various rhythmic electrical oscillations during wakefulness and sleep. Based on these observations, in 1925, Hans Berger in Jena, Germany, used a galvanometer more sensitive than those used by Caton and Beck, with which he managed to record different brain electrical rhythms through electrodes placed on the scalp of his son; the results obtained were published in 1929 (Berger 1929). This was the first electroencephalographic (EEG) record from a human.

Initially, sleep was considered as a unitary phenomenon, defined behaviorally as a state reversible of perceptual disengagement from and unresponsiveness to the environment. Sleep is typically accompanied by postural recumbence, behavioral quiescence and closed eyes (Carskadon and Dement, 2011). But in the middle of 1950, Eugene Aserinsky and Nathaniel Kleitman, in the University of Chicago, in the United Sates, reported the existence of a second state of sleep, which involved the presence of rapid eye movements (REMs), thus, rapid eye movement (REM) sleep was discovered. Interestingly, this sleep stage is also called paradoxical sleep because brain activity during this stage is similar during to that of wakefulness (Jouvet 1969). REM sleep is defined by EEG activation, muscle atonia, and episodic bursts of REMs (Carskadon and Dement, 2011). During this phase, dreams occur, which are a series of subconscious images and can include sounds, sights, and other sensations. Dreams occur most frequently during REM sleep (National Sleep Foundation, 2019). Therefore, in sleep we can appreciate two great states -sleep and dream- which show different degrees of depth and specific physiologic and behavioral processes (Jiménez-Anguiano 1992).

Historical Review of Neuroanatomical Sleep Substrates

The first observations that showed the relationship of the brain in sleep regulation were published in 1930, by Constantine Von Economo, based in studies performed in brains of patients with *encephalitis lethargica*, a variant of the Spanish influenza, an epidemic that at the end of the World War I (1914-1918) caused the death of more people (between 25 and 30 millions) than those who died in the war itself (9 millions). During the course of the disease, individuals reported drowsiness or insomnia. Subsequently, in post-mortem studies, Von Economo observed that the brains of patients who manifested insomnia presented lesions in the anterior hypothalamus (AH) and in the preoptic basal forebrain area, while those who slept excessively showed lesions in the posterior region of the hypothalamus (PH), located at the base of the diencephalon and in the mesencephalic reticular formation (MRF). Based on these observations, he proposed the existence of a sleep center located in the AH and an antagonist center of wakefulness, located in the PH. Subsequent studies with hypothalamus lesions carried out in animals confirmed that AH lesions produce insomnia and that PH lesions produce sleep (Nauta 1946).

Shortly thereafter, Ranson in 1939 confirms that AH lesions of monkeys produce insomnia. However, lesions in the PH showed no effect. Years later Nauta in 1946, confirms the results obtained by Von Economo by demonstrating that lesions in the AH of rats produced insomnia and lesions at the level of the MRF and the PH produce sleep. These results suggested the existence of two neuroanatomical centers responsible for the generation of sleep and wakefulness: on one hand the PH as a wakefulness center which actively inhibits the sleep center located in the AH (Von Economo 1930).

Hess in 1944 reported that electrical stimulation of the intermediate mass of the thalamus in the cat produces a behaviorally asleep animal. Later, Monnier in 1963 demonstrated that electrical stimulation of the intralaminar nuclei of the thalamus induces EEG synchronization similar to what he called: "orthodox sleep." These works evidenced an important participation of the thalamus in the generation of the sleep. However, Andersen in 1968 demonstrated that lesions in the thalamus did not *produce insomnia*, so he determined that this brain structure would only be participating in sleep modulation.

Currently, there is a great deal of experimental evidence using various techniques that have fully supported these observations (Saper 2001), from which, the participation of AH in sleep regulation and PH in wakefulness has been highlighted. Still, the idea of the presence of unique centers responsible for sleep and wakefulness regulation has been discarded. After many years of research, we know that during sleep various brain nuclei are activated, which facilitate and/or inhibit chemical substances in order for sleep or wakefulness to occur.

Another of the important works in the investigation of the cerebral mechanisms involved in the regulation of sleep, was carried out by the Belgian researcher Frederick Bremer. In 1935, he worked with two preparations carried out in the brain of the cat under general anesthesia. On the one hand by separating the anterior region of the brain from the brain stem he prepared the isolated brain or *"cerveu isolé."* When recording the EEG, a permanent cortical synchronization was observed, which he interpreted as an animal constantly asleep. On the other hand, by sectioning the brain of the spinal cord, he performed the preparation isolated encephalon or "*encéphale isolé*"; under this preparation he observed that the sleep-wake cycle of the cat was not modified. Bremer concluded that sleep was the product of the insufficient arrival of sensory stimuli capable of awakening the animal and that wakefulness was the opposite case: the number of stimuli received were important to keep the animal awake. From his observations Bremer conceived the sleep as a passive phenomenon and postulated for wakefulness an active mechanism dependent on specific visual, auditory and somatic afferences.

After the Word War II (1939-1945), the research of basic brain mechanisms involved in the regulation of the sleep-wake cycle continued. In the early 1950s, Guiseppi Moruzzi and Horace Magoun in the Northwestern University in Evanston, Illinois, contributed two outstanding discoveries that ended up banishing the idea that the sleep was a passive state. They considered the existence of an intermediate structure between the two sections responsible for generating this effect. First, they showed that if they eliminated the sensory pathways of the brainstem they did not modify the amount of wakefulness or sleep. Second, they observed that lesions in a portion of the brain stem – the reticular formation (RF) – caused a sleep-like state. From these results, Moruzzi and Magoun concluded that the tonic activity of the RF, driven by sensory afferences, keeps the anterior portion of the brain awake, and that the reduction in RF activity produces sleep.

Interestingly, these researchers observed that the stimulation of RF produced EEG synchronization and suggested that through a system called the Ascending Reticular Activating System (ARAS), the RF exerts a tonic activating influence on subcortical and cortical structures, increasing and maintaining the waking state. Yet, in order for sleep to appear, it is necessary that the activity in the RF must decrease. The existence of this system was demonstrated by lesions from the RF which produced an immediate and chronic decrease in tonic cortical activity (Lindsley 1950).

From the work of Moruzzi and Magoun it was demonstrated that sleep is a condition as active as wakefulness. Subsequent evidence has shown that electrical stimulation of MRF induces cortical activation in rats and cats that have been anesthetized (Starzl 1951). Records of the unitary activity of neurons of the medial RF (mRF) in the cat brain show that most of the neurons that comprise this nucleus are more active during wakefulness than in the Slow Wave Sleep (SWS). The local microinjection of kainic acid in the MRF causes cortical activation and a behavior associated with activation in the cat (Kitsikis and Steriade, 1981). Non-invasive studies such as neuroimaging in humans have shown that the activity of the MRF is higher during wakefulness than in SWS (Kajimjura 1999). Subsequently, the involvement of other brain nuclei involved in the modulation of wakefulness and sleep was evidenced.

Modulation of Wakefulness and SWS

Noradrenergic Cells of the Locus Coeruleus (LC)

Wakefulness is modulated by multiple cerebral mechanisms, among which the noradrenergic cells of Locus Coeruleus (LC) stand out. These neurons are known to have afferences to the cerebral cortex, hippocampus, amygdala, and other subcortical areas such as the thalamus, hypothalamus, and basal forebrain (Maeda 1972). Findings from unitary neuronal activity and gene expression studies suggest that these cells participate in cortical activation and thus in alerting behavior. LC cells, which are noradrenergic, have a characteristic discharge pattern throughout the sleep-wake cycle. These reach a maximum firing rate when the predominant behavior is wakefulness and as sleep appears, the firing rate decreases. Thus, as SWS and REM sleep appears, these cells are no longer active and remain silent (Siegel 1977).

Maximum Noradrenaline (NA) release levels are reached in the forebrain and cortex when wakefulness is present (Park 2002). In addition, either activation or inactivation of LC cells lead to increased or decreased wakefulness, respectively. Experimental infusions of noradrenergic drugs directly into the thalamo-cortical, hypothalamus-cortical and basal-cortical activator systems facilitates the activity in cortical regions promoting wakefulness in turn (Panksepp 1973). Also, stress conditions such as forced wakefulness are able to activate the neurons of LC (Nakane 1994). Other studies using mice that cannot produce NA showed less insomnia than normal animals after being subjected to mild stress and low-dose administrations of amphetamines (Hunsley 2004).

The evidence described above shows an important participation of the noradrenergic neurons in the LC as well as in other regions in the maintenance of wakefulness. Naturally, this is not the only system that is carrying out the massive task beneath the generation, maintenance and promoting mechanisms of wakefulness. More experimental evidence will be covered in the following paragraphs.

Serotoninergic Neurons of the Raphé Nuclei

The Raphé Nuclei (RN) (mainly raphé dorsalis (RD) and raphé medianus), are known to have the largest source of Serotonin (5-HT) in the brain and are involved in the control of sleep-wake cycle. The afferences held by this group of neurons are practically the same towards which the LC projects, however, the participation of the serotonergic neurons of the RN is not as clear as those of the LC neurons (Steinbusch 1981). There is evidence of the involvement of these cells in both sleep and wakefulness behavior. Works of raphé neurons unitary activity have shown that the maximum level of discharge occurs in wakefulness and diminishes progressively as SWS appears, to the point of being completely silent during REM sleep. This initial work led to the conclusion that the RN was involved in the promotion of wakefulness (Jacobs 1981).

However, subsequent studies showed that serotonergic neurons exhibit differences with respect to noradrenergic neurons, since it has been shown that the activity of RN neurons seems to participate in SWS modulation, in addition to RN lesions and inactivations of these neurons, suppress SWS (Jouvet 1969). Also, it has been shown that the microinjection of 5-HT in the medium Preoptic Area (mPOA) decreases wakefulness and increases SWS (Denoyer 1989). Similarly, the administration of serotonergic agonists, exert an inhibitory action on the cholinergic neurons of the basal forebrain, which are promoters of wakefulness, thus suppressing cortical activation. As we can see, the pharmacological evidence points to a participation of serotonergic neurons of RD in the modulation of wakefulness and sleep (Rancillac 2016).

Currently, from a vast experimental contribution, it has been shown that the main brain structures that regulate the SWS are located in the anterior portion of the brain.

Orexins/Hypocretins

These neuropeptides were identified in the Lateral Hypothalamus (LH) simultaneously by two independent research groups and have the name of Orexins (OXs) because they were initially related in the regulation of food intake and Hypocretins due to its location in the hypothalamus and by the homology with the hormone Secretin (De Lecea 1998; Sakurai 1998). For the purpose of this chapter, we will use the term OXs in this section.

There are two types of OXs: Orexin (OX-A) and Orexin-B (OX-B). made up of 33 and 28 amino acids that bind to two distinct G- protein coupled receptor subtypes termed the OX-1 and 2 receptor (OX_1R, OX_2R) (Bourgin 2000). Orexinergic neurons are located predominantly in three contiguous hypothalamic regions: LH, Perifornical area (PFA) and dorsomedial hypothalamus (DMH) (Siegel 2004), from where they innervate virtually the entire brain, they project to major arousal centres in the brain, such as basal forebrain (BF) (Sakurai 2005), brain stem and spinal cord (De Lecea 1998). OXs have been recognized for their diverse functions in sleep/wakefulness/arousal, appetite regulation and may play provocative roles in stress response. The activity of the orexinergic neurons during the sleep is differential. It has been shown, that these neurons are active during waking, decrease firing rate during SWS and cease the activity during REM sleep (Lee 2005).

OX-A, unlike OX-B, has been associated to the regulation of arousal mechanisms of more potent manner (España 2001). OX-A stimulates arousal in rats and mice when administrated intracerebroventricular (ICV) a dose-dependent. The highest dose produced a major enhancement of arousal which was accompanied by a marked reduction in REM sleep, SWS and the latency to the first occurrence of REM sleep was also prolonged (Pipper 2000). In addition, the blocked of OX-$_1$R by use of selective antagonist administered ICV reversed the effect induced by OX-A (Smith 2003).

OX-A administration in specific cerebral nuclei has similar effect to ICV injections. In this sense, it has been shown that the injection of OX-A into the medial parvocellular subdivision of nucleus paraventricular (PVN), in basal forebrain (BF) and in the lateral preoptic area (lPOA) produced a reduction of sleep and increased the wakefulness (España 2001) via the OX$_2$R (Eggermann 2001). In the BF are present orexinergic receptors (Marcus 2001). Still more, the perfusion of OX-A into the histaminergic tuberomammillary nucleus (TMN) of rats promptly increased wakefulness for 2 hours after starting the perfusion, concomitant with a reduction in REM sleep and SWS (Huang 2001).

As far as the participation of OX-A in several nucleus of brain stem, it has been shown that the application of OX-A into the noradrenergic neurones of the LC increases cell firing (Hagan 1999) and the local administration of OX-A but not OX-B suppressed REM sleep in a dose-dependent manner and increased wakefulness at the expense of SWS. These effects were blocked with an antibody that neutralizes OX binding to OX$_1$R (Bourgin 2000). Likewise, OX-A was microinjected into the laterodorsal tegmental nucleus (LDT) of chronic, unanesthetized cats and they found a significant increase in the time spent in wakefulness and a significant decrease in the time spent in REM sleep (Xi 2001).

In addition, it has been shown that knockout and transgenic overexpression studies have implicated orexins neurons in arousal and sleep. Mice lacking OX peptide display increases in REM and SWS and a decrease in awake time during the active period of normal rodents (Chemelli 1999). In humans, Nishino and colleagues (2000) found that seven of nine patients with narcolepsy had undetectable OX-A in cerebrospinal fluid (Nishino 2000)

These evidences indicate that orexinergic processes, in particular OX-A, play an important role in both of the promotion of wakefulness and the suppression of SWS and REM sleep. Moreover, to the fact that orexins have been strongly linked with the pathological basis of narcolepsy-cataplexy.

Regulation of REM Sleep

As for the substrates responsible for the generation of REM sleep, it has been evidenced that they are located in the brain stem, since previous structures of the brain do not exhibit any pattern of activity characteristic of this state when separated from the brain stem (Figure 1) (Prospéro-García 1999; Reinoso-Suárez 2001). The appearance of Ponto-Geniculo-Occipital (PGOs) spikes and muscular atony, characteristic of REM sleep state, are generated in the brain stem and the participation of the pons is specifically highlighted, as one of the most important regions for the regulation of the REM sleep state, since transects carried out at the middle level of the brain do not prevent the periodic appearance of this state. While transects through the medulla oblongata eliminate muscular atony, they do not prevent the periodic appearance of phasic activity, such as PGO spikes and REMs (Siegel 2002). The results of these studies indicated the importance of the Pontine Tegment in the generation of phasic and tonic activation, as well as the inhibitory processes of REM sleep (Jouvet 1962) Subsequently, it was demonstrated that extensive lesions of the pontine segment produce a total decrease in REM sleep state, including PGO spikes, REMs and muscular atony (Jones 2005; Ramírez-Salado 2014).

Cholinergic Neurons of the Pedunculopontine Tegmental Nucleus (PPT)

The pedunculopontine tegmental (PPT) nucleus is one of the main cholinergical nuclei of the CNS, which participates in the modulation of wakefulness and REM sleep. PPT and LDT constitute the Peribraquial region of the brain stem. (Datta and Hobson, 1994).

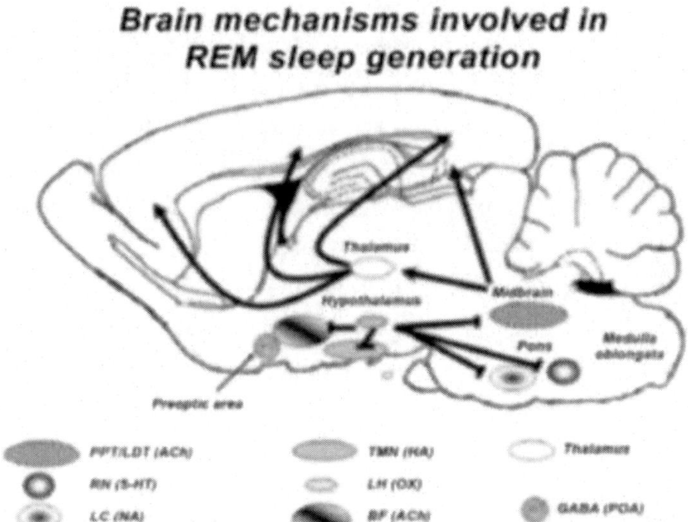

Figure 1. Schematic representation of the main brain nuclei involved in the regulation of REM sleep. As shown in the scheme, the main cholinergic nuclei of PPT/LDT send projections towards the Thalamus and Cerebral Cortex. The inhibitory projections from the Lateral Hypothalamus (LH) (Orexinergic nucleus) towards the brain stem, specifically towards the RD and LC (serotonergic and noradrenergic nucleus, respectively) are illustrated. Finally, the participation of regions of the anterior brain is depicted. Specifically, the functional connectivity between the basal forebrain (BF) (cholinergic nucleus), the medial preoptic area (POA) (GABAergic nucleus) and the Tuberomamilar Nucleus (TMN, Histaminergic Nucleus) is shown in association to REM sleep generation.

It possesses cholinergic neurons and various neurotransmitters and peptides, such as: glutamate (Glu), gamma-aminobutyric acid (GABA) and nitric oxide (NO) (Spann 1992). It is known that electrical and chemical stimulation of PPT in cats and rats promotes wakefulness. In addition, it has been evidenced through studies of cellular registry in rats in free movement, that at least 60% of the neurons that PPT has activity both in wakefulness and REM sleep (Datta and Siwek, 1997). It has been demonstrated that the PPT nucleus of the cat, increases its firing frequency one minute before the first signs of REM sleep appear (Datta 1991). Therefore, it has been proposed that within the PPT region there are 2 types of neuronal triggers: one that correlates with the initiation and maintenance of REM sleep (REM-on neurons); and another that is tonically active during wakefulness and inhibits REM sleep (REM-off neurons) (Steriade and McCarley, 2005). Notably, within the REM-on area are two populations of glutamatergic neurons, the first of which projects to the BF and regulates EEG components of REM sleep and the second of which projects to the ventromedial medulla and spinal cord and regulates atonia during REM sleep (Saper and Fuller, 2017).

Moreover, it has been shown that the firing patterns of PPT neurons in rats showed 3 types of activity: the first neuronal population showed a higher firing frequency during REM sleep (12.86%), in relation to SWS and wakefulness; a second neuronal population was more active during wakefulness and REM sleep (60%), in relation to SWS; and a third group of neurons showed no changes in function of behavioral status (27. 14%). From these findings we can realize that PPT neurons participate in the regulation of the mechanisms involved in the maintenance of wakefulness and in the generation of REM sleep (Datta and Siwek, 1997).

In addition, it has been reported that lesion with kainic acid of the PPT and LDT nucleus, produce a decrease in REM sleep, which is proportional to the number of cholinergic cells destroyed (60%) (Webster and Jones 1988, 458). Other studies have shown that stimulation of PPT/LDT induces the release of Acetylcholine (ACh) and increases latency for the appearance of excitatory potsynaptic potentials (EPSPs) (Lydic and Baghdoyan, 1993).

Molecular cloning has identified five muscarinic cholinergic receptors (mAChRs) and studies of hybridization *in situ* have shown that mRNA for M2 receptors predominate in the brain stem, including PPT and LDT (M2 > M3 = M4, not M1 or M5) (Mash and Potter 1986; Vilaro 1994). Also, injections of M2 muscarinic agonists in the medial pontine reticular formation (mPRF) promote the release of ACh and increase the appearance of REM sleep, while antagonist of mAChRs reduce the release of ACh and decrease the amount of sleep in rats, cats and humans (Gillin 1993; Velázquez-Moctezuma 1998; Baghdoyan and Lydic, 1999).

From the evidence described above we can summarize that the nuclei of PPT-LDT increase their activity during the REM sleep state, associated with an increase in the release of ACh in the mPRF (Prospéro-García 1999). In addition, the interaction of PPT-

LC has been corroborated using RNAi technology *in vivo,* showing that NA from LC neurons inhibited PPT neurons and reduces REM sleep (Khandai 2016). Recently, it has been shown the interaction between the pons and the ventral portion of the mesencephalon through GABAergic modulation in the sleep-wake cycle. They found that inhibitory neurons regulate wakefulness by suppressing dopaminergic neurons, and the activation of GABA-ergic neurons induces SWS and reduces wakefulness (Takata 2018).

CONCLUSION

Throughout this chapter, we have reviewed that during sleep, there is an enormous activation of various brain nuclei, highlighting the pontine region of the brain stem, in which we locate the cholinergic mesopontines of the nuclei of PPT/LDT, in addition to the nuclei of RD and LC, and hypothalamic nuclei, which produce various neurotransmitters, such as ACh, 5-HT, NA, DA and Histamine, respectively. Recently, through the advent of new techniques, especially of molecular biology, have been identified the interaction with other substances, such as GABA, Glu, NO and OXs, among others; which modulate the activity of cholinergic, serotonergic, dopaminergic and noradrenergic neurons, described above. However, we still have to determine to a large extent, the neural networks and biochemical mechanisms involved in the genesis of waking, sleeping and dreaming under normal conditions and in various sleep disorders.

REFERENCES

Andersen, P., Andersson, S. A., Junge, K. and Sveen, O. (1968). Patterns of spontaneous barbiturate spindle activity within the thalamus. *Electroencephalogr Clin Neurophysiol,* 24(1):90-95.

Baghdoyan, H. and Lydic, R. (1999). M2 muscarinic receptor subtype in the feline medial pontine reticular formation modulates the amount of rapid eye movement sleep. *Sleep,* 22(7):835-847.

Berger, H. (1929). About of the Electroencephalogram. *Arch Psychiatr Nervous,* 87:527-570.

Bourgin, P., Huitrón-Reséndiz, S., Spier, A. D., Fabre, V., Morte, B., Criado, J. R., Sutcliffe, J. G., Henriksen, S. J. and De Lecea, L. (2000). Hypocretin-I modulates rapid eye movement sleep through activation of locus coeruleus neurons. *J. Neurosci,* 20: 7760- 7765.

Bremer, F. (1935). *"Cerveau isolé"* et physiologie du sommeil." *C. R. Soc. Biol,* 118: 1235-1241.

Carskadon, M. A. and Dement, W. C. (2011). Monitoring and staging human sleep. In. *Principles and practice of sleep medicine*. Edited by M. H. Kryger, T. Roth and W. C. Dement, 16-26. St. Louis: Elsevier Saunders.

Chemelli, R. M., Willie, J., Sinton, C., Elmquist, J., Scammell, T., Lee, C., Richardson, J., Williams, S., Xiong, Y. and Kisanuki, Y. (1999). Narcolepsy in orexin knockout mice: molecular genetics of sleep regulation. *Cell,* 98, 437-–451.

Datta, S. and Hobson, A. (1994). Neuronal activity in the caudo-lateral peribrachial pons: relationship to PGO waves and rapid eyes movements. *J Neurophysiol,* 71: 95- 109.

Datta, S. and Siwek, D. (1997). Excitation of the brain stem pedunculopontine tegmentum cholinergic cells induces wakefulness and REM sleep. *J. Neurophysiol*, 77: 2975- 2988.

Datta, S., Calvo, J. M., Quattrochi, J. and Hobson, A. (1991). Long-term enhancement of REM sleep following cholinergic stimulation. *Neuroreport,* 2: 619-622.

De Lecea, L., Kilduff, T. S., Peyron, C., Gao, X. B., Foye, P. E., Danielson, P. E., Fukuhara, C., Battenberg, E. L., Gautvik, V. T., Bartlet, F. S., Frankel, W. N., Van Den Pol, A. N., Bloom, F. E., Gautvik, K. M. and Sutcliffe, J. G. (1998). The hypocretins: hypothalamus- specifc peptides with neuroexcitatory activity. *Proc Natl Acad Sci USA,* 95: 322-327.

Denoyer, M., Sallanon, M., Kitahama, K., Aubert, C. and Jouvet, M. (1989). Reversibility of para-chlorophenylalanine-induced insomnia by intrahypothalamic microinjection of L-5-hydroxytryptophan. *Neurosci,* 28(1): 83-94.

Eggermann, E., Serafin, M., Bayer, L., Machard, D., Saint-Mleux, B., Jones, B. E. and Muhlethaler, M. (2001). Orexins/hypocretins excite basal forebrain cholinergic neurons. *Neurosci*, 108: 177–181.

Espana, R. A., Baldo, B. A., Kelley, A. E. and Berridge, C. W. (2001). Wake promoting and sleep-suppressing actions of hypocretin (orexin): basal forebrain sites of action. *Neurosci*, 106: 699–715.

Gillin, J. C., Salin-Pascual, R., Velazquez-Moctezuma, J., Shiromani, P. and Zoltoski, R. (1993). Cholinergic receptor subtypes and REM sleep in animals and normal controls. *Prog Brain Res,* 98:379-87.

Hagan, J., Leslie, R. A., Patel, S., Evans, M. L, Wattam, T. A., Holmes, S., Benham. C. D., Taylor, S. G., Routledge, C., Hemmati, P., Munton, R. P., Ashmeade, T. E., Shah, A. S., Hutcher, J. P., Hatcher, P. D., Jones, D. N. C., Smith. M. I., Piper, D. C., Hunter, A. J., Porter, R. A. and Upton, N. (1999). Orexin A activates Locus Coeruleus cell firing and increases arousal in the rat. *Proc Natl Acad Sci USA,* 96:10911-10916.

Hess, W. R. (1944). Das Schlafsyndrom als Folge diencephaler Reizung [Sleep syndrome as a result of diencephalic irritation]. *Helv Physiol Pharmacol Acta,* 2: 305-344.

Huang, Z., Qu, W., Li, W., Mochizuki, T., Eguchi, N., Watanabe, T., Urade, Y. and Hayaishi, O. (2001). Arousal effect of orexin A depends on activation of the histaminergic system. *Proc Natl Acad Sci* USA, 14 (98):9965-70.

Hunsley, M. S. and Palmiter, R. D. (2004). Altered sleep latency and arousal regulation in mice lacking norepinephrine. *Pharmacol Biochem Behav*, 78(4):765-73.

Jacobs, B. L., Heym, J. and Trulson, M. E. (1981). Behavioral and physiological correlates of brain serotoninergic unit activity. *J Physiol,* 77(2):431-436.

Jiménez-Anguiano, A., Merchant-Nancy, H. and Drucker-Colín, R. (1992). Gastropancreatic peptides in the regulation of REM sleep. In *Sleep and aging*, edited by Salvatore Smirne, 5-25. Italy, Raven Press.

Jones, B. (2005). From waking to sleeping: neuronal and chemicals substrates. *Trends Phamacol Science,* 26: 578-586.

Jouvet, M. (1962). Reacherches sur les structures nerveuses et les mécanismes responsables des diffé-rentes phases du sommeil physiologique [Research on the nervous structures and mechanisms responsible for the different phases of physiological sleep]. *Arch Ital Biol,* 100: 125-206.

Jouvet, M. (1969). Biogenic amines and the states of sleep. *Science,* 163: 32-41.

Kajimura, N., Uchiyama, M., Takayama, Y., Uchida, S., Uema, T., Kato, M., Sekimoto, M., Watanabe, T., Nakajima, T., Horikoshi, S., Ogawa, K., Nishikawa, M., Hiroki, M., Kudo, Y., Matsuda, H., Okawa, M. and Takahashi, K. (1999). Activity of midbrain reticular formation and neocortex during the progression of human non-rapid eye movement sleep. *J. Neurosci,* 19 (22):10065-10073.

Khanday, M. J., Somarajan B. I., Mehta, R. and Mallic, B. (2016). Noradrenaline from Locus Coeruleus neurons acts on Pedunculo-Pontine Neurons to prevent REM sleep and induces its loss-associated effects in rats. *eNeuro*, 2-19. DOI:http://dx.doi.org/10.1523/ENEURO.0108-16.2016.

Kitsikis, A. and Steriade, M. (1981). Immediate behavioral effects of kainic acid injections into the midbrain reticular core. *Behav Brain Res,* 3:361-380.

Lee, M. G., Hassani, O. K. and Jones, B. E. (2005). Discharge of identified orexin/hypocretin neurons across the sleep-waking cycle. *J. Neurosci,* 25, 6716–6720.

Lindsley, D. B., Schreiner, L. H., Knowles, W. B. and Magoun, H. W. (1950). Behavior and EEG changes following chronic brain stem lesions in the cat. *Electroencephalogr Clin Neurophysiol,* 2: 483-498.

Lydic, R. and Baghdoyan, H. (1993). Pedunculopontine stimulation alters respiration and increases ACh release in the pontine reticular formation. *Am J Physiol,* 264: R544-R554.

Maeda, T. and Shimizu, N. (1972). Ascending projections from the locus coeruleus and other aminergic pontine neurons at the level of the rat prosencephalon. *Brain Res,* 36(1):19-35.

Marcus, J. N., Aschkenasi, C. J., Lee, C. E., Chemelli, R. M., Saper, C. B., Yanagisawa, M. and Elmquist, J. K. (2001). Differential expression of orexin receptors 1 and 2 in the rat brain._*J. Comp. Neurol,* 435, 6–25.

Mash, D. and Potter, L. (1986). Autogradiographic localization of M1 and M2 muscarinic receptors in the rat brain. *Neurosci,* 19: 551-564.

Monnier, M., Koller, T. and Graber, S. L. (1963). Humoral infuences of induced sleep and arousal upon electrical brain activity of animals with crossed circulation. *Exp Neurol*, 23: 264-277.

Moruzzi, G. and Magoun, H. (1949). Brain stem reticular formation and activation of the EEG. *Electroenceph Clin Neurophysiol,* 1: 455-473.

Nakane, H., Shimizu, N. and Hori, T. (1994). Stress-induced norepinephrine release in the rat prefrontal cortex measured by microdialysis. *Am J Physiol,* 267(6):R1559-66.

National Sleep Foundation, 2019. www.sleep.org

Nauta, H. (1946). Hypothalamic regulation of sleep in rats." An experimental study. *J Neurophysiol,* 9: 285-316.

Nishino, S., Ripley, B., Overeem, S., Lammers, G.J. and Mignot, E. (2000). Hypocretin (orexin) deficiency in human narcolepsy. *Lancet,* 355:39–40.

Panksepp, J., Jalowiec, J. E., Morgane, P. J., Zolovick, A. J. and Stern, W. C. (1973). Noradrenergic pathways and sleep-waking states in cats. *Exp Neurol,* 41(2):233-45.

Park, S. P. (2002). *In vivo* microdialysis measures of extracellular norepinephrine in the rat amygdala during sleep-wakefulness. *J Korean Med Sci,* 17(3):395-9.

Piper, D. C., Upton, N., Smith, M. I. and Hunter, A. J. (2000). The novel brain neuropeptide, orexin-A, modulates the sleep-wake cycle of rats. *Eur J Neurosci,* 12:726-730.

Prospéro-García, O., Navarro, L., Murillo-Rodríguez, E., Sánchez-Alavez, M., Guzmán-Marín, R. M., Méndez-Díaz, M., Goméz-Chavarín, M., Jiménez-Anguiano, A. and Drucker-Colín R. (1999). Cellular and molecular changes ocurring during REM sleep. In: *Rapid eye movement sleep*, edited by Birenda Mallick B. and Shojiro Inoué, 367-381. New Delhi, Narosa Publishing House.

Ramírez-Salado, I., Rivera-García, A. P., Velázquez-Moctezuma, J. Jiménez-Anguiano, A. and Pellicer, F. (2014). $GABA_A$ receptor agonist at the caudo-lateral peribrachial area suppresses ponto-geniculo-occipital waves and its related states. *Pharmacol, Biochem & Behav,* 124: 333–340.

Rancillac, A. (2016). Serotonin and sleep-promoting neurons. *Oncotarget,* 7(48): 78222–78223.

Ranson, S. (1939). Somnolence caused by hypothalamus lesion in the monkey. *Arch Neurol Psychiat,* 41:1-23.

Reinoso-Suárez, F., de Andrés, I., Rodrigo-Angulo, M. and Garzón, M. (2001). Brain structures and mechanisms involved in the generation of REM sleep. *Sleep Med,* 5: 63-77.

Sakurai, T., Amemiya, A., Ishii, M., Matsuzaki, I., Chemelli, R. M., Tanaka, H. Williams, S. C. Richardson, J. A. Kozlowski, G. P. and Wilson, S. (1998). Orexins and orexin receptors: a family of hypothalamic neuropeptides and G protein-coupled receptors that regulate feeding behavior. *Cell,* 92:573–585.

Sakurai, T., Nagata, R., Yamanaka, A., Kawamura, H., Tsujino, N., Muraki, Y., Kageyama, H., Kunita, S., Takahashi, S., Goto, K., Koyama, Y., Shioda, S. and Yanagisawa, M. (2005). Input of orexin/hypocretin neurons revealed by a genetically encoded tracer in mice. *Neuron*, 46, 297–308.

Saper, C. B., Chou, T. C., Scammell, T. E. (2001). The sleep switch: hypothalamic control of sleep and wakefulness. *Trends Neurosci,* 24: 726–731.

Saper, C. B. and Fuller, P. M. (2017). Wake-sleep circuitry: an overview. *Curr Opin Neurobiol,* 44:186-192.

Siegel, J. (2002). *The neural control of sleep and waking*. New York, Springer-Verlag.

Siegel, J. M. and McGinty, D. J. (1977). Pontine reticular formation neurons: relationship of discharge to motor activity. *Science,* 196 (4290):678-80.

Siegel, J.M. Hypocretin (orexin): role in normal behaviour and neuropathology. (2004). *Annu. Rev. Psychol,* 55, 125-148.

Smith, M., Piper, D., Duxon, M. and Upton, N. (2003). Evidence implicating a role for orexin-1 receptor modulation of paradoxical sleep in the rat. *Neurosci Lett,* 341: 256-258.

Spann, B. and Grofova, I. (1992). Cholinergic and non cholinergic neurons in the rat pedunculopontine nucleus. *Anat Embryol,* 186: 215-227.

Starzl, T. E. Taylor, C. W. and Magoun, H. W. (1951). Ascending conduction in reticular activation system, with special reference to the diencephalon. *J Neurophysiol,* 14: 461-477.

Steinbusch, H. W. (1981). Distribution of serotonin-immunoreactivity in the central nervous system of the rat-cell bodies and terminals. *Neurosci,* 6(4):557-618.

Steriade, M. and McCarley, R. W. (2005). Afferent and efferent conenections of brainstem and forebrain modulatory systems. In *Brainstem control of wakefulness and sleep*, edited by Mircea Steriade and Robert McCarley, 1-728. New York: Plenum Publishers.

Takata, Y., Oishi, Y., Zhou, X.Z., Hasegawa, E., Takahashi, K., Cherasse, Y., Sakurai, T. and Lazarus M. (2018). Sleep and wakefulness are controlled by ventral medial midbrain/pons GABAergic neurons in mice. *J. Neurosci,* 38 (47) 10080-10092.

Velázquez-Moctezuma, J., Jiménez-Anguiano, A. and Vázquez-Palacios, G. (1998). Participation of cholinergic receptors subtypes in the regulation of REM sleep. *Wakefulness-Sleep.* 10: 46-53.

Vilaro, M. T., Palacios, J. M. and Mengod, G. (1994). Multiplicity of muscarinic autoreceptor subtypes? Comparison of the distribution of cholinergic cells and cells containing mRNA for fve subtypes of muscarinic receptors in the rat brain. *Molecular Brain Res,* 21: 30–46.

Von Economo, C. (1930). Sleep as a problem of localization. *J Nerv Ment Dis* 71: 249-259.

Webster, J. and Jones, B. (1988). Neurotoxic lesions of the dorsolateral pontomesencephalic-tegmentum-cholinergic cell area in the cat. Effects upon sleep-waking states. *Brain Res,* 458: 285-302.

Xi, M. C., Morales, F. R. and Chase, M. H. (2001). Effects on sleep and wakefulness of the injection of hypocretin-1 (orexin-A) into the laterodorsal tegmental nucleus of the cat. *Brain Res,* 901, 259-264.

In: Sleep, Cognition and Emotion
Editors: I. Ramírez Salado et al.

ISBN: 978-1-53617-561-5

Chapter 2

NEUROENDOCRINOLOGY OF SLEEP

Anabel Jiménez-Anguiano*
Área de Neurociencias,
Departamento de Biología de la Reproducción,
Universidad Autónoma Metropolitana-Iztapalapa, Mexico City, Mexico

ABSTRACT

The sleep-wake cycle is a complex and neurobiological process of enormous significance for various animal species. The sleep is not a unitary state, we can distinguish two great states: sleep and dream. Sleep includes light sleep and Slow Wave Sleep (SWS), and dream involves the Rapid Eye Movements (REM) sleep state. During the two states of sleep-REM and no-REM we can see different patterns of hormonal secretion.

The neuroendocrine system regulates a wide variety of processes, which cooperate in the maintenance of the homeostatic equilibrium of the organism. Sleep is no exception, as it has been observed that there is an important correlation between the sleep-wake cycle and the secretion of various hormones. These correlations have been attributed, in part, to a close relationship and circadian synchronization between hormone secretion and sleep. Nowadays it is known that the amount of hormones involved in sleep regulation is enormous, and they establish a complex interaction with several neurotransmitter systems, neuropeptides and/or steroids. Sex and age have also been described as factors that affect the endocrine regulation of sleep. In this sense, it has been shown that the secretion of the Cortisol in young patients is higher in women than men, in such a way that for Growth Hormone (GH), men show only one peak of secretion, and *in women* can be registered *several*, besides the fact that *along the life the peaks* of liberation *change*.

It has been shown that in sleep disorders such as sleep apnea-hypopnea syndrome (SAHS), insomnia, and narcolepsy, among others, exist significant alterations in hormone

* Corresponding Author's Email: aja@xanum.uam.mx.

levels with respect to healthy patients. Hormonal correlations have provided valuable information on the endocrine regulation of sleep.

In this chapter, a brief review was made of the neurohumoral secretions that have been shown a prominent role in sleep modulation, with the purpose of elucidating its possible mechanisms of action under normal conditions and in several disorders of sleep.

Keywords: sleep, neuroendocrinology, hormones

INTRODUCTION

> "One generation succeeds the other, and each one repeats the acts of the previous one. Just a moment before the death, they discover it was possible to dream and to draw the world in their own way, to then wake up and start a different drawing."
>
> Elena Garro (1916-1998), Recollections of things to come

The central nervous system (CNS) and the endocrine system work together to regulate various physiological processes in the body, including the sleep-wake cycle. It has been shown that there is an important circadian correlation in the secretion of various hormones during sleep (Smith and Mong, 2019). It has been known for more than 25 years of sleep research that the maximum liberation peaks of pituitary hormones: GH and prolactin (PRL), occur during sleep (Antonijevic 1999). For GH, maximum release occurs during the onset of SWS in various mammalian species. For PRL, it has been examined that the highest levels in humans are observed after sleep onset, but in contrast to GH, the concentration of PRL in the plasma reaches peak values during midnight. The most relevant data of these protein nature hormones in sleep regulation are analyzed below.

GROWTH HORMONE (GH)

GH, also known as somatotropin, is a hormone formed by a protein chain with a molecular weight of 22,005 kD. Its functions include stimulating protein synthesis by promoting the transport of amino acids through cell membranes and increasing the nuclear transcription of deoxyribonucleic acid (DNA) to form ribonucleic acid (RNA); increasing the mobilization of fatty acids from adipose tissue; increase the amount of free acids in the blood and encourage the use of them as a source of energy, which promotes the "protein savings"; which, by not being degraded, cause an increase in lean body mass; finally, the amount of glucose used throughout the body decreases. These functions together, enable the growth of the body (Ganong 2010). GH is an essential anabolic hormone for growth with a circadian rhythm that has been clearly linked to sleep pattern

(Sassin 1972). Its secretion is modulated by a hormone that stimulates it, the growth hormone releasing hormone (GHRH) produced in the hypothalamus, and another that inhibits it, somatostatin (Ganong 2010; Leng 2018).

GH secretion in mammals (Kimura and Tsai 1984) is related to the SWS stage (Garnier 1988, 133), specifically to the first episode of the SWS (Hindmarch 1987). In humans, the main secretion of GH in the 24-hour cycle occurs during the first part of the night and coincides with the appearance of delta sleep (Takahashi 1968). Various human studies have shown that the SWS precede the release of GH (Pawel 1972; Holl 1991).

In healthy individuals, the average daily values of GH for adults (6-3 ng/ml) and children (6 ng/ml) (Leal-Guadarrama 2003). GH has quiet periods and pulses of secretion up to 20 or 30 μg/l (Bidlingmaier and Strasburger 2007). It has also been observed that sleep deprivation (SD) in young males prevents GH secretion (Takahashi 1968; Davidson 1991), and that if sleep onset is delayed by about three hours, a delay in GH secretion also occurs (Mendelson 1987). However, other reports have shown that, although the peak of GH is lost during the first few hours of sleep following SD, total secretion over 24 hours of GH secretion occurs during the first 100 minutes of sleep. In men, GH secretion decreases gradually from the age of 30, and after age 50 it has completely disappeared. In the case of women, GH release is associated with delta sleep, but GH secretion is lower compared to men (Jarrett 1990), and during menopause a significant decrease is observed (Van Cauter 2000). Adults with a deficiency in GH secretion show a decrease in SWS (Astrom and Lindholm, 1990).

Pharmacological studies have shown that when the action of GH releasing hormone (GHRH) is blocked, delta sleep is significantly decreased which, in turn the administration of GHRH promotes delta sleep in some mammalian species, including humans (Steiger 1998). In addition, transgenic animals with GHRH receptor deficiency have been shown to have a significant decrease in delta sleep (Obal 2001). Based on the experimental evidence described above, the strong association between SWS and the mechanisms regulating GH secretion has been supported. It is striking that during the REM sleep state the clearly distinguished peaks of secretion during SWS are not appreciated (Jarrett 1990).

On the other hand, several studies have been shown the effects of GH in the CNS based on animal models in which GH protects the brain and spinal cord from different types of degeneration and is able to promote the survival of neurons that have suffered a hypoxic-ischemic insult. It is postulated that one of the mechanisms of neuroprotection of GH is more related to the decrease of apoptosis rather than the increase of neurogenesis and it is precisely this effect that suggests that the decrease of this hormone with age or due to other circumstances can affect the brain and contribute to its deterioration, as well as the loss of its functional capacity (Tresguerres 2010). Other possible neuroprotective effect is the proliferation of glial cells that play a fundamental role in the exchange of ions between cells in the cortex and intercellular space. Whatever the effect by which GH acts as a neuroprotector, it is not known exactly whether it is by direct action of GH or through insulinic growth factor type 1 (IGF-1), which is directly modulated by GH, located mainly in fibroblasts, chondrocytes, osteoblasts and renal cells (Leal-Guadarrama 2003; Tresguerres 2010), as the glomerulus and in some segments of the distal and proximal tubule, as well as in the collecting ducts. It should be noted that the somatotroph hormone stimulates the differentiation of various cell types while the factor IGF1 acts as a mediator for mitogenic and multiplier effects.

PROLACTIN

Prolactin (PRL) secretion levels have a circadian cycle because they vary throughout the day. The secretion of PRL occurs in the early hours of the morning, although a slight increase is also observed at the beginning of sleep (Sassin 1972). It has been noted that in women the secretion of PRL is higher, although both sexes have the same pattern of secretion.

The secretion of PRL is accentuated at the end of puberty and in pregnancy. The maximum peak of PRL secretion occurs during SWS (Van Cauter 1982). On the other hand, PRL increases in the nap and remains high in sleep; this pattern of secretion is reversed when awakening begins. Studies in subjects with SD have shown a decrease in PRL levels (Treuer 1996), as well as in people with continuous waking's at night (Spiegel 1994). It has also been recorded that in women with premenstrual syndrome who are in treatment with SD of the first part of the night, the secretion of PRL is delayed and in SD of the second part of the night it is advanced with respect to SWS (Parry 1996).

Moreover, it has been observed that the dopaminergic pathway and particularly the serotonergic participates in the control of PRL secretion. As for the dopaminergic system, methysergide decreases PRL concentrations at night and produces a minimal effect on EEG traces (Mendelson 1975). While serotonergic receptor subtypes regulate the secretion of PRL at night, zimelidine and indalipine, inhibitors of serotonin (5-HT) reuptake, decrease REM sleep, as does buspirone, a compound with 5-HT agonist properties. In addition, intracerebroventricular (i.c.v.) administration of PRL increases REM sleep in rats. This effect is also observed in rats with hyperprolactinemia (Jiménez-Anguiano 2010).

ADRENOCORTICOTROPIC HORMONE (ACTH) AND CORTISOL

A close relationship has been shown between the activity of the Hypothalamus-Hypophysis-Adrenal (HHA) axis and sleep. The main components of this axis involve the Corticotrophin Releasing Factor (CRF) synthesized in the paraventricular nucleus of the hypothalamus and secreted to the pituitary portal vein. The CRF promotes the secretion of ACTH in the anterior pituitary or adeno-hypophysis, stimulating the secretion of cortisol in the adrenal cortex, specifically in the fasciculate area (Figure 1). Cortisol inhibits the secretion of CRF and ACTH by the hypothalamus and anterior pituitary, respectively, under physiological conditions (Ganong 2010).

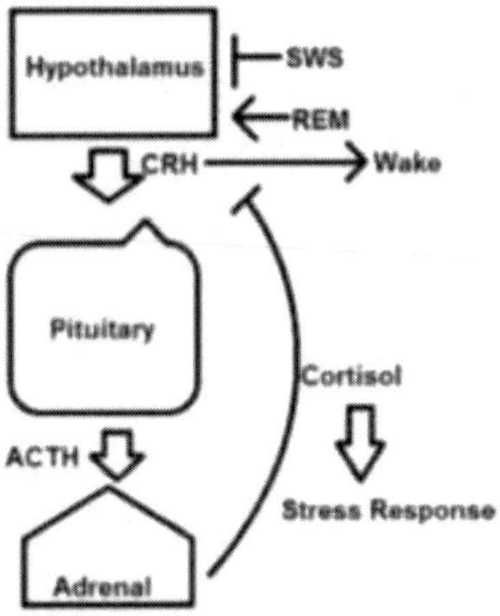

Figure 1. Schematic of sleep impacts on the HPA axis. Slow-wave sleep is an inhibitor of the HPA axis stress pathway, while REM sleep stimulates cortisol production. Corticotropin releasing hormone (CRH) has been shown to stimulate wake, while exogenous glucocorticoids have been shown to counteract this effect and stimulate sleep by feedback inhibition (Taken from Smith and Mong, 2019).

The pulsating secretion of ACTH promotes an increase in cortisol secretion hours before waking up. ACTH and cortisol also have circadian secretion in sleep. The highest peaks of cortisol release occur in the early morning hours, about two to three hours before awakening, and the lowest are recorded in the early hours of sleep (Krieger 1977; Weitzman 1971). In contrast, ACTH secretion has two release patterns: circadian and ultradian (Fehm and Born 1991). It has also been shown that cortisol release peaks are not always associated with ACTH secretion (Carnes 1988; Krieger and Glick, 1974), and that SD for short (one day) or long (about one week) periods does not affect cortisol secretion at night (Poland 1972; Salín-Pascual 1988). It has also been observed that crossing areas from east to west with very long flying hours produces serious alterations in the endocrine system and sleep, as it takes some time for the HHA axis to adapt. This phenomenon has been called time zone change syndrome, also known as Jet Lag (Mendelson 1987). Acute stress, on the other hand, activates the HHA axis by stimulating the hypothalamus to release CRF, resulting in increased plasma levels of ACTH and cortisol.

The intense rhythm of life that we are living has gradually increased the incidence of stress in many countries such as Mexico, which has a direct impact on people's health (Pozos-Padilla 2016; Jiménez-Anguiano 2018). One of the first researchers to study stress was Hans Selye in 1950. Selye identified it as "*nonspecific response of the body to any demand*" and a generalized effort by the organism to adapt to new conditions that could be compared to other general defense reactions. He called it the general adaptation syndrome and divided it into three phases: the alarm reaction, the resistance stage, and the stage of exhaustion. Most of the characteristic manifestations of the first stage are hypoglycemia, gastrointestinal disorders and hemoconcentration, which disappear or are reversed during the resistance stage, but reappear in the exhaustion stage (Selye 1950).

The maintenance of homeostasis is vital for individuals, so it is necessary, among other healthy activities: to eat, drink and sleep well. Various evidences indicate that the modification of the Cortisol rhythm is associated with sleep disorders, depression and metabolic alterations (Debono and Ross, 2011). It is currently suggested that if there is an alteration in the morning levels of free cortisol in plasma or saliva, then there is a possible risk factor for acquiring this type of ailment (Maldonado and Carranque, 2004).

In the case of depression, it has been pointed out that alterations of the sleep-wake cycle occur (Dallaspezia and Benedetti, 2011; Jiménez Anguiano and Velázquez Moctezuma, 2000). Recent research indicates that chronic depression is intimately related to dementia, as there are important physical and metabolic variations during this condition, such as altered glucocorticoid levels, atrophy of the hippocampus, increased deposition of β-amyloid plaques, inflammatory changes and deficit of nervous growth factors (Byers and Yaffe, 2011). Likewise, compared to diagnostic methods based on psychological and sometimes physical questionnaires, it has recently been shown that the ratio cortisol to dehydroepiandrosterone sulphate (DHEA) in saliva offers a more objective and rapid diagnostic indicator. Although high baseline cortisol values have been shown to be a characteristic of depressive illness, high cortisol concentration does not always lead to depression (Campagne 2003). In addition, it has been shown that the physiological response to stress induces an increase in cortisol secretion that could become the emergence of insulin resistance and metabolic syndrome (Álvarez 2010).

GONADOTROPINS (LUTEINIZING HORMONE AND FOLLICLE STIMULATING HORMONE)

Luteinizing hormone (LH) and follicle stimulant hormone (FSH) are secreted in the adenohypophysis. LH shows a pattern of secretion during the night, with a rapid increase and a slow decrease, while FSH, less sensitive to the gonadotropin releasing hormone (GnRH) and with a much longer half-life, has fewer fluctuation (Leng 2018).

The relationship between gonadotropins and the different states of sleep varies according to the age of the individuals; during puberty, in both genders, there is no congruent relationship between LH secretion and the sleep-wake cycle. In adults and children it has been shown that there are no variations in blood measurements of LH, during sleep and waking, although there is a small increase in LH during REM sleep, when compared to other sleep states (Rubin and Poland, 1982). SD causes a reduction in nocturnal LH concentrations (Strassman 1991) although this effect has not been constant (Gonzalez-Santos 1989). However, it is not clear whether the decrease in LH concentrations, caused by SD, is due to a smaller amount of sleep or some states of sleep.

The investigations carried out in normal women, during the follicular phase of the menstrual cycle, indicate that there is an association between the decrease in the frequency of the secretion pulse and the LH concentration. Both in normal conditions and in an inverted sleep cycle, LH secretion and concentration are decreased by 13% in the first hours of sleep, in a 24-hour cycle (Sitaram and Gillin, 1982).

STEROIDS OF GONADAL ORIGIN

The influence of gonadal hormones on the sleep-wake cycle has been fully confirmed in the estrous cycle, during the stages of pregnancy and birth, and as a result of menopause (Carrier 2017; Branchey and Branchey, 1970). Initial observations were described in rabbits after sexual behavior in which the females passed through a sleep stage (Sawyer 1969). The same effect was subsequently observed in male hamsters, sexual behavior produces an increase in SWS (Jiménez-Anguiano 2003). It was later shown that during the proestro phase, at which point that estrogens are increased, a total of REM sleep suppression occurs. It was also reported that estrogen administration produced a considerable decrease in the REM sleep state (Colvin 1968; Colvin 1969). While in ovariectomized rats, in addition to having decreased REM sleep also presented a reduction in SWS (Branchey and Branchey, 1973).

These changes of sleep pattern in females were not observed in castrated males and the possibility was some level of sexual dimorphism for the sleep pattern (Branchey 1971). On the other hand, progesterone has been postulated as an anesthetic because of its stabilizing effects on the membrane. In turn, it has been shown that in normal conditions Progesterone induces sleep in particular SWS because it decreases neuronal excitability (Andersen 2006) and megestrol acetate, an antagonist of Progesterone receptors significantly reduces the amount of REM sleep (Weidemann 1994). It has also been shown that progesterone decreases the number of apneas episodes in women with obesity due to its respiratory stimulating capacity (Andersen 2005). For Testosterone (T), no notable and repeatable effects have been shown directly on the sleep pattern; however, given the complex metabolic structure of steroids, it is possible that T exerts an indirect effect through its metabolites (Carrier 2017). It has also been shown the prevalence of insomnia is lower in men than in women (Wang 2016).

NEUROSTEROIDS

Neurosteroids are compounds synthesized and accumulated in the nervous system from cholesterol or other steroid precursors. They have been involved in different functions such as: food intake, memory, learning and sleep. Knowledge about steroids of cerebral origin has increased considerably in recent years (Baulieu 1991). In this way it has been shown that the two main steroids that are synthesized from the glial tissue are pregnenolone (PNE) and DHEA. In addition, it has been shown that neurosteroids are synthesized *de novo* directly in the brain from cholesterol (Baulieu 1997). It has been shown that both exert their effects on the nervous system by acting through of the modulation of GABA-A receptors, by way off an allosteric transmission mechanism (Rupprecht and Holsboer 1999). 3-α-21-dihydroxy-5-α-pregnan-20-one (Alopregnanolone) and 3-α-21-dihydroxy-5-β-pregnan-20-one (Pregnanolone) act in membrane sites activating conductance of the chlorine channels that are linked to GABA. The mechanism of action of neurosteroids on the GABA-A receptor is different from that of barbiturates and benzodiazepines (Paul and Purdy, 1991).

In the case of Alopregnanolone (ALO) it has been shown that its administration in rats produces an electroencephalographic effect similar to that of barbiturates and is also capable of inducing sleep (Mendelson 1987). It has also been shown that the administration of ALO significantly modifies the sleep pattern, since it increases the total time of the SWS and shortens its latency, at the expense of a reduction in the amount of REM sleep and wakefulness. Additionally they showed that ALO increases both the SWS and an intermediate phase between the SWS and the REM sleep phase, the pre-REM (Damianisch 2001).

Likewise, it has been observed that ALO administered in the basalis nucleus, a cholinergic nucleus, reduces REM sleep and increases SWS (Darnaudéry 1999a). At the same time, it has been seen that the PNA, like the ALO, also produces an increase in the SWS (Dale 1997). From the evidence described above, it can be seen that the, ALO and PNA are stimulatory neurosteroids of the GABA-A receptor that modify the sleep architecture, by preferentially increasing the amount of SWS and pre-REM, at the expense of a reduction in sleep REM and wakefulness.

On the other hand, the PNE -GABA-A receptor inhibitor neurosteroid-, unlike the ALO and PNA stimulators, has opposite effects on the sleep pattern. Thus, it has been observed that PNE applied in the basalis nucleus, a cholinergic nucleus, increases the amount of REM sleep (Darnaudéry 1999b). These results suggested the possible participation of the cholinergic system in the increased sleep produced by these two neurosteroids (Darnaudéry 1999a). Recently, this idea has been supported by administering the PNE in the Pedunculopontine Tegmental (PPT) nucleus, since it also produces an increase in the amount of REM sleep (Darbra 2004). From these results it has been proposed that the cholinergic system maintains an important interaction with endogenous neurosteroids such as PNE, for the regulation of REM sleep.

In addition, it has been shown that PNE and DHEA administered in rats are capable of producing a proconvulsive effect. DHEA, which is considered an excitatory neurosteroid, causes an increase in REM sleep in young men, also causing an increase in the power spectrum of the α, β, and δ bands through a direct action on the receptors GABA (Fries 1995). In humans it has been shown that PNE causes electroencephalographic changes that resemble those induced by some benzodiazepine agonists. Its effect on sleep specifically is to increase the amount of delta sleep (Steiger 1993). On the other hand, the DHEA and the PNE do not produce any change on the secretion of GH or on the release of cortisol, as to suggest an effect mediated by the participation of these hormones (Lancel 1992).

MELATONIN

Melatonin (MEL) (N-acetyl-5-methoxytryptamine) is synthesized in the Pineal Gland from the 5-Hidroxitriptamine (5-HT) or Serotonin (Hardeland 2006). Before talking about MEL, the most important components that generate and keep circadian rhythms, such as temperature, hibernation, migration, reproductive behaviors and sleep-wake cycle in mammals, are the suprachiasmatic nuclei (SCN) and the Pineal Gland (Murphy and Cambell, 1996). Anatomically, Pineal is located on the upper zone of the III ventricle below the posterior boundary of the corpus callosum and is connected by a stem to the posterior commissure and the habenular commissure. Several works detected as long as the SCN could be the generator of the mammalian circadian rhythms, the Pineal gland could be responsible for the communication of the synchronization and rhythmicity signal to the rest of the organism through MEL, which is secreted to the blood flow and to the CSF. The two sites, the SCN and the Pineal, are neurally connected by polysynaptic efferent pathways that arrives to the upper cervical ganglion; whose noradrenergic sympathetic fibers innervate the pinealocite, where the enzymes that limit MEL metabolism are regulated by α and β-adrenergic receptors. It has been shown that MEL is also synthesized in the retina, Harder's gland, extraorbital lacrimal glands, portions of the gastrointestinal tract and in certain types of blood cells. The synthesis of MEL in the Pineal gland shows a cyclic activity that is inhibited by ambient light and stimulated by darkness (Hardeland 2006). MEL has a cyclic activity that is transmitted from mothers to infants and its effect can be observed after six weeks of age (Sadeh 1997).

MEL is known for its sedative and sleep-promoting effects in cats (Marczynski 1964), rats (Holmes and Sugden 1982), chickens (Hishikawa 1969) and humans (Vollrath 1981). Without going into greater detail, a single oral, intravenous or intranasal dose of

MEL improves sleep efficiency in humans, by reducing sleep latency and the number of awakenings during the total sleep period. It also decreases stage N1 and increases the stage N2 (Waldhauser 1990). However, it does not appear that MEL participates directly in the regulation of sleep, but can indirectly affect it by altering the circadian pacemaker phase. In order that, MEL has some of the important characteristics to be considered as a possible candidate for the treatment of various sleep disorders such as insomnia, in the alterations of sleep in Alzheimer's disease, in the syndrome due to change of time zones (jet-lag) and in the synchronization of altered sleep-wake cycles (Arendt 1988; Cruz-Aguilar 2018; Van Maanen 2017). Concerning the therapeutic properties of MEL, recent research demonstrates its effectiveness in alterations such as jet-lag and sleep disorder due to rotating schedules, when used together with other synchronizers for exposure to intense light and total darkness (Burgess 2002; Auger 2015).

CONCLUSION

Throughout this brief review we have realized the relevance and complexity of the various hormonal events involved in regulating sleep. More and more people are talking about the close relationships that exist between the sleep cycle and the endocrine system, as well as its relationship with other systems, especially with the immune system in some diseases.

As we have seen, experimental methods have too much to contribute to the study of the neuroendocrine regulation of sleep; undoubtedly, elucidating their mechanisms will open up a large number of therapeutic possibilities, both for diseases limited to the endocrine system and for various sleep disturbances and stress.

REFERENCES

Álvarez, A. M., González, R. M. and Marrero, M. A. (2010). Role of Testosterone and Cortisol in metabolic syndrome. *Cuban Journal Endocrinol,* 21(1):80-90.

Andersen, M. L., Bittencourt, L., Antunes I. and Tufik, S. (2006). Progesterone on Sleep: A Possible Pharmacological Treatment for Sleep-Breathing Disorders?. *Current Medicinal Chemistry,* 13, 3575-3582.

Antonijevic, I. A., Murck, H., Frieboes, R. M., Holsboer, F. and Steiger A. (1999). On the gender differences in sleep-endocrine regulation in young normal humans. *Neuroendocrinol,* 70: 280–287.

Arendt, J., Aldhous, M. and Wright, J. (1988). Synchronization of a disturbed sleep-wake cycle in a blind man by melatonin treatment. *Lancet,* 1: 772-773.

Astrom, C. and Lindholm, J. (1990). Growth hormone-deficient young adults have decreased deep sleep. *Neuroendocrinol,* 51:82-84.

Auger, R. R., Burgess, H. J., Emens, J. S., Deriy, L. V., Thomas, S. M. and Sharkey, KM. (2015). Clinical practice guideline for the treatment of intrinsic circadian rhythm sleep-wake disorders: Advanced Sleep-Wake Phase Disorder (ASWPD), Delayed Sleep-Wake Phase Disorder (DSWPD), Non-24-Hour Sleep-Wake Rhythm Disorder (N24SWD), and Irregular Sleep-Wake Rhythm Disorder (ISWRD). An update for 2015: An American Academy of Sleep Medicine Clinical Practice Guideline. *J Clin Sleep Med*, 11:1199-2.

Baulieu, E. (1991). Neurosteroids: a new function in the brain. *Biol Cell,* 71:3-10.

Baulieu, E. E. (1979). Neuroesteroids: of the nervous system, by the nervous system, for the nervous system. *Recent Prog Horm Res*, 52:1-32.

Bidlingmaier, M. and Strasburger, C. J. (2007). Growth hormone assays: current methodogies and their limitations. *Pituitary,* 10:115–119.

Branchey, L., Branchey, M. and Nadler R. (1973). Effects of sex hormones on sleep patterns of female rats gonadectomized in adulthood and in the neonatal period. *Physiol Behav,* 11:609-611.

Branchey, L., Branchey, M. and Nadler, R. (1971). The influences of sex hormones on brain activity in male and female rats. In: *Influence of hormones on the nervous system,* edited by Ford D. H. 334-340. Karger Basel.

Branchey, M. and Branchey, L. (1970). Sleep and wakefulness in female rats during pregnancy. *Physiol Behav,* 5: 365-368.

Burgess, H., Sharkey, K. and Eastm, C. H. (2002). Bright light, dark and melatonin can promote circadian adaptation in night shift workers. *Sleep Med Rev,* 6 (5): 407-420.

Byers, A. L. and Yaffe, K. (2011). Depression and risk of developing dementia. *Nat Rev Neurol,* 7(6):323-31.

Campagne, D. M. (2003). Medir la depresión con la razón cortisol/DHEA. *Psiq Biol,* 10(6):200-202.

Carnes, M., Kalin, N. and Lent, S. (1988). Pulsatile ACTH secretion: variation with time of day and relationship to cortisol. *Peptides,* 9: 325-378.

Carrier, L., Semba, K., Deurveilher, S., Drogos, L., Cyr-Cronier, J., Lord, C. and Sekerovick, Z. (2017). Sex differences in age-related changes in the sleep-wake cycle. *Front Neuroendocrinol,* 47; 66-85.

Colvin, G., Whitmoyer, D. and Sawyer, C. (1969). Circadian sleep- wakefulness patterns in rats after ovariectomy and treatment with estrogen. *Exp Neurol,* 25:616-625.

Colvin, G., Whitmoyer, D., Lisk, R., Walter, D. and Sawyer, C. (1968). Changes in sleep-wakefulness in female rats during circadian and estrous cycles. *Brain Res,* 7:173-181.

Cruz-Aguilar, M. A., Ramírez-Salado, I., Guevara M. A., Hernández-González, M. and Benítez-King, G. (2018). Melatonin effects on EEG activity during sleep onset in mild-to-moderate Alzheimer's disease: A pilot study. *J Alzheimer's Dis Reports,* 2: 55–65.

Dale, M., Wesley, F., Seidel, K., Lan, N. G., Xia, H., Hawkinson, J., Wieland, S., Carter, R. and Wood, P. (1997). An orally bioavailable analog of the endogenous neuroactive steroid, pregnanolone, demonstrates potent sedative hypnotic actions in the rat. *Journal of Pharmachol and Experim Therapeutics,* 282:420-429.

Dallaspezia, S. and Benedetti, F. (2011). Chronobiological therapy for mood disorders. *Expert Rev Neurother,* 11(7):961-70.

Damianisch, K., Rupprecht, R. and Lancel, M. (2001). The Influence of subchronic administration of the neuroesteroid allopregnanolone on sleep in the rat. *Neuropsychopharmacol,* 24(4):576-584.

Darbra, S., George, O., Bouyer, J. J., Piazza, P. V., Le Moal M. and Mayo, W. (2004). Sleep-wake states and cortical synchronization control by pregnenolone sulfate into the pedunculopontine nucleus. *J. Neurosci Res,* 76:742-747.

Darnaudéry, M., Pallarés, M., Bouyer, J. J., Le Moal M. and Mayo W. (1999a). Infusion of neuroesteroids into the rat nucleus bassalis affects paradoxical sleep in accordance with their memory modulating properties. *Neurosci,* 2:583-588.

Darnaudéry, M., Pallarés, M., Bouyer, J. J., Moal, M. and Mayo W. (1999a). The promnesic neuroesteroid pregnenolone sulfate increases paradoxical sleep in rats. *Brain Res,* 818:492-498.

Davidson, J., Moldofsky, H. and Lúe, F. (1991). Growth hormone and cortisol secretion in relation to sleep and wakefulness. *Psychiatr Neurosci,* 16:96-102.

Debono, M. and Ross, R. J. (2011). Optimal glucocorticoid therapy. *Endocr Dev,* 20:173-80.

Fehm, H. and Born, J. (1991). Evidence for entertainment of nocturnal cortisol secretion to sleep processes in human beings. *Neuroendocrinol,* 53:171-176.

Friess, E., Trachsel, L., Guldner, J., Schier, T., Steiger, A. and Holsboer, F. (1995). Dehydroepiandrosterone increases rapid-eye-movement sleep and EEG power in the spindle frequency range. *Am J Physiol,* 268:E107-113.

Ganong, F. (2010). *Fisiología Médica*. 23ª Edición. Mexico City. Mac. Graw Hill.

Garnier, P. and Raymond F. Tob, J. (1988). Growth hormone secretion during sleep. *Hormone Res,* 29:133-139.

González-Santos, M., Gajá-Rodríguez, V. and Alonso-Uriarte, R. (1989). Sleep deprivation and adaptative hormonal responses of healthy men. *Arch Androl,* 22:203-207.

Hardeland, R., Pandi-Perumal, S. R. and Cardinali D. P. (2006). Melatonin. *Inter J Biochem & Cell Biol,* 38: 313–316.

Hindmarch, P., Smith, P. J., Brook, C. G. and Matthew, D. R. (1987). The relationship between velocity and growth hormone secretion in short prepubertal children. *Clin Endocrinol,* 27:581-91.

Hishikawa, Y., Cramer, H. and Kuholo, W. (1969). Natural and melatonin-induced sleep in young chickens-a behavioral and electrographic study. *Exp Brain Res,* 7:84-94.

Holl, R., Hartman, M. and Veldhuis, J. (1991). Thirty-second sampling of plasma growth hormone in man: correlation with sleep stages. *J Clin Endocrinol Metab,* 72:854-861.

Holmes, S. and Sugden, D. (1982). Effects of melatonin on sleep and neurochemistry in the rat. *Br J Pharmaco*l, 76:95-101.

Jarrett, D. Greenhouse, J. and Miewald, J. (1990). A reexamination of the relationship between growth hormone secretion and slow wave sleep using delta wave analysis. *Biol Psychiatr,* 27:497-509.

Jiménez Anguiano, A. and Velázquez Moctezuma, J. (2000). Depression and sleep. In *Biological Psyquiatry Topics,* edited by Torner Aguilar, C. and Velázquez Moctezuma J. Mexico City: Mexican Society of Biological Psyquiatry and Metropolitan Autonomous University.

Jiménez Anguiano, A., Jiménez Vázquez, F. and Velázquez Moctezuma, J. (2018). Neuroendocrine control of sleep and stress. In: The Nature of the dreams, edited by Alejandra Rosales-Lagarde, José L. Díaz, Markus Mueller, and Anabel Jiménez-Anguiano, A. 185-201. Mexico City, Ed. Herder.

Jiménez Anguiano, A., Arteaga Silva M. and Velázquez Moctezuma J. (2003). Masculine sexual activity affects slow wave sleep in golden hamsters. *Brain Res Bull,* 59: 429-432.

Jiménez-Anguiano, A., Arana-Lechuga, Y., Esqueda-León, E., Terán-Pérez, G. and Velázquez-Moctezuma J. (2010). Neurohumoral sleep regulation. In: *Sleep disorders.* Edited by Matilde Valencia, Rogelio Pérez-Padilla, J. and Rafael Salín-Pascual. Mexico City, McGraw Hill.

Kimura, F. and Tsai, C. (1984). Ultradian rhythm of growth hormone secretion and sleep in the adult male rat. *J Physiol,* 353:305-3315.

Krieger, D. (1977). Regulation of circadian periodicity of plasma ACTH levels. *Ann NY Acad Sci,* 297:561-567.

Krieger, D. and Glick, S. (1974). Sleep EEG stages and plasma growth hormone concentration in states of endogenous hypercortisolemia or ACTH elevation. *J Clin Endocrinol Metab,* 39:986-1000.

Lancel, M., Cronlein, T., Muller-Preuss, P. and Holsboer, F. (1992). Pregnenolone enhances EEG delta activity during non- rapid eye movement sleep in the rat, in contrast to midazolam. *Brain Res,* 575:63-68.

Leal-Guadarrama, L. I., Ochoa-Rojas, M. C. and Méndez, J. D. (2003). Clinical importance of growth factors similar to insulin. *Mex Med Gazette,* 139, 589-599.

Leng, G. (2018). *The heart of the brain.* The Hypothalamus and its hormones. Cambridge, MA: MIT Press. 265.

Maldonado, E. F. and Carranque, G. (2004). Influence of subjetive sleep quality on morning Cortisol levels. *C. Med. Psicosom,* 69:9-14.

Marczynski, T., Yamaguchi, N., Ling, G. and Grodzinska, L. (1964). Sleep induced by the administration of melatonin (5-metoxy-N-acetyltriptamine) to the hypothalamus of unrestrained cats. *Experientia,* 20:436-437.

Mendelson, W. (1987) *Human sleep: research and clinical care.* New York: Plenum PubHshing.

Mendelson, W. (1987). *Human sleep: research and clinical care.* New York: Plenum PubHshing 129-179.

Mendelson, W., Jacobs, L. and Reichman, J. (1975). Supression of sleep-related prolactin secretion and enhancement of sleep-related growth hormone secretion. *J Clin Invest,* 56:690-697.

Mendelson, W. B. (1987). Human sleep: research and clinical care. New York: Plenum.

Jarrett, D., Greenhouse, J. and Miewald, J. (1990). A reexamination of the relationship between growth hormone secretion and slow wave sleep using delta wave analysis. *Biol Psychiatr,* 27:497-509.

Murphy, P. and Cambell, S. (1996). Physiology of the circadian system in animals and humans. *J Clin Neurophysiol,* 13:2-16.

Obal, F., Fang, J. and Taishi, P. (2001). Deficiency of growth hormone releasing hormone signaling is associated with sleep alteration in the rat. *J Neurosci,* 21:2912-2918.

Parry, B., Hauger, R., LeVeau, B., Mostofi, N., Cover, H., Clopton, P. and Gillin, C. (1996). Circadian rhythms of prolactin and thyroid-stimulating hormone during the menstrual cycle and early versus late sleep deprivation in premestrual dysphoric disorder. *Psychiatry Res,* 62 (2):147-160.

Paul, S. and Purdy, R. (1991). Neuroactive steroids. *Faseb J,* 6:2311- 2322.

Pawel, M., Sassin, J. and Weitzman, D. (1972). The temporal relation between HGH release and sleep stage changes at nocturnal sleep onset in man. *Life Sci,* 11: 587-593.

Poland, R., Rubin, R., Clark, B. and Gouin, P. (1972). Circadian patterns of urine 17-OHCS and VMA excretion during sleep deprivation. *Dis Ner System,* 33:456-458.

Pozos-Radillo, E., Preciado-Serrano, L., Plascencia-Campos, A., Váldez-López and Morales-Fernández, A. (2016). Psychophysiological Manifestations associated with stress in students of a public University in Mexico. *J Child Adolesc Psychiatr Nurs,* 29(2):79-84.

Rubin, R. and Poland, R. (1982). The chronoendocrinology of endogenous depression. In*: Neuroendocrine perspectives,* edited by Müller E, Macleod R. Amsterdam: Elsevier-Biomedical Press.

Rupprecht, R. and Holsboer, F. (1999). Neuroactive steroids: mechanism of action and neuropsychopharmacological perspectives. *Trend Neurosci,* 22:410-416.

Sadeh, A. (1997). Sleep and melatonin in infants: a preliminary study. *Sleep,* 20; 185-191.

Salín-Pascual, R. Ortega-Soto, H. and Huerto-Delgadillo, L. (1988). The effect of total sleep deprivation on plasma melatonin, cortisol in healthy human volunteers. *Sleep,* 11: 362-369.

Sassin, J., Frantz, A. and Weitzman, E. (1972). Human prolactin: 24-hour pattern with increased released during sleep. *Science,* 177:1205-1207.

Sawyer, C. (1969). Some effects of hormones on sleep. *Exp Med Surg,* 27:177-186.

Selye, H. (1950). Stress and the general adaptation syndrome. *Br Med,* 1(4667):1383-92.

Sitaram, N. and Gillin, C. (1982). The effect of naloxone on normal human sleep. *Brain Res,* 244:387-392.

Smith, P. C. and Mong, J. A. (2019). Neuroendocrine Control of Sleep. In: *Curr Top Behav Neurosci.* Springer Berlin Helderberg.

Spiegel, K., Follenius, M., Simon, C. and Saini, J. (1994). Prolactin secretion and sleep. *Sleep,* 17: 20-27.

Steiger, A., Antonijevic, I. A. and Bohlhalter, S. (1998). Effects of hormones on sleep. *Horm Res,* 49:125-130.

Steiger, A., Trachsel, L., Guldner, J., Hemmeter, U., Rothe, B., Rupprecht, R., Vedder, H. and Holsboer, F. (1993). Neurosteroid pregnenolone induces sleep-EEG changes in man compatible with inverse agonistic GABAa-receptor modulation. *Brain Res,* 615:267-274.

Strassman, R., Qualls, C., Lisansky, E. and Peake, G. (1991). Sleep deprivation reduces LH secretion in men independently of melatonin. *Acta Endocrinol,* 124:646-651.

Takahashi, Y., Kipnis, D. and Daughaday, W. (1968). Growth hormone secretion during sleep. *J Clin Invest,* 47: 2079-2090.

Tresguerres, A. F., Kireev, R., Salazar, V. and Ariznavarreta, C. (2010). *Effects of Growth hormone and Melatonin on the Central Nervous System.* Annals of the Royal National Academy of Pharmacy. Action of hormones at the brain level 1054: 241-282. Accessed October 25, 2019. doi. http://analesranf.com/index.php/article/view/1054/1053.

Treuer, K., Norman, T. and Armstrong, S. (1996). Overnight human plasma melatonin, cortisol, prolactin, TSH, under conditions of normal sleep, sleep deprivation, and sleep recovery. *J Pineal Res,* 20:7-14.

Van Cauter, E., Leproult, R. and Plat, L. (2000). Age-related changes in slow wave and REM sleep and relationship with Grow Hormone and cortisol levels in healthy men. *JAMA,* 284:861-868.

Van Cauter, E., Dresser, D. and Rafetoff, S. (1982). The relationship between episodic variations of plasma prolactin and REM-non-REM cyclicity is an artefact. *J Clin Endocrinol Metab,* 54:70-75.

Van Maanen, A., Meijer, AM., Smits, M. G., Van der Heijden, K. B. and Oort, F. J. (2017). Effects of melatonin and bright light treatment in childhood chronic sleep onset insomnia with late melatonin onset: A randomized controlled study. *Sleep,* 40:1-11.

Vollrath, L., Sernm, P. and Gammel, G. (1981). Sleep induction by intranasal application of melatonin. *Bioscience,* 29:327-329.

Waldhauser, F., Saletu, B. and Trinchard-Lugan, L. (1990). Sleep laboratory investigations on hypnotic properties of melatonin. *Psychopharmacol,* 100: 222-226.

Wang, Y. M., Chen, H. G., Song, M., Xu, S. J., Yu, L. L., Wang, L., Wang, R., Shi, L., He, J., Huang, Y. Q., Sun, H. Q., Pan, C. Y., Wang, X. Y. and Lu, L. (2016). Prevalence of insomnia and its risk factors in older individuals: a community-based study in four cities of Hebei Province, China. *Sleep Med,* 19:116–122.

Weidemann, K., Seier, F., Hirchmann, M., Lauer, C. and Holsboer, F. (1994). Sleep endocrine effects of megestrelone acetate and mifepristone, a mixed glucocorticoid and progesterone-receptor antagonist. *Neuropsychopharmacol,* 10:2-21.

Weitzman, D., Fukushima, D. and Nogeire, C. (1971). Twenty-four hour pattern of the episodic secretion of cortisol in normal subjects. *J Clin Endocrinol Metab,* 33:14-22.

In: Sleep, Cognition and Emotion
Editors: I. Ramírez Salado et al.
ISBN: 978-1-53617-561-5

Chapter 3

EFFECTS OF MENOPAUSE ON SLEEP-WAKING CYCLE

Mónica Dafne García-Granados[1,2,*], Nadia Alejandra Torres-Marcial[1] and Dulce Alondra Guadalupe Islas-Jiménez[3]

[1]Laboratorio de Cronobiología y Sueño, Dirección de Investigaciones en Neurociencias, Instituto Nacional de Psiquiatría, Ramón de la Fuente Muñiz, Ciudad de México, México
[2]Posgrado en Ciencias Biológicas, Universidad Nacional Autónoma de México, México
[3]Facultad de Psicología, Universidad Nacional Autónoma de México, Ciudad de México, México

ABSTRACT

Given the changes that occur during the biological life cycles of menstruation, pregnancy, and menopause, the understanding of sleep in women is an important topic to study. Research shows that female sex hormones play a key role in regulating sleep, and thus, each phase of a woman's life —from adolescence to menopause— uniquely increases the risk of sleep disturbances. Menopause is an important stage in women's lives, which is marked by drastic hormonal fluctuations, causing disturbances in normal sleep patterns or sleep behaviors.

In this chapter, we review some of the severe symptoms that have been identified as risk factors for sleep disturbances during the menopausal transition, such as vasomotor symptoms (e.g., hot flashes and sweating) and mood symptoms (e.g., depression). Also, we present scientific findings on the sleep disorders that are more prevalent as women go

* Corresponding Author's Email: mgdafne@gmail.com.

through menopause, such as insomnia, sleep-disordered breathing, restless legs syndrome, and periodic limb movement disorder.

Keywords: menopausal transition, female sex hormones, sleep-waking cycle, sleep disorders, sleep disturbances

INTRODUCTION

The sleep–wake cycle is a neurobiological phenomenon in which intervals of activity alternate with restfulness, which occurs with a periodicity approximating the 24-hour day-night cycle. The sleep–wake cycle is disturbed during the menopausal transition, causing some aberrant features known as sleep disorders. Menopause occurs between 50 and 52 years of age. It is defined as the cessation of menstruation due to the degeneration of ovaries and follicles along with changing ovarian hormone levels (Parra et al. 2018). According to the World Health Organization's (1996) scientific group on menopause research, 1.5 million women worldwide enter menopause each year, and this is expected to increase to 1.2 billion by the year 2030. Sleep disturbances and sleep disorders increase in prevalence during the menopausal transition, and they have a far-reaching impact on quality of life, mood, productivity, and physical health, especially for women with severe sleep disturbances that are associated with functional impairment (Joffe et al. 2010; Polo-Kantola et al. 2011). The menopausal transition begins 2 years before cessation of menses (Harlow et al. 2012). It is associated with fluctuating hormone levels as well as the emergence of physiological and psychological symptoms, which include hot flashes, night sweats, insomnia, mood changes, fatigue, and excessive daytime sleepiness (Kravitz et al. 2008; Tom et al. 2010; Archer et al. 2011).

In this chapter, the hormonal physiology of the woman's menstrual cycle is first explained in order to understand the basic characteristics of menopause, which is characterized by the loss of ovarian follicular activity, meaning that there is a total cessation of menstruation. Therefore, the first part of this chapter gives an overview of the endocrinology of both the menstrual cycle and the reproductive stages of women, including the menopausal transition and postmenopause. Then, the effect of female sex hormones on sleep is explained, and a brief report about sleep in women across the life span is given. Lastly, the chapter reviews some severe symptoms that are identified as risk factors for sleep disturbances in the menopausal transition, such as vasomotor symptoms (e.g., hot flashes and sweating) and mood symptoms (e.g., depression). We also present scientific findings about the sleep disorders that are more prevalent as women go through menopause, such as insomnia, sleep-disordered breathing, restless legs syndrome, and periodic limb movement disorder. All of these sleep disturbances and disorders can have long-term effects on health and wellbeing during several years of the menopausal transition.

Defining Menopause

To understand the basic characteristics of menopause, the hormonal physiology of the woman's menstrual cycle needs to be addressed, given that menopause is characterized by the loss of ovarian follicular activity, meaning that there is a total cessation of menstruation (Parra et al. 2018). Therefore, this section will discuss the endocrinology of both the menstrual cycle and the reproductive stages of women, including the menopausal transition and postmenopause.

The Menstrual Cycle

A normal menstrual cycle is a coordinated serial event that repeats from month to month, which occurs every 28 ± 7 days. The first days of the menstrual cycle represent the menstrual phase. Menstruation is the monthly discharge of up to 80 mL of endometrial tissues through the vagina, which lasts from 3 to 8 days. The follicular phase, or preovulatory phase, begins at the end of the menstrual phase. The first two weeks of the cycle constitute this phase, at the end of which ovulation occurs (approximately on day 14 of the cycle). Later during the remaining weeks, the luteal phase occurs (Solomon, Berg, and Martin 2008; Barriga-Pooley and Brantes-Glavic 2018).

A normal menstrual cycle results from the coordination of stimulatory and inhibitory influences that determine the release of a mature oocyte from a large population of primordial oocytes. At the beginning of each cycle, the hypothalamus is involved along with the secretion of gonadotropin-releasing hormone (GnRH), stimulating the pituitary gland which, in turn, is responsible for secreting follicle stimulating hormone (FSH) and luteinizing hormone (LH). The ovary then responds to those hormones, recruiting a dominant follicle and secreting estradiol (E2) and inhibin A (potential marker of follicular maturity and luteal function). The E2 produced by the follicles exerts negative feedback on the FSH. FSH is regulated, in part, by the inhibin-B peptide that is secreted by granulosa cells, thereby creating a negative feedback system. During the menstrual cycle, the production of inhibin B by the cohort of growing follicles reaches its maximum level in the follicular phase. This mechanism determines the onset of FSH suppression during this phase of the cycle, which favors the selection of a single dominant follicle and serves as a regulatory mechanism that prevents super ovulation. For its part, the anti-mullerian hormone (AMH) is not only responsible for suppressing the development of the Müllerian ducts in the sexual differentiation of a male embryo, but it also plays a role in recruitment and follicular selection, inhibiting FSH-dependent follicular growth. The expression of the anti-müllerian hormone reflects the transition from resting primary follicles to growing follicles, and their plasma concentration reflects the number of antral and pre-antral follicles (pre-maturation phase of the follicles) in the ovary. In addition, its measurement makes it possible to quantify the ovarian follicular reserve as well as inhibin B. Estradiol continues to increase due to the growth of the dominant follicle, until it reaches its peak, triggering the discharge of LH, which is responsible for ovulation and posterior secretion of progesterone by the corpus luteum. Through ovulation, an oocyte is released from the ovary, where follicular cells decrease estrogen production. The luteal phase begins immediately after ovulation, and is preceded by a significant increase in LH. This stimulates the development of the corpus luteum, which secretes a large amount of progesterone, estrogen, and inhibin. These are key hormones for pregnancy preparation. If there is no embryo implantation, the endometrium detaches, resulting in menstrual flow (Solomon, Berg and Martin 2008; Valenzuela et al. 2016; Barriga-Pooley and Brantes-Glavic 2018). (Figure 1). Except for pregnancy, the fluctuation in female sex hormones is stable during the menstrual cycle in women of reproductive ages. Nevertheless, upon reaching a certain age, the body naturally begins to deteriorate, and so fertility begins to decrease and the menstrual cycle changes, leading to menopause.

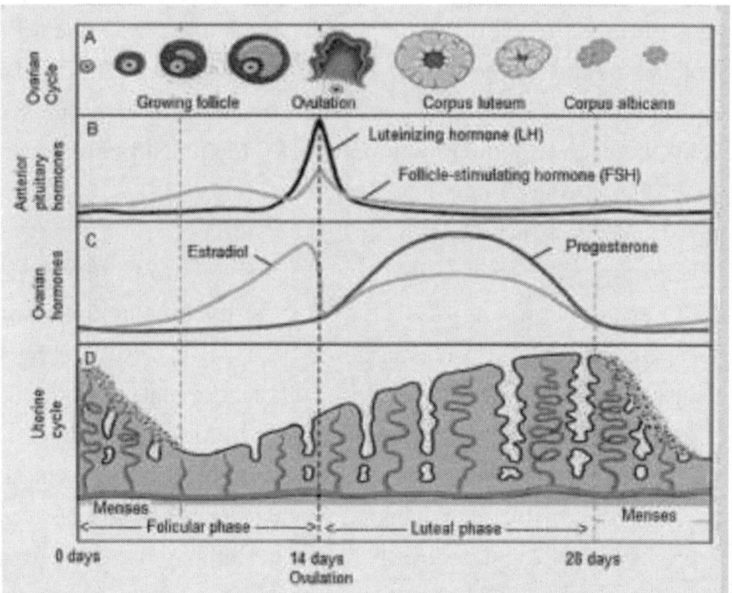

Figure 1. (A) Follicular growth and the involution of the corpus luteum during a normal menstrual cycle. (B and C) Hormonal concentration levels during the normal menstrual cycle of women in the reproductive stage (luteinizing hormone, follicle stimulating hormone, progesterone, and estradiol). (D) Proliferation of the endometrial glands of the uterus. (Image taken from https://www.nzgp-webdirectory.co.nz/).

MENOPAUSAL TRANSITION AND POSTMENOPAUSE

A current classification system for women's reproductive aging was proposed by the STRAW+10 expert workshop. The three main Stages of Reproductive Aging are: reproductive, menopausal transition, and postmenopause. These in turn are subdivided according to variability and regularity in the duration of menstrual cycles and sex hormones (Harlow et al. 2012) (Table 1).

Table 1. The Stages of Reproductive Aging Workshop +10 staging system for reproductive aging in women

Menarche (at start of stage -5); FMP (0) (between stages -1 and +1a)

Stage	-5	-4	-3b	-3a	-2	-1	+1a +1b	+1c	+2
Terminology	**REPRODUCTIVE**				**MENOPAUSAL TRANSITION**		**POSTMENOPAUSE**		
	Early	Peak	Late		Early	Late	Early		Late
					Perimenopause				
Duration	*variable*				*variable*	1-3 years	2 years (1+1)	3-6 years	*Remaining lifespan*
PRINCIPAL CRITERIA									
Menstrual Cycle	Variable to regular	Regular	Regular	Subtle changes in Flow/ Length	Variable Length Persistent ≥7- day difference in length of consecutive cycles	Interval of amenorrhea of >=60 days			
SUPPORTIVE CRITERIA									
Endocrine									
FSH			Low	Variable*	$\uparrow$ Variable*	$\uparrow$ >25 IU/L**	$\uparrow$ Variable	Stabilizes	
AMH			Low	Low	Low	Low	Low	Very Low	
Inhibin B				Low	Low	Low	Low	Very Low	
Antral Follicle Count			Low	Low	Low	Low	Very Low	Very Low	
DESCRIPTIVE CHARACTERISTICS									
Symptoms						Vasomotor symptoms *Likely*	Vasomotor symptoms *Most Likely*		*Increasing symptoms of urogenital atrophy*

* Blood draw on cycle days 2-5 $\uparrow$ = elevated

**Approximate expected level based on assays using current international pituitary standard (Abbott Laboratories FSH 2007-10; Roche Diagnostics FSH 2007-11; Stricker et al. 2006).

(Image taken from Harlow et al. 2012 with permission).

Reproductive Stage

The reproductive stage is divided into three periods: early (-5), medium (-4), and late (-3). For the purpose of this chapter, we will only focus on the late reproductive stage, since variability in the menstrual cycle begins during this stage, which marks the beginning of the menopausal transition. According to Valenzuela et al. (2016), the late reproductive stage (-3) is characterized by a decrease in fertility, where women can begin to notice changes in their menstrual cycles. Since hormonal concentrations are altered, stage -3 is subdivided into two stages: -3b and -3a. In stage -3b, the regularity and duration of the menstrual cycle is maintained and FSH is stable. However, the anti-müllerian hormone concentration, antral follicle count, and perhaps inhibin B concentrations are found to be low. In stage -3a, changes in the cycles begin and FSH varies between high and normal measurements. Given the increase in FSH, a greater number of follicles are activated during each cycle, resulting in an increase in the production of estrogen and a shortening of the follicular phase due to accelerated folliculogenesis and early elevated estradiol. This can possibly result in symptoms of hyperestrogenism, such as mastalgia and mucorrhoea.

Mastalgia is breast pain without an adjacent breast pathology, which is classified as cyclic mastalgia, non-cyclic mastalgia, and chest pain. Cyclic mastalgia affects over 40% of patients before menopause, mainly after 30 years of age. Pain is severe in approximately 8% of these women and can affect their daily activities. It can continue for many years, and usually disappears after menopause (Rosolowich, Saettler and Szuck 2006; Secretary of Health, 2009). With regard to mucorrhoea, this refers to an asymptomatic vaginal secretion that occurs around ovulation, as an expression of the estrus estrogen peak (Rigol and Santiesteban 2014).

Menopausal Transition

Menopausal transition is characterized by irregular menstrual cycles associated with elevated and variable concentrations of FSH in the follicular phase. A progressive decrease in fertility occurs during the menopausal transition. During this period, symptoms arise such as temperature dysregulation (hot flashes), disturbed sleep, altered circadian rhythms, insomnia, pain, depression, poor sensory processing, and cognitive dysfunction. The symptoms tend to appear due to a disruption in multiple estrogen-regulated systems. According to the STRAW +10 classification, the menopausal transition is divided into early menopausal transition (-2) and late menopausal transition (-1) (Nelson 2008; Brinton et al. 2015; Valenzuela et al. 2016; Scheyer et al. 2018). Stage -2 is known to have more variability in the periodicity of menstrual cycles (short or long cycles), with a difference of ≥ 7 days in the duration of consecutive and persistent cycles. With regard to the hormonal profile, high plasma concentrations of FSH are observed, due to the low production of inhibin B and a progressive decrease in the circulating anti-müllerian hormone and antral follicle count. Stage -1 is marked by the presence of amenorrhea ≥ 60 days. In this phase, menstrual cycles become more variable in duration, since at the endocrinological level, a diminished ovarian reserve results in fluctuations in hypothalamic, pituitary, and ovarian hormone levels, and notably, a decrease in ovarian secretion of estrogen and progesterone. This stage has been described as lasting from 1 to 3 years, on average (Harlow et al. 2012; Valenzuela et al. 2016; Scheyer et al. 2018). When studying a group of women in different stages of reproductive aging, more than 60% of the cycles in the late menopausal transition were found to be anovulatory, and only 25% were ovulatory, with cycles longer than 60 days (Valenzuela et al. 2016). In addition, large variability on the day of ovulation has been found. For example, a study by O'Connor et al. (2009) carried out with 108 women over 5 years reported that, in the late menopausal transition phase, ovulation occurred on the 27th day of the cycle, on average. The luteal out-of-phase (LOOP) event explains the high concentrations of estradiol observed in some menstrual cycles during the menopausal transition. FSH hyperstimulation occurs, which leads to recruitment of a second follicle during the luteal phase (Hale et al. 2009). This results in an increase in estradiol beginning in the middle of the luteal phase, and consequently, menstruation occurs with high estradiol and low FSH. The cycle that starts will be abnormally short if the LOOP event results in ovulation, or abnormally long if the LOOP event does not result in ovulation. In the late menopausal transition, "out-of-phase" luteum events occur in 37% of the cycles.

In addition, Santoro and Sutton-Tyrrell (2011) demonstrated that this late transition stage coincides with climacteric symptoms and physiological changes in multiple systems. Thus, vasomotor symptoms are more frequent, the rate of psychological disorders increases, sleep disorders increase, loss of bone mineral density accelerates significantly, and the adventitia or outer layer of the carotid arterial wall has been found to be thicker in stage -1 (Bromberger et al. 2001; Kravitz et al. 2003; Gold et al. 2006; Finkelstein et al. 2008; Wildman et al. 2008). Menopause is established at the end of this transition and after 12 months of amenorrhea.

Menopause is defined as the time of a woman's life when menstrual cycles cease, which marks the end of the woman's fertile life (Parra et al. 2018). Menopause is caused by reduced secretions of the ovarian hormones (estrogen and progesterone). This occurs at an average age of 51 years, with a range of 40-58 years. While many symptoms have been attributed to menopause, only vasomotor dysfunction and vaginal dryness are consistently associated with this condition. Other symptoms include mood changes, sleep disturbances (insomnia and night sweats), urinary incontinence, cognitive changes, somatic complaints, sexual dysfunction (decreased libido), fatigue, joint pain, and reduced quality of life (Nelson 2008; Santoro, Epperson and Mathews 2015). Also during this period, the body begins to use energy differently, fat cells change, and women may gain weight more easily (National institute of aging 2017).

Postmenopause

Postmenopause starts immediately after the menopausal transition, and is characterized by estrogen depletion with definitive amenorrhea and high levels of FSH. Symptoms (e.g., mood changes and sleep disturbances) may persist during this stage (Duval et al. 2010; Brinton et al. 2015). As in the other stages, postmenopause is divided into an early stage (+1) and late stage (+2) (Valenzuela et al. 2016). Stage +1 lasts for approximately 5 to 8 years. Since the greatest decrease in ovarian function occurs during this stage, it is further subdivided into 3 stages. The first is stage +1a, defined as the first 12 months after the last menstrual bleeding. The next stage is +1b, which lasts roughly 12 months, during which FSH and estradiol concentrations change. These phases together (+1a and +1b) last approximately two years, and it is during this period that vasomotor symptoms may appear. The third stage is +1c, during which high FSH and low estradiol have stabilized. This stage lasts between 3 and 6 years. With regard to stage +2, the main medical concern during this period is the impact of menopause on urogenital health. In addition, a variable proportion of women continue to experience vasomotor symptoms (Valenzuela et al. 2016).

NEUROENDOCRINOLOGY OF MENOPAUSE

Before diagnosing menopause, there is a decrease in inhibin B levels (a hormone that regulates the low synthesis of FSH, resulting in normal or slightly low estradiol levels). This shortens the estrogen-dependent follicular phase, thereby shortening menstrual cycles. As a result, FSH levels increase due to the low inhibin B levels, and this increases the recruitment of follicles. Consequently, the production of estrogens in the ovary initially rises along with an increase in the rate of follicles that try to reach a stage of mature follicle, in order to release their respective oovocyte. However, only one can reach ovulation, and thus the other "hyperstimulated" follicles will become atrial follicles (undergo a process of atrophy), generating a greater loss of the follicular reserve. As ovarian dysfunction increases, the elevation of FSH and estrogens is lost, giving way to the subsequent effect of hypoestrogenemia (low levels of estradiol). When the amount of estrogen (mainly estradiol) in the bloodstream decreases, the hypothalamus releases more GnRH so that the adenohypophysis receives the signal to release more FSH, but the problem is that there are no longer enough follicles for production. The decrease in estradiol levels produces irregular follicular maturation with ovulatory and anovulatory cycles (Rueda-Beltz et al. 2016; Torres-Jiménez and Torres-Rincón 2018).

In the longitudinal Study of Women's Health Across the Nation (SWAN), conducted with 3,302 menstruating women 42 to 52 years of age, Randolph et al. (2011) determined the patterns of change in serum E2 and FSH throughout the menopausal transition, in relation to age at final menstrual period. These researchers found that during the menopausal transition, an increase in the FSH pattern began 6.10 years before the final menstrual period, it accelerated 2.05 years before the final menstrual period, decelerated beginning 0.20 years before the final menstrual period, and reached stable levels 2.00 years after the final menstrual period, independent of age at the final menstrual period or ethnicity. The mean E2 concentration did not change until 2.03 years before the final menstrual period, when it began decreasing. It reached its maximal rate of change at the final menstrual period, then decelerated, and stabilized 2.17 years after the final menstrual period (Figure 2).

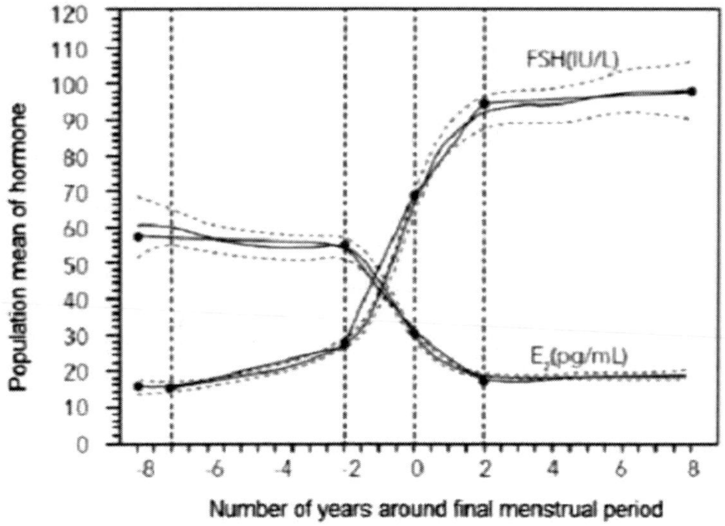

Figure 2. Variation in FSH and estradiol in population means (n = 1,215) around the last menstruation. (Image taken from Randolph et al. 2011 with permission).

POLYSOMNOGRAPHY

Sleep is defined as a reversible behavioral and physiological state of perceptual disconnection and lack of response to the environment (Carskadon et al. 2011). Polysomnography (PSG) is a multi-parametric test that is used to record multiple parameters during sleep, and as a diagnostic tool in sleep medicine. To this end, clinicians must obtain two major groups of parameters: those that are essential to determining the sleep stage and those that demonstrate the related physiology. A routine overnight PSG simultaneously records multiple physiological characteristics during sleep. In general, patients spend one entire night in the sleep laboratory with the goal of capturing a typical night's sleep. The study assesses wakefulness and sleep stages, respiration, cardiopulmonary function, and body movements. Electroencephalography (EEG), electro-oculography (EOG), and chin muscle electromyography (EMG) channels are used to stage sleep. Airflow and respiratory effort channels are used to score sleep-disordered breathing (Vaughn and Giallanza 2008).

EEG activity changes as the depth of sleep increases, with EEG recordings becoming progressively dominated by the activity of high voltage and low frequency waves. Based on these electroencephalographic changes, sleep can be divided into two stages: rapid eye movement sleep (REM) and non-REM sleep (NREM). The latter is further divided into three stages: N1, N2, and N3 (Data and Maclean 2007). N1 NREM sleep is characterized by relatively low voltage, mixed frequency activity (3–7 Hz), and acute vertex waves in the EEG. N2 NREM sleep is characterized by slow oscillation (less than 1 Hz) with distinctive sleep spindles (increase and decrease of 12–14 Hz waves lasting between 0.5–1.0 seconds) and the K-complex. In the deepest stages of sleep, N3 NREM (also called SWS), there is only low frequency wave activity (Data and Maclean 2007; Carskadon et al. 2011; Hachul et al. 2014). In control individuals, NREM sleep predominates during the first half of the night, while REM sleep is more frequent during the second half of the night (Hachul et al. 2014). REM sleep is characterized by low-amplitude synchronization of fast oscillations in the cortical EEG, atonia in the EMG (particularly observed in antigravity muscles, whereas the diaphragm and extra-ocular muscles retain substantial tone), and singles and clusters of rapid eye movements (REMs) in the EOG. In addition to these polysomnographic signs, other REM sleep-specific physiological signs are myoclonic twitches (most apparent in the facial and distal limb musculature), pronounced fluctuations in cardio-respiratory rhythms and core body temperature, and penile erection and clitoral tumescence (Data and Maclean 2007).

The sleep architecture consists of a collection of quantitative sleep measures that are measured with PSG, including: sleep latency (time it takes to fall asleep, which is normally less than 30 minutes), total sleep time (TST) (variable within the person), sleep onset latency (SOL), degree of fragmentation, total waking time, sleep efficiency (more than 85% of TST), and sometimes disruptive sleep events such as spontaneous arousals or apnea (Rechtschaffen et al. 1968; Kristal et al. 2008). The architecture of dreaming is very important because it is directly related to the quality of the dream. Furthermore, the architecture of sleep is highly susceptible to being altered by the environment as well as by endogenous factors such as hormones and age.

Effect of Female Sex Hormones on Sleep

Estrogen

Estrogen has been shown to decrease sleep latency, the number of awakenings after sleep occurs, and the number of cyclic spontaneous arousals, as well as increase total sleep time and REM cycles (Scharf et al. 1997). In premenopausal women, the number of arousals doubles during the luteal phase, when both estrogen and progesterone levels are low (Manber and Armitage 1999). Estrogen is also related to regulating body temperature. In fact, low estrogen levels raise both peripheral and central temperatures (hot flashes), a common symptom that characterizes menopause. Freedman and Woodward (1996) reported that hot flashes are also characterized by bursts of catecholamines and surges in luteinizing hormone. Hormonal therapy has shown that estrogen replacement is associated with increases in the amount of both slow wave sleep and REM (Antonijevic et al. 2000), which improves sleep quality during the menopausal transition. The effect that a high concentration of estrogens has on regulating temperature has been proposed as one reason why sleep is deeper during the follicular phase in premenopause. Since estrogen is a thermoregulatory hormone that helps regulate low body temperature at night, an incorrect regulation of estrogen could cause a lower drop in temperature (Freedman and Woodward 1996), which would result in more arousability and lighter sleep. In addition, estrogen may have a direct impact on mood through central nervous system receptors in the modulation of neurotransmitters. Estrogen is involved in norepinephrine (NE) and serotonin (5 HT) regulation. It increases 5 HT postsynaptic responsivity as well as both the number of 5 HT receptors and their uptake, and it tends to be a 5 HT agonist. Estrogen also selectively increases NE activity in the brain due to decreased monoamine oxidase activity, and it has other mixed effects on NE. These effects imply that it would have an antidepressant effect on the central nervous system (Halbreich 1997).

Progesterone

Progesterone has sedative qualities by stimulating benzodiazepine receptors that in turn stimulate the production of NREM associated gamma-aminobutyric acid (GABA) receptors (Manber and Armitage 1999). As a GABA agonist, progesterone is anxiolytic (Schweizer 1995). During a normal menstrual cycle, progesterone has been reported to drop in the premenstrual phase, which is associated with sleep difficulties and increased arousal. Progesterone also affects breathing, given that this hormone is a respiratory stimulant, which has been used to treat mild obstructive sleep apnea (Empson and Purdie 1999). There is remarkably little obstructive sleep apnea during pregnancy, and considering the amount of weight gain that typically occurs during this period, the high progesterone levels that are characteristic of pregnancy are considered to function as a respiratory stimulant. Similarly, less sleep apnea is seen prior to menopause, when progesterone levels are naturally higher.

Brief Report About Sleep in Women across the Lifespan

In the female population, sleep is constantly altered throughout a woman's life since age is highly related with the fluctuation of female sex hormones (Pengo et al. 2018) (Figure 3). Next, we will give a brief report about sleep in women across the lifespan.

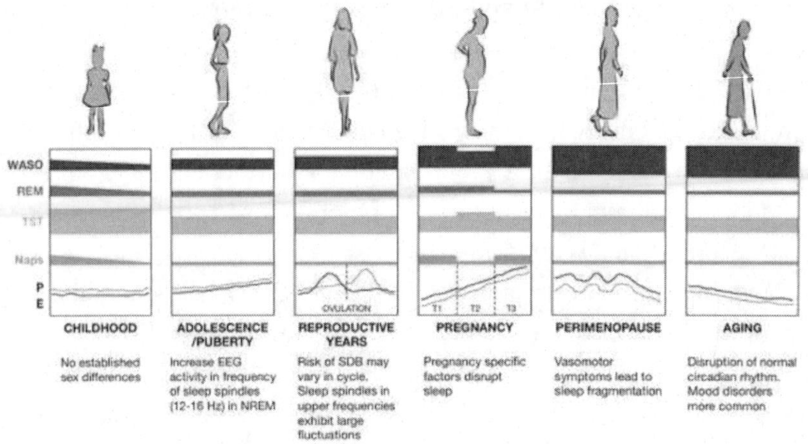

Figure 3. Sleep in women across the lifespan. E: estrogen, NREM: non-rapid eye movement, P: progesterone, REM: rapid eye movement, SDB: sleep-disordered breathing, T: trimester, TST: total sleep time, WASO: wake after sleep onset. (Image taken from Pengo et al. 2018 with permission).

Sleep during Menstruation

The first significant differences in sleep architecture between men and women appear with the onset of menarche. These changes are related to increases in ovarian function and hormonal release (estradiol and progesterone) in women (Bernhardt et al. 2016; Rebollar et al. 2017). The most dramatic change during the menstrual cycle is increased EEG activity in the frequency range of sleep spindles (12-16 Hz) during NREM sleep. This occurs in the post-ovulatory luteal phase of the menstrual cycle when progesterone and estradiol are relatively elevated (Baker et al. 2007). Sleep measurements taken by PSG show that sleep quality is generally better for women than for men (Carrier et al. 2017).

Sleep during Women's Reproductive Years

The menstrual cycle occurs monthly from puberty to menopause, and is regulated by different female sex hormones, including estradiol, progesterone, luteinizing hormone (LH), and follicle stimulating hormone (FSH) (Rebollar et al. 2017). Romans et al. (2015) evaluated 76 women (18 to 42 years old) who reported their daily sleep quality for 6 months. Although subjective sleep quality was significantly lower in the premenstrual phase than during menses and at the end of the follicular phase, the effect of the menstrual cycle phase was small and not significant when controlling for perceived stress and social support (Romans et al. 2015; Carrier et al. 2017). Driver et al. (1996) recorded the sleep of 9 women every other night throughout an ovulatory cycle, for a total of 138

PSG recordings. Researchers observed no significant changes during the menstrual cycle in terms of subjective sleep ratings and objective sleep measurements (total sleep time, sleep efficiency, sleep latency, REM sleep latency, and SWS). They also did not find any significant changes in sleep architecture during the menstrual cycle (Chuong et al. 1997; Baker et al, 2001).

Sleep during Pregnancy

Sleep disturbances during pregnancy are caused by common symptoms such as anxiety, urinary frequency, backache, fetal movement, general abdominal discomfort, breast tenderness, leg cramps, heartburn, and reflux (Nowakowski and Meers 2019).

Sleep during the Menopausal Transition and Postmenopause

The main changes reported in perimenopause are deterioration in the quality of subjective sleep, fragmentation of sleep, increased awakening, poor sleep quality, and insomnia (Pengo et al. 2018). PSG data have not found differences in total sleep time, sleep fragmentation, and sleep efficiency in peri- and postmenopausal women compared to younger women (Xu et al. 2011). Other studies have reported that reduced estrogen levels are related to high percentages of SWS and TST in post-reproductive women (Sowers et al. 2008). And it has been reported that high levels of FSH result in an increase in WASO (de Zambotti et al. 2015). It is also important to consider that different mechanisms could affect the quality of sleep during the last part of a woman's life, such as vasomotor symptoms, hormonal changes, comorbidity conditions, and a medical history of diseases.

Sleep Quality Reported in Women during Menopausal Transition and Postmenopause

Many established scales of menopausal symptoms include items that assess sleep quality symptoms. The most commonly used scales of menopause symptoms are the Greene Climacteric Scale, which includes the item "difficulty sleeping" (Greene 1998), and the Menopause Rating Scale, which includes the item "sleep problems" (difficulty falling asleep, difficulty sleeping through, and waking up early) (Schneider et al. 2000). Items that assess sleep quality have been included in these types of scales based on clinical observations of women's complaints of disturbed sleep during the menopausal transition. Epidemiological studies show that vasomotor symptoms (e.g., hot flashes) and mood symptoms (e.g., depression) are frequently identified as risk factors for sleep disturbances in the menopausal transition. Also, the prevalence of sleep disorders (e.g., sleep-related breathing, restless legs syndrome, and periodic limb movement disorder) increase as women go through menopause. Clinicians' observations have been supported by results from early research studies that have reported more sleep disturbances in perimenopausal and postmenopausal women (Ballinger 1979).

Several studies have shown that women who are transitioning through menopause and postmenopause are more likely to report sleep difficulties, with prevalence rates of self-reported sleep difficulties ranging between 40% and 56%, compared to premenopausal women in the late-reproductive stage who have rates of 31% (Kravitz et al. 2003). There is convincing evidence from several cross-sectional studies (Kravitz et al. 2003; Young et al. 2003; Shin et al. 2005; Cheng et al. 2008; Hung et al. 2014) and longitudinal studies (Kravitz et al. 2008; Berecki-Gisolf, Bergum, and Dobson 2009; Tom et al. 2010; Woods and Mitchell 2010) that the prevalence of perceived sleep disturbances increases during the menopausal transition, even after controlling for age. For example, a meta-analysis of cross-sectional data from 24 studies reported higher odds of experiencing sleep disturbance in peri-menopause (1.60), post-menopause (1.67), and surgical menopause (2.17) relative to premenopause (Xu and Lang 2014). And a longitudinal analysis of 8 years of data from 3,045 women who participated in the Study of Women's Health Across the Nation (SWAN) showed that sleep difficulties occurred due to frequent awakenings, after adjusting for demographics and health-related factors (Kravitz et al. 2008). This research also showed that difficulty falling asleep increased during the menopausal transition, whereas early morning awakening decreased from late perimenopause to postmenopause. In addition, a longitudinal study conducted by Tom et al. (2010) showed the relevance of considering the severity of sleep difficulties, with findings of a stronger relationship between the menopausal stage and difficulty sleeping for women with severe vs moderate sleep difficulties. Both studies confirmed cross-sectional data that indicate that surgical menopause (when both ovaries are removed) without hormone therapy had the greatest association with sleep difficulties.

While most studies have examined associations between sleep and menopausal stages based on bleeding patterns, some have considered a relationship between hormone levels and sleep difficulties. In the Seattle Midlife Women's Health Study cohort, Woods and Mitchell (2010) reported that higher FSH and lower estrone levels were associated with more severe nighttime awakenings, although a multivariate analysis did not find this relationship to be significant. In a cross-sectional study of 160 peri- and postmenopausal women, no relationship was found between estradiol (E2) levels or follicle-stimulating hormone (FSH) levels and diary-reported sleep measures, although wake time was strongly associated with the menopausal stage (Moreno-Frias 2014). However, longitudinal studies are better suited for tracking the relationship between changes in hormone levels and sleep. SWAN found that decreasing E2 and increasing FSH levels were associated with a higher odds of frequent awakenings during follow-up (Kravitz et al. 2008), and a higher rate of change in FSH was associated with poorer sleep quality (Sowers 2008). Other studies suggest that declining E2 levels are associated with worsening sleep quality even in the late reproductive stage (Hollander et al. 2001), and are associated with more severe sleep problems (Dennerstein et al. 2007). This evidence implies that a faster change in hormones is associated with a poorer sleep profile.

Polysomnographic Studies during the Menopausal Transition and Postmenopause

Polysomnography (PSG), including electroencephalographic (EEG) analysis, provides detailed information about biological sleep and its regulation, and is also a powerful tool for examining relationships between aspects of sleep and other biological measures, such as hormone levels and respiratory functions. Importantly, PSG is necessary for diagnosing sleep disorders such as obstructive sleep apnea syndrome, restless legs syndrome, and periodic limb movement disorder, although it is not required for an insomnia diagnosis.

Even though there is strong evidence of declining self-reported sleep quality during the menopausal transition, PSG studies have generally not found a corresponding negative change in sleep architecture (Kaplan et al. 2017), and psychological state may influence sleep quality judgments by affecting the sleep assessment process rather than sleep itself (Krystal et al. 2008). In addition, PSG studies have not consistently shown evidence that sleep disturbances are directly linked with menopausal stages, which could be explained by findings by epidemiological studies that multiple factors can influence sleep difficulties and their measurement during the menopausal transition (e.g., presence of preexisting sleep complaints, severity of sleep complaints, symptoms such as hot flashes, hormone fluctuations, and the influence of other conditions related with midlife). The use of different methodologies could also contribute to the lack of consistency that has been found between the stages of menopause and the architecture of sleep. Most PSG studies have been observational and include small samples based on only one or a few nights, which capture a snapshot of sleep. Furthermore, most studies have not controlled for confounding factors such as age, health status, medications, and presence of a sleep disorder, leading to contradictory results and challenges when trying to compare studies.

Although some studies have not found any differences in sleep architecture between pre- and postmenopausal women, a few have reported some differences in sleep architecture between perimenopausal, postmenopausal, and pre-menopausal women. The Wisconsin Sleep Cohort Study of 589 midlife women showed that peri- and postmenopausal women had better PSG sleep, with slower wave sleep (SWS) than premenopausal women (Young et al. 2003). While more SWS could be interpreted as reflecting a better sleep pattern, alternatively it could reflect a recovery response to sleep deprivation. Other studies have shown that postmenopausal women had a higher apnea-hypopnea index and lower arterial oxyhemoglobin saturation (Hachul et al. 2015). Meanwhile, SWAN did not find any differences in PSG measures according to menopausal status (Campbell et al. 2011). And in another study, late-perimenopausal and postmenopausal women had more high-frequency beta EEG activity, suggesting greater cortical hyperarousal during sleep for those women than for premenopausal and early perimenopausal women, an effect partially explained by a higher frequency of self-reported hot flashes (Campbell et al. 2011). A limited number of studies have investigated the association between PSG measures and FSH, an endocrine marker of the menopausal transition. In this regard, SWAN found that a more rapid rate of change in FSH in women in the menopausal transition was associated with a higher amount of SWS and longer total sleep time (TST), but less favorable self-reported sleep quality (Sowers et al. 2008). A cross-sectional study by de Zambotti, Colrain, and Baker (2015) carried out with women between the ages of 43 and 52 years (in the early menopausal transition) without sleep complaints found that higher FSH concentrations were associated with more wakefulness after sleep onset (WASO) and with more awakenings and arousals, after adjusting for age and body mass index. However, in women with menopausal insomnia, PSG measures did not correlate with FSH, but rather, they were associated with anxiety and depression symptoms. A few other studies have evaluated the associations between sleep and hypothalamic-pituitary-ovarian axis hormones in a group of pre- and postmenopausal women diagnosed with depression, finding that FSH levels were positively associated with WASO and negatively associated with SWS (Antonijevic et al. 2003). Most PSG studies have been cross-sectional, and therefore, they are unable to track changes in PSG measures over the course of the menopausal transition. Furthermore, women may begin to experience menopausal symptoms as early as their 40s, when they are still premenopausal, which could confound cross-sectional comparisons of the different stages. Currently, there is only one longitudinal study of changes in PSG measures throughout the menopausal transition. Lampio et al. (2017) designed a prospective 6-year follow-up study to evaluate changes in sleep stages and sleep continuity in women during menopausal transition. They studied 60 women (mean= 46 years, S.D. = 0.9) at a premenopausal baseline visit and then 6 years later. At follow-up, the women had less TST and more WASO, after adjusting for vasomotor symptoms, body mass index, and mood. These changes in sleep were linked with advancing age rather than increased FSH levels. Increasing FSH was associated with a greater proportion of SWS but not with slow wave EEG activity, which the authors proposed as reflecting an adaptive change to counteract age-related sleep fragmentation.

According to scientific studies, PSG studies have shown few consistent effects of the menopausal stage on sleep architecture, independent of age. Since all but one study were cross-sectional, they were unable to track changes in sleep throughout the transition. And few have used more complex indicators of sleep disruption, such as quantitative EEG analysis. Given that between-women variability in hormone trajectories, speed of transitioning menopause, and life experiences as women go through menopause can all interact with sleep, further longitudinal studies of midlife women are needed in order to determine whether PSG measures fluctuate along with the changing hormone environment during the menopausal transition. It is also important to consider that subgroups of women might be more susceptible to sleep disturbances, particularly those with severe vasomotor symptoms, depression, or sleep disorders.

SLEEP DISTURBANCE AS A CONSEQUENCE OF HOT FLASHES AND SWEATING

Sleep disturbance is defined as subjectively perceived sleep problems that do not necessarily meet the criteria for a clinical disorder but are bothersome to the individual (Joffe et al. 2010). Hot flashes (HFs) and sweating are conditions that can potentially affect normal sleep. HFs and sweating are vasomotor symptoms defined by transient sensations of heat, sweating, flushing, anxiety, and chills lasting for 1 to 5 min, although some flashes last up to 60 minutes (Freedman 2014). HFs can cause considerable distress, especially when severe and frequent (de Zambotti et al. 2014). More than 80% of women experience HFs and sweating during the menopause transition (Archer et al. 2011), and data from SWAN show that the median duration of HFs is 7.4 years (vis NE et al. 2015). While vasomotor symptoms may occur at any time (Freedman 2014), particularly at night they cause sudden perspiration during sleep and require changing clothing and bed linen, which may result in sleep disturbance, although not every episode of hot flash/night sweat awakens women (Moe 2004).

Vasomotor symptoms are partly due to estrogen depletion during menopause (Core 1995). Elevated central sympathetic activation, mediated through α2-adrenergic receptors, is one factor responsible for narrowing the thermoneutral zone. Some studies with animal models show that injecting norepinephrine into the preoptic hypothalamus causes heat dissipation responses followed by a decline in body temperature (Brück and Zeisberger 1990). Also, serotonergic mechanisms (Archer et al. 2011), hypothalamic kisspeptin, neurokinin B, and dynorphin neurons (Rance et al. 2013) have been associated with regulation of HFs.

There is evidence that the presence of self-reported HFs is consistently associated with poorer self-reported sleep quality and chronic insomnia (Kravitz et al. 2003, 2008;

Ohayon 2006; Blumel et al. 2012), indicating that women link HFs with nocturnal awakenings. Studies that have investigated the relationship between reported vasomotor symptoms and PSG measures have produced conflicting results, with early studies showing no relationship (Young et al. 2003) and more recent studies finding an association between HFs and disrupted sleep (Joffe et al. 2010, 2013; de Zambotti et al. 2014; Thurston et al. 2014). Longitudinal data from SWAN show that women with moderate to severe HFs are nearly three times more likely to report problems of frequent nocturnal awakenings compared to women without HFs (Kravitz et al. 2011). HF-associated bother and interference over and above frequency appear to be important constructs to consider. It has been reported that bothersome HFs but not HFs alone are associated with sleep difficulties (Xu et al. 2012). Studies showing that the effective treatment of HFs with HTs are associated with improved sleep quality (Polo-Kantola 2011) also support evidence of HFs as a cause of poor sleep. A group of researchers who analyzed clinical trial data in women with frequent moderate to severe HFs (more than 7 per day) quantified the effect of treatment to reduce the number and severity of HFs on quality of life and sleep problems. They found that a reduction of 5 HFs per day is associated with a clinically meaningful decrease in sleep problems (Pinkerton et al. 2016). A serious problem is that HFs can emerge in a person's late 30s and 40s, and are associated with poor sleep even at this late-reproductive stage (Hollander et al. 2001). Given that in some women HFs can continue to persist for several years postmenopause, they may be an important source of sleep disturbance well after midlife (Freeman et al. 2014). With objectively measured HFs, evidence shows that symptomatic postmenopausal women have more awakenings and lower sleep efficiency than asymptomatic postmenopausal women (Woodward and Freedman 1994), or no differences in PSG measures (Freedman et al. 2004). Studies consistently show a low incidence of HFs during REM sleep (Freedman et al. 2004; de Zambotti et al. 2014), possibly owing to the inhibition of thermo-regulatory responses during this stage. However, studies do not consistently indicate an association between HF and waking, with reports of awakenings more likely to occur before an HF than after an HF (Freedman et al. 2004), HFs occurring before an awakening only in the first half of the night (Freedman and Roehrs 2006), or HFs coinciding with awakenings in the majority of events, with no differences in the first and second parts of the night (de Zambotti et al. 2014). A study by de Zambotti et al. (2014) found that roughly 80% of HFs in perimenopausal women interfered with sleep. Differences among studies might relate with the classification of HFs in association with awakenings, as well as with between-women variability in the impact of HFs on sleep. In an experimental model of new-onset HFs in young premenopausal women treated with a gonadotropin-releasing hormone agonist that simulates menopause, HFs were linked with more PSG awakenings, more WASO, and more N1 sleep (Joffe et al. 2013). Bianchi et al. (2016) found that WASO time increased by 6.3 min for each additional HF measured, on average. Furthermore, the majority (66%) of HFs occurred during wake or stage N1 sleep (few HFs occurred during REM sleep), with no difference between the first and second halves of the night. In an analysis of the overall impact of HFs on sleep architecture, de Zambotti et al. (2014) reported that wake time attributed to HFs was responsible for 27% of objective WASO, on average, although the impact of hot flashes varied widely among the women. They also reported that the occurrence of an awakening coincided with the majority (69%) of HFs. Thus, these data provide support for a relationship between HFs and disturbed sleep.

An interesting study performed by Buchanan et al. (2017) used objective assessments of sleep to compare the effects of yoga and aerobic exercise with usual activity in midlife women. These researchers analyzed a randomized controlled trial called Menopause Strategies: Finding Lasting Answers for Symptoms and Health (MsFLASH). The analysis was conducted with 186 late transition and postmenopausal women aged 40 to 62 years with hot flashes. Study subjects had an average of 7.3 to 8 HFs per day. Participants were randomized to 12 weeks of yoga, supervised aerobic exercise, or usual activity. Sleep measures were evaluated using wrist actigraphy, and bedtimes and rise times were determined primarily from the participants' sleep diaries. Linear regression models were used to compare the mean and coefficient of variation of change in the actigraph sleep measurements obtained from each intervention group with those obtained from the usual activity group. For the entire sample, Buchanan et al. (2017) found that mean baseline values of the primary sleep measures were 407.5 ± 56.7 min for TST, 54.6 ± 21.8 min for WASO, 37.7 ± 18.7 for coefficient of variation for WASO, and 81.5 ± 46.9 for coefficient of variation for number of long awakenings > 5 min. Changes in the actigraphic sleep outcomes from baseline to weeks 11 and 12 were small, with no differences among groups. Nevertheless, in a secondary exploratory analysis, women with a baseline Pittsburgh Sleep Quality Index over 8 had a significantly lower TST coefficient of variation following yoga compared with usual activity. This finding shows that yoga and aerobic exercise improve sleep parameters in menopausal women with HFs.

DEPRESSION AND SLEEP DISTURBANCE

Epidemiological studies show a rise in self-reported sleep difficulties as women go through menopause. These data reflect how women are feeling, and depression is frequently identified as a risk factor for sleep disturbance during the menopausal transition (Cohen et al. 2006; Bromberger et al. 2007, 2011). It is hypothesized that fluctuations and an eventual drop in estrogen is responsible for impaired mood (Borrow et al. 2014), and women vulnerable to developing mood symptoms during the premenstrual stage of the menstrual cycle or postpartum are at greater risk for developing depressive symptoms during the menopausal transition (Studd 2011). Campbell,

Dennerstein, Finch, and Szoeke (2017) carried out a longitudinal study that used a scale and self-reporting to measure how depressive symptoms and negative moods (e.g., loneliness, helplessness, impatience, depression, hopelessness, withdrawal, unhappiness, confusion, stress, and feelings of insignificance) may be related with women's reproductive stages. They found that negative moods and depressive symptoms were highest during the menopausal transition and lowest in late postmenopause. Increasing age was associated with a reduction in depressive symptoms and negative moods. The findings by this study suggest that women have a higher risk of depressive symptoms and higher levels of negative moods during the menopausal transition and early postmenopause than during late postmenopause. While this may be due to a range of bio-psycho-social risk factors, changes in FSH levels that occur during perimenopause may contribute to an elevated risk (Campbell, Dennerstein, Finch, and Szoeke 2017).

The relationship between depressive symptoms and subjective sleep disturbances has also been documented in women in the menopausal transition. One longitudinal study of 309 women going through menopause reported that depressive symptoms were unrelated to menopausal status or annual change in E2, but were associated with hot flashes and sleep disturbance (Avis et al. 2001). However, 13 years of longitudinal SWAN data on midlife women showed that the presence of subjective sleep problems at baseline was a significant predictor of persistent and recurrent major depressive disorder at follow-up (Bromberger et al. 2016).

Studies of the relationship between PSG and depressive symptoms are controversial. On the one hand, Toffol et al. (2014) found that more depressive symptoms were associated with lower sleep efficiency and shorter TST in perimenopausal women, and with a higher percentage of REM sleep in postmenopausal women. On the other hand, Kravitz et al. (2011) found that mood symptoms were not independently related with sleep architecture, and that anxiety symptoms were related to longer sleep onset latency and lower sleep efficiency, but only in women who also reported vasomotor symptoms. In addition, recent studies have shown that depression is associated with difficulty falling asleep and waking up earlier than desired (Vousoura et al. 2015). And in an intervention study by Joffe et al. (2011) of perimenopausal women with depressive symptoms, improvement in depression was predicted by improved sleep and increased E2.

SLEEP DISORDERS DURING MENOPAUSAL TRANSITION

Sleep disorders are a group of conditions in which the normal sleep pattern or sleep behavior is disturbed. These disorders can present as complaints of either insufficient sleep, perceived excessive amount of sleep, or abnormal movements during sleep (Sateia 2014). In the case of women in menopause, there is an increase in the prevalence of sleep disorders, such as insomnia, sleep disordered breathing, restless legs syndrome, and periodic limb movement disorder.

Insomnia Disorder during the Menopausal Transition

More than 30% of adults experience transient insomnia at some point in their lives. According to the American Academy of Sleep Medicine (2014), insomnia can develop into a more chronic and persistent condition in roughly 40% of cases. The diagnosis of insomnia is made when the patient reports dissatisfaction with sleep, such as sleep-onset or sleep-maintenance insomnia, as well as other daytime symptoms (e.g., sleepiness, impaired attention, mood disturbances) for at least 3 nights per week, lasting for more than 3 months and causing significant distress and daytime consequences (American Academy of Sleep Medicine 2014). In the context of menopause, classic clinical features of insomnia —such as rumination, anxiety, and generalized hyperarousal— overlap with aspects of insomnia that are specific to menopause, such as vasomotor symptoms (Shaver et al. 2002; Riemann et al. 2015). Ohayon (2006) reported that 26% of a sample of 1,000 perimenopausal women met the DSM-IV diagnosis of insomnia, with difficulty staying asleep being the most common symptom. While some women had preexisting insomnia symptoms before entering the menopausal transition, 31.8% reported that their insomnia symptoms were related to the development of menopause. In addition, researchers have shown that the likelihood of having chronic insomnia symptoms increases with the severity of hot flashes. Lampio et al. (2016) found that other factors are associated with increased risk of insomnia disorder during the menopausal transition, including: personality factors such as neuroticism, past depression, a history of severe premenstrual symptoms, and chronic stress.

Studies of PSG measures of insomnia disorder during the menopausal transition have shown shorter TST, more WASO, and poorer sleep efficiency. This disruption matches subjective reports of poor sleep quality in this group relative to women without insomnia who were also in menopausal transition. In fact, nearly 50% of the insomnia group had short sleep duration of less than 6 hours (Baker et al. 2015). PSG findings were also reported to be consistent with sleep diary data collected over a 2-week period. Women with insomnia were more likely to have objectively measured hot flashes, and the presence of hot flashes predicted the number of PSG-awakenings per hour of sleep. A comparative study of premenopausal, perimenopausal, and postmenopausal women with insomnia (ages between 40 and 59 years) showed that subjective measures, including insomnia severity index scores, Beck depression inventory scores, and sleep diary measures, were similar among premenopausal, perimenopausal, and postmenopausal women, while peri- and postmenopausal women with insomnia had a longer PSG-defined total wake time and lower sleep efficiency (peri and postmenopausal= 81.8%, premenopausal= 86.0%). This suggests that PSG measures of sleep quality are more strongly impacted in peri and postmenopausal women than in premenopausal women with insomnia (Xu et al. 2011). In addition, it is now known that chronic stress could add a layer of complexity to PSG sleep insomnia disorder in midlife women. Women with more chronic stress exposure over a 9-year period had greater PSG-measured WASO and were more likely to have insomnia than participants with moderate stress exposure (Hall et al. 2015). And other researches have reported that women in the menopausal transition with insomnia showed less within-night recovery in autonomic vagal functioning following pre-sleep acute stress exposure (de Zambotti et al. 2016). In this sense, it is unclear whether insomnia disorder that develops in the context of the menopausal transition has a similar impact on long-term health as insomnia that develops in other circumstances, and how comorbidity factors, such as hot flashes or depression, might interact with sleep difficulties to contribute to impairment.

Sleep Disordered Breathing

Sleep apnea is a primary sleep disorder that is characterized by snoring, upper airway obstruction, inspiratory flow limitation, and excessive daytime sleepiness (Shneerson 2000). Sleep apnea can be diagnosed with PSG, in which the severity of sleep apnea is quantified by the number of respiratory events per hour of sleep. Along with clinical symptoms, at least 5 events per hour (Apnea-Hypopnea Index: AHI $\geq$ 5) are required for a diagnosis of sleep apnea (Iber et al. 2007). In the case of obstructive apnea, this is defined as a cessation of airflow for at least 10 seconds, and results from the collapse of the upper away during sleep. In the general middle-aged population, moderate to severe sleep apnea can be found in roughly 30% to 50% of men and 11% to 23% of women (Heinzer et al. 2015; Bjornsdottir 2016). Clinical symptoms most often include loud snoring, choking/gasping, apneas witnessed by the bed partner, excessive sleepiness and fatigue, and morning headache. However, the prevalence of sleep apnea in women increases following menopausal transition (Young et al. 2003; Polesel et al. 2015; Mirer et al. 2017). In a sample of midlife women with sleep complaints, Freedman and Roehrs (2007) reported that 53% had Sleep Disordered Breathing (SDB). It has been proposed that changes in body habitus during menopausal transition affect breathing during sleep. Postmenopausal women tend to have a higher body mass index and central adiposity (Davis et al. 2012), both of which are strongly associated with sleep disordered breathing.

In the Wisconsin Sleep Cohort, postmenopausal women were 2.6 times more likely to have an AHI $\geq$ 5 per hour and 3.5 times more likely to have an AHI $\geq$ 15 per hour, compared with premenopausal women, after adjusting age, body mass index, and smoking (Young et al. 2003). Recently, a longitudinal analyses by Mirer et al. (2017) found that AHI increases from premenopause to peri- and postmenopause, independent of age and changes in body habitus. The greater prevalence of SDB after menopause might be partly due to loss of the protective effects of female reproductive steroid hormones, especially progesterone (Young et al. 2003). The decline in estrogen and progesterone after menopause likely plays a role in the increased risk of SDB, given that progesterone increases the ventilatory drive, affects upper airway dilatory muscles (Kapsimalis and Kryger 2002), and partly mediates its effects through estrogen-dependent receptors. Hence, the menopausal decline in both these hormones could impact respiration. In addition, hormonal therapy is associated with a lower prevalence of SDB (Young et al. 2003), although clinical trials of hormonal therapy have produced conflicting results (Polo-Kantola 2008).

Restless Legs Syndrome and Periodic Limb Movement Disorder

Restless legs syndrome (RLS) is characterized by an uncomfortable sensation leading to an urge to move the limbs, which occurs or worsens while at rest, with consistent evening predominance. It is associated with dysesthesia and is partially relieved by physical activity. Patients often describe the sensation as “creeping, crawling tingling,” shock-like feelings, or simply indescribable discomfort. Over the course of the disease, the sensations can spread to the arms or trunk. One of the major characteristics of RLS is its worsening in the evening and at night, which results in difficulty initiating sleep, as patients often get up and pace around the room to relieve the discomfort. In turn, poor sleep often leads to fatigue and daytime sleepiness. With regard to Periodic Limb Movement Disorder (PLMD), this may be diagnosed when the frequency of limb movement (as defined by the American Academy of Sleep Medicine scoring manual) is - greater than 15 per hour in adults. Periodic limb movements must be accompanied by sleep disturbance to establish this diagnosis. RLS and PLMD frequently co-occur. The latter is present in 80% to 90% of patients diagnosed with RLS. The presence of PLMD also supports the diagnosis of RLS. According to the American Academy of Sleep Medicine (2014), RLS is one of the most common sleep-related movement disorders, affecting about 15% of adults. The prevalence of RLS and PLMD increases with age, and RLS is more common in women (Harmell and Ancoli-Israel 2011). Freedman et al. (2007) found PLMD and apneas to be the best predictors of poorer sleep efficiency in peri- and postmenopausal women who report sleep disturbances. However, in a group of asymptomatic postmenopausal women, the incidence of PLMD was unrelated to E2 or FSH levels (Polo-Kantola 2001), suggesting that the increase in prevalence of RLS and PLMD after menopause may be related more to aging than to hormonal changes.

CONCLUSION

Sleep difficulties increase in prevalence during the menopausal transition, and they have a far-reaching impact on quality of life, mood, and physical health, especially in women for whom sleep disturbances are severe and are associated with functional impairment. Nocturnal hot flashes and sweating are an important component of sleep difficulties for women who are going through menopause. Evidence shows that sleep disturbances in perimenopausal and menopausal women are associated with self-reports of sleep disturbances as well as PSG-measured wakefulness. However, since not all women who have menopause-related sleep problems complain of hot flashes, other factors that could disturb sleep must be considered, such as insomnia, sleep disordered breathing, movement disorders, and depression.

Given that menopausal transition is a period that affects the quality of life of midlife women, women need to know about the physiological and behavioral changes that occur during this period of life. As women approach menopause, this knowledge would allow them to consult specialists in order to prevent or counter the symptoms that are characteristic of this stage. The American Academy of Sleep Medicine suggests that symptoms occurring during this period of life can usually be treated with a combination of hormonal therapy, antidepressants, aerobic exercise, diet, application of sleep hygiene measures, and cognitive or supportive therapy.

REFERENCES

Abbott Laboratories FSH. 2007–10. *Package Insert: Architect Analyzers;* 6C24, 34–4680/R1.

Abe, T., K. Ogawa, H. Nittono and T. Hori. 2008. "Neural generators of brain potentials before rapid eye movements during human REM sleep: a study using sLORETA." *Clinical Neurophysiology* 119(9):2044-53. doi: 10.1016/j.clinph.2008.05.008.

Achermann, P. and A. A. Borbely.1997. "Low-frequency (<1 Hz) oscillations in the human sleep electroencephalogram." *Neuroscience* 81:213–22.doi: 10.1016/s0306-4522(97)00186-3.

American Academy of Sleep Medicine. 2014. *International Classification of Sleep Disorders: Diagnostic and Coding Manual*. Darien: American Academy of Sleep Medicine.

Antonijevic, I. A., G. K. Stalla., and A. Steiger. 2000. "Modulation of sleep electroencephalogram by estrogen replacement in postmenopausal women." *American Journal of Obstetrics and Gynecology* 182:277-82. doi: https://doi.org/10.1016/S0002-9378(00)70211-0.

Antonijevic, I. A., H. Murck, R. M. Frieboes, M. Uhr, and A. Steiger. 2003. "On the role of menopause for sleep-endocrine alterations associated with major depression." *Psychoneuroendocrinology* 28:401–418. doi: 10.1016/s0306-4530(02)00031-8.

Arnardottir, Erna S., Erla Bjornsdottir, Kristin A Olafsdottir, Bryndis Benediktsdottir, and Thorarinn Gislason. 2016. "Obstructive sleep apnoea in the general population: highly prevalent but minimal symptoms." *European Respiratory Journal* 47(1):194–202. doi: 10.1183/13993003.01148-2015.

Avis, N. E., S. Crawford, R. Stellato, and C. Longcope. 2001. "Longitudinal study of hormone levels and depression among women transitioning through menopause." *Climacteric* 4:243–249. doi: https://doi.org/10.1080/cmt.4.3.243.249.

Baker, Fiona C., Tracey L. Kahan, John Trinder and Ian M. Colrain. 2007. "Sleep quality and the sleep electroencephalogram in women with severe premenstrual syndrome." *Sleep* 30(10):1283-1291. doi:10.1093/sleep/30.10.1283.

Baker, Fiona C., Adrian R. Willoughby, Stephanie A. Sassoon, Ian M Colrain and Massimilianode de Zambotti. 2015. "Insomnia in women approaching menopause: Beyond perception." *Psychoneuro-endocrinology* 60:96–104. doi: 10.1016/ j.psyneuen.2015.06.005.

Ballinger, C. B. 1976. "Subjective sleep disturbance at the menopause." *Journal of Psychosomatic Research* 20:509-513. doi: 10.1016/0022-3999(76)90015-5.

Barriga-Pooley, P. and S. Brantes-Glavic. 2018. "Normal menstrual cycle." In *Menstrual Cycle,* edited by Olena Ivanivna Lutsenko, IntechOpen. doi: 10.5772/intechopen.79876.

Bianchi, Matt T., Semmie Kim, Thania Galvan, David P. White, and Hadine Joffe. 2016. "Nocturnal hot flashes: relationship to objective awakenings and sleep stage transitions." *Journal of Clinical Sleep Medicine* 12(7):1003–1009. doi: http://dx.doi.org/10.5664/jcsm.5936.

Blümel, Juan E., Antonio Canoa, Edward Mezones-Holguín, Germán Barón, Ascanio Bencosme, Zully Benítez, Luz M. Bravo, Andrés Calle, Daniel Flores, María T. Espinoza, Gustavo Gómez, José A. Hernández-Bueno, Fiorella Laribezcoa, Mabel Martino, Selva Lima, Alvaro Monterrosa, Desiree Mostajo, Eliana Ojeda, William Onatra, Hugo Sánchez, Konstantinos Tserotas, María S. Vallejo, Silvina Witis, María C. Zúniga and Peter Chedraui. "A multinational study of sleep disorders during female mid-life." 2012. *Maturitas* 72(4):359–366. doi: 10.1016/j.maturitas.2012.05.011.

Borrow, A. P. and N. M. Cameron. 2014. "Estrogenic mediation of serotonergic and neurotrophic systems: implications for female mood disorders." *Progress in Neuropsychopharmacology and Biologycal Psychiatry* 54:13–25. doi: https://doi.org/10.1016/j.pnpbp.2014.05.009.

Brinton, R. D., J. Yao, F. Yin, W. J. Mack, and E. Cadenas. 2015. "Perimenopause as a neurological transition state." *Nature Reviews Endocrinology* 11:393–405. doi: 10.1038/nrendo.2015.82.

Bromberger, J. T., and H. M. Kravitz. 2011. "Mood and menopause: findings from the Study of Women's Health Across the Nation (SWAN) over 10 years." *Obstetrics and Gynecology Clinics of North America* 38: 609–625. doi: 10.1016/j.ogc.2011.05.011.

Bromberger, J. T., H. M. Kravitz, A. Youk, L. L. Schott, and H. Joffe. 2016. "Patterns of depressive disorders across 13 years and their determinants among midlife women: SWAN mental health study." *Journal of Affective Disorders* 206:31–40. doi: https://doi.org/10.1016/j.jad.2016.07.005.

Bromberger, J. T., K. A. Matthews, L. L. Schott, Sarah Brockwell, Nancy E. Avis, Howard M. Kravitz, Susan A. Everson-Rose, Ellen B. Gold, MaryFran Sowers, and John F. Randolph, Jr. 2007. "Depressive symptoms during the menopausal transition: the Study of Women's Health Across the Nation (SWAN)." *Journal of Affective Disorders* 103:267–272. doi: 10.1016/j.jad.2007.01.034.

Bromberger, J. T., P. M. Meyer, H. M. Kravitz, B Sommer, A. Cordal, L. Powell, P. A. Ganz, and K. Sutton-Tyrrell. 2001. "Psychologic distress and natural menopause: a multiethnic community study." *American Journal of Public Health* 91:1435-1442. doi: https://doi.org/10.2105/AJPH.91.9.1435.

Brück, K., and E. Zeisberger. 1990. *Thermoregulation: Physiology and Biochemistry.* New York: Pergamon.

Buchanan, D. T., Carol A. Landis, Chancellor Hohensee, Katherine A. Guthrie, Julie L. Otte, Misti Paudel, Garnet L. Anderson, Bette Caan, Ellen W. Freeman, Hadine Joffe, Andrea Z. LaCroix, Katherine M. Newton, Susan D. Reed, and Kristine E. Ensrud. 2017. Effects of Yoga and Aerobic Exercise on Actigraphic Sleep Parameters in Menopausal Women with Hot Flashes. *Journal of clinical sleep medicine:* JCSM: official publication of the American Academy of Sleep Medicine, 13(1):11–18. doi:10.5664/jcsm.6376.

Burgess, N, E. A. Maguire, and J. O'Keefe. 2002. "The human hippocampus and spatial and episodic memory." *Neuron* 35:625– 641. doi: 10.1016/s0896-6273(02)00830-9.

Campbell, I. G., J. T. Bromberger, D. J. Buysse, M. H. Hall, K. A. Hardin, H. M. Kravitz, K. A. Matthews, M. O. Rasor, J. Utts, and E. Gold. 2011 "Evaluation of the association of menopausal status with delta and beta EEG activity during sleep." *Sleep* 34:1561–1568. doi: 10.5665/sleep.1398.

Campbell, Katherine E., Lorraine Dennerstein, Sue Finch, and Cassandra E. Szoeke. 2017. "Impact of menopausal status on negative mood and depressive symptoms in a longitudinal sample spanning 20 years." *Menopause* 24:490-496. doi: 10.1097/GME.0000000000000805.

Carrier Julie, Kazue Semba, Samuel Deurveilher, Lauren Drogos, Jessica Cyr-Cronier, Catherine Lord, and Zoran Sekerovick. 2017. "Sex differences in age-related changes in the sleep-wake cycle." *Frontiers in Neuroendocrinology* 47:66-85. doi: doi.org/10.1016/j.yfrne.2017.07.004.

Carskadon, Mary A., and William C. Dement. 2011. "Monitoring and staging human sleep." In *Principles and practice of sleep medicine*, 5th edition, ed. Kryger MH, Roth T and Dement WC. St. Louis: Elsevier Saunders.

Cheng, Ming-Huei, Chung-Yao Hsu, Shuu-Jiun Wang, Shin-Jung Lee, Peng-Hui Wang, and Jong-Ling Fuh. 2008. "The relationship of self-reported sleep disturbance, mood, and menopause in a community study." *Menopause* 15:958–962. doi: 10.1097/gme.0b013e318160dafa.

Cohen, L. S., C. N. Soares, A. F. Vitonis, M. W. Otto, and B. L. Harlow. 2006. "Risk for new onset of depression during the menopausal transition: the Harvard study of moods and cycles." *Archives of General Psychiatry* 63:385–390. doi: 10.1001/archpsyc.63.4.385.

Cornwell, B. R., L. L. Johnson, T. Holroyd, F. W. Carver, and C. Grillon. 2008. "Human hippocampal and parahippocampal theta during goal-directed spatial navigation predicts performance on a virtual Morris Water Maze." *Journal Neuroscience* 28(23): 5983–5990. doi: 10.1523/JNEUROSCI.5001-07.2008.

Datta, Subimal, and Robert MacLean. 2007. "Neurobiological mechanisms for the regulation of mammalian sleep-wake behavior: Reinterpretation of historical evidence and inclusion of contemporary cellular and molecular evidence." *Neuroscience and Biobehavioral Reviews* 31(5): 775-824. doi: 10.1016/ j.neubiorev.2007.02.004

Davis, Susan R., C. Castelo-Branco, P. Chedraui, M. A. Lumsden, R. E. Nappi, D. Shah, and P. Villaseca. 2012. "Understanding weight gain at menopause." *Climacteric* 15(5):419-429. doi: 10.3109/13697137.2012.707385.

de Zambotti Massimiliano, David Sugarbaker D, John Trinder, Ian M. Colrain, and Fiona C. Baker. 2016. "Acute stress alters autonomic modulation during sleep in women approaching menopause." *Psychoneuroendocrinology* 66:1–10. doi:10.1016/j.psyneuen.2015.12.017.

de Zambotti, Massimiliano, Ian M Colrain, and Fiona C Baker. 2015. "Interaction between reproductive hormones and physiological sleep in women." *Journal of Clinical Endocrinology and Metabolism* 100(4):1426-1433. doi: 10.1210/jc.2014-3892.

de Zambotti, Massimiliano, Ian M. Colrain, Harold S. Javitz, and Fiona C. Baker. 2014. "Magnitude of the impact of hot flashes on sleep in perimenopausal women." *Fertil Steril* 102(6):1708–1715.

Dennerstein, Lorraine, Philippe Lehert, Henry Burger, Janet Guthrie. 2007. "New findings from non-linear longitudinal modelling of menopausal hormone changes." *Human Reproduction* 13:551–557. doi: https://doi.org/10.1093/humupd/dmm022.

Driver, Helen S, Derk-Jan Dijk, Esther Werth, Kurt Biedermann, and Alexander a Borbély. 1996. "Sleep and the sleep electroencephalogram across the menstrual cycle in young healthy women." *Journal of Clinical Endocrinology and Metabolism* 81(2):728–35. doi: 10.1210/jcem.81.2.8636295.

Duval, Fabrice, M. A. Martine Jautz-Duval, Felix González, and Hassen Rabia. 2010. "Neurobiological bases of psychiatric vulnerability throughout the stages of the hormonal women`s life." *Chilean Journal of Neuro-Psychiatry* 48:292-306. doi: https://dx.doi.org/10.4067/S0717-92272010000500005.

Empson, Jacob A. C., and David W. Purdie. 1999. "Effects of sex steroids on sleep," *Annals of Medicine* 31(2):141-145. doi:10.3109/07853899708998790.

Finkelstein, J. S., S. E. Brockwell, V. Mehta, G. A. Greendale, M. R. Sowers, B. Ettinger, J. C. Lo, J. M. Johnston, J. A. Cauley, M. E. Danielson, and R. M. Neer. 2008. "Bone mineral density changes during the menopause transition in a multiethnic cohort of women." *The Journal of Clinical Endocrinology and Metabolism* 93:861-868. doi: https://doi.org/10.1210/jc.2007-1876.

Freedman, Robert R, Dena Norton, Suzanne Woodward, and Germaine Cornélissen. 1995. "Core body temperature and circadian rhythm of hot flashes in menopausal women." *The Journal of Clinical Endocrinology and Metabolism* 80(8):2354-2358. doi:10.1210/jcem.80.8.7629229.

Freedman, Robert R., and Roehrs Timothy A. 2007. "Sleep disturbance in menopause." *Menopause* 14(5):826–829. doi:10.1097/GME.0b013e3180321a22.

Freedman, Robert R., and Suzanne Woodward. 1996. "Core body temperature during menopausal hot flushes." *Fertility and Sterility* 65(6): 1141-1144. doi.org/10.1016/S0015-0282(16)58328-9.

Freedman, Robert R., and Timothy A. Roehrs. 2004. "Lack of sleep disturbance from menopausal hot flashes." *Fertility and Sterility* 82(1):138–144. doi: https://doi.org/10.1016/j.fertnstert.2003.12.029.

Freedman, Robert R., and Timothy A. Roehrs. 2006. "Effects of REM sleep and ambient temperature on hot flash-induced sleep disturbance." *Menopause* 13(4):576–583. doi: 10.1097/01.gme.0000227398.3192.bc.

Gold, E. B., A. Colvin, N. Avis, J. Bromberger, G. A. Greendale, L. Powell, B. Sternfeld, and K. Matthews. 2006. "Longitudinal analysis of the association between vasomotor symptoms and race/ethnicity across the menopausal transition: study of women's health across the nation." *American Journal of Public Health* 96:1226-1235. doi: 10.2105/AJPH.2005.066936.

Greene, J. G. 1998. "Constructing a standard climacteric scale." *Maturitas* 29:25-31. doi: 10.1016/s0378-5122(98)00025-5.

Hachul, H., C. Frange, A. G. Bezerra, C. Hirotsu, G. N. Pires, M .L. Andersen, L. Bittencourt, and S. Tufik. 2015. "The effect of menopause on objective sleep parameters: data from an epidemiologic study in Sao Paulo, Brazil." *Maturitas* 80:170–178. doi: 10.1016/j.maturitas.2014.11.002.

Halbreich, U. 1997. "Role of estrogen in postemenopausal depression." *Neurology* 48(5):16S-20S. doi: https://doi.org/10.1212/WNL.48.5_Suppl_7.16S.

Hale, G. E., C. L. Hughes, H. G. Burger, D. M. Robertson, and I. S. Fraser. 2009. "Atypical estradiol secretion and ovulation patterns caused by luteal out-of-phase (LOOP) events underlying irregular ovulatory menstrual cycles in the menopausal transition." *Menopause* 16:50-59. doi: 10.1097/GME.0b013e31817ee0c2.

Hall Martica H., Melynda D. Casement, Wendy M. Troxel, Karen A. Matthews, Joyce T. Bromberger, Howard M. Kravitz, Robert T. Krafty, and Daniel J. Buysse. 2015. "Chronic stress is prospectively associated with sleep in midlife women: the SWAN sleep study." *Sleep* 38(10):1645–1654. doi: https://doi.org/10.5665/sleep.5066.

Harlow, S. D., M. Gass, J. E. Hall, R. Lobo, P. Maki, R. W. Rebar, S. Sherman, P. M. Sluss, and T. J. de Villiers. 2012. "Executive summary of the Stages of Reproductive Aging Workshop + 10: addressing the unfinished agenda of staging reproductive aging." *Menopause* 19:387-395. doi: 10.1097/gme.0b013e31824d8f40.

Harmell, A. and S. Ancoli-Israel. 2011. Handbook of sleep medicine. Philadelphia: Lippincott Williams and Wilkins.

Heinzer, Raphaël, Sopharat Vat, Pedro Marques-Vidal, Helena Marti-Soleret, Daniela Andries, Nadia Tobback, Vincent E. Mooser, Martin Preisig, Atul Malhotra, Gérard Waeber, Peter Vollenweider, Mehdi Tafti, and José Haba-Rubio. 2015. "Prevalence of sleep-disordered breathing in the general population: The HypnoLaus study." *Lancet Respiratory Medicine* 3(4):310–318.doi:10.1016/S2213-2600(15)00043-0.

Health Secretary. 2009. Diagnosis and treatment of benign breast pathology in the first and second level of care. Clinical practice guide Mexico: CENETEC.

Hollander, Lorin E., Ellen W. Freeman, Mary D. Sammel, Jesse A. Berlin, Jeane Grisso, and Michelle Battistini. 2001. "Sleep quality, estradiol levels, and behavioral factors in late reproductive age women." *Obstetrics and Gynecology* 98(3):391–397. doi:10.1016/s0029-7844(01)01485-5.

Hung, Hao-Chang, Feng-Hwa Lu, Horng-Yih Ou, Jin-Shang Wu, Yi-Ching Yang, and Chih-Jen Chang. 2014. "Menopause is associated with self-reported poor sleep quality in women without vasomotor symptoms." *Menopause* 21:834–839. doi: 10.1097/GME.0000000000000183.

Hutchison, I. C, and S. Rathore. 2015. "The role of REM sleep theta activity in emotional memory.*" Frontiers in Psychology* 6:1439. doi:10.3389/fpsyg.2015.01439.

Iber, Conrad, Sonia Ancoli-Israel, Andrew Chesson, and Stuart F. Quan. 2007. The AASM Manual for the Scoring of Sleep and Associates Events: Rules, Terminology and Technical Specifications. Westchester: American Academy of Sleep Medicine.

Jensvold, Margarte, Uriel Halbreigh, and Jean A. Hamilton. 1996. *Psychopharmacology and women: sex, gender, and hormones.* Washington: American Psychiatric Press.

Joffe, Hadine, Laura Fagioli Petrillo, Alexia Koukopoulos, Adele Viguera, April Hirschberg, Ruta Nonacs, Brittny Somley, Erica Pasciullo, David White, Janet Hall, and Lee Cohen. 2011. "Increased estradiol and improved sleep, but not hot flashes, predict enhanced mood during the menopausal transition." *The Journal of Clinical Endocrinology and Metabolism* 96:E1044–1054. doi: https://doi.org/10.1210/jc.2010-2503.

Joffe, Hadine, Anda Massler, and Katherine M. Sharkey. 2010. "Evaluation and management of sleep disturbance during the menopause transition." *Seminars in Reproductive Medicine* 28(5):404–421.

Joffe, Hadine, David P. White, Sybil L. Crawford, Kristin E. McCurnin, Nicole Economou, Nicole Economou, and Janet E. Hall. 2013. "Adverse effects of induced hot flashes on objectively recorded and subjectively reported sleep: results of a gonadotropin-releasing hormone agonist experimental protocol." *Menopause* 20(9):905–914. doi: 10.1097/GME.0b013e31828292d1.

Kahn, D., E. F. Pace-Schott, and J. A. Hobson. 1997. "Consciousness in waking and dreaming: the roles of neuronal oscillation and neuromodulation in determining similarities and differences." *Neuroscience* 78:13–38. doi:10.1016/s0306-4522(96)00550-7.

Kantola-Polo, Päivi. 2011. "Sleep problems in midlife and beyond." *Maturitas* 68(3):224–232. doi: 10.1016/j.maturitas.2010.12.009.

Kantola-Polo, Päivi, Esa Rauhala, Risto Erkkola, Kerttu Irjala, and Olli Polo. 2001. "Estrogen replacement therapy and nocturnal periodic limb movements: a randomized controlled trial." *Obstetrics and Gynecology* 97(4):548–554. doi: https://doi.org/10.1016/S0029-7844(00)01191-1.

Kantola-Polo, Päivi. 2008. "Dealing with menopausal sleep disturbances." *Sleep Medicine Clinics* 3:121–131.

Kaplan, K. A., P. P. Hardas, S. Redline, J. M. Zeitzer, and Sleep Heart Health Study Research Group. 2017. "Correlates of sleep quality in midlife and beyond: a machine learning analysis." *Sleep Medicine* 34:162–167. doi: 10.1016/j.sleep.2017.03.004.

Kapsimalis, Fotis and Meir H. Kryger. 2002. "Gender and obstructive sleep apnea syndrome, part 2: mechanisms." *Sleep* 25(5):499–506. doi: https://doi.org/10.1093/sleep/25.5.497.

Krause, Adam J., Eti Ben Simon, Bryce A. Mander, Stephanie M. Greer, Jared M. Saletin, Andrea N. Goldstein-Piekarski, and Matthew P. Walker. 2017. "The sleep-deprived human brain." *Nature Reviews Neuroscience* 18(7):404–418. doi: 10.1038/nrn.2017.55.

Kravitz, Howard M., Imke Janssen, Natette Santoro, Joyce T. Bromberger, Miriam Schocken, Susan A. Everson-Rose, Kelly Karavolos, MA, and Lynda H. Powell. 2005. "Relationship of day-to-day reproductive hormone levels to sleep in midlife women." Archives of Internal Medicine 165(20):2370–2376. doi:10.1001/archinte.165.20.2370.

Kravitz, Howard M., Patricia A. Ganz, Joyce Bromberger, Lynda H. Powell, Kim Sutton-Tyrrell, and Peter M. Meyer. 2003. "Sleep difficulty in women at midlife: a community survey of sleep and the menopausal transition." *Menopause* 10(1):19–28. doi: 10.1097/00042192-200301000-00005.

Kravitz, Howard M., Xinhua Zhao, Joyce T. Bromberger, Ellen B Gold, Karen A Matthews, Martica H Hall, and MaryFran R. Sowers. 2008. "Sleep disturbance during the menopausal transition in a multi-ethnic community sample of women." *Sleep* 31(7):979–990. doi: https://doi.org/10.5665/sleep/31.7.979.

Krystal, Andrew, and Jack Edinger. 2008. "Measuring sleep quality." *Sleep Medicine* 9:S10–17. doi: https://doi.org/10.1016/S1389-9457(08)70011-X.

Lampio, L., P. Polo-Kantola, S.L. Himanen, S. Kurki, E. Huupponen, J. Engblom, O.J. Heinonen, O. Polo, and T. Saaresranta. 2017. "Sleep during menopausal transition: A six-year follow-up." *Sleep* 40. doi: 10.1093/sleep/zsx090.

Lampio, Laura, Tarja Saaresranta, Janne Engblom, Olli Polo, and Päivi Polo-Kantola. 2016. "Predictors of sleep disturbance in menopausal transition." *Maturitas* 94:137–142. doi: 10.1016/j.maturitas.2016.10.004.

Manber, Rachel, and Roseanne Armitage. 1999. "Sex, steroids, and sleep: a review." *Sleep* 22(5):540-55.

Mirer, Anna G., Terry Young, Mari Palta, Ruth M. Benca, Amanda Rasmuson, and Paul E. Peppard. 2017. "Sleep-disordered breathing and the menopausal transition among participants in the Sleep in Midlife Women Study." *Menopause* 24(2):157–162. doi: 10.1097/GME.0000000000000744.

Moreno-Frias, Carmen, Nicté Figueroa-Vega, and Juan Manuel Malacara. 2014. "Relationship of sleep alterations with perimenopausal and postmenopausal symptoms." *Menopause* 21:1017–1022. doi: 10.1097/GME.0000000000000206.

National Institute on Aging. 2017. *What Is Menopause.* Last modified June 27. https://www.nia.nih.gov/health/what-menopause.

Nelson, H. D. 2008. "Menopause." *Lancet* 371:760–770. doi: 10.1016/S0140-6736(08)60346-3.

Nielsen, T., and R. Levin. 2007. "Nightmares: a new neurocognitive model." *Sleep Medicine Reviews* 11(4): 295–310. doi: 10.1016/j.smrv.2007.03.004.

Nowakowski, Sara, and Jessica Meers. 2019. "Cognitive Behavioral Therapy for Insomnia and Women's Health. Sex as a Biological Variable." *Sleep Medicine Clinics* 14:185-197. doi: doi.org/10.1016/j.jsmc.2019.01.002.

O'Connor, K. A., R. Ferrell, E. Brindle, B. Trumble, J. Shofer, D. J. Holman, and M. Weinstein. 2009. "Progesterone and ovulation across stages of the transition to menopause." *Menopause* 16:1178-1187. doi: 10.1097/gme.0b013e3181aa192d.

Ocampo, Ana, Francisco Menéndez, and Montserrat Conde. 2017. "Comparison of affect changes during the ovulatory phase in women with and without hormonal contraceptives." *Heliyon* 3(4):1-18. doi: doi.org/10.1016/j.heliyon.2017.

Ogawa, K, H. Nittono, and T. Hori. 2005. "Brain potentials before and after rapid eye movements: an electrophysiological approach to dreaming in REM sleep." *Sleep* 28:1077–82. doi: https://doi.org/10.1111/j.1479-8425.2008.00358.x.

Ohayon, Maurice M. 2006. "Severe hot flashes are associated with chronic insomnia." *Archives of Internal Medicine* 166(12):1262–1268. doi:10.1001/archinte.166.12.1262.

Parra, M., N. Lagos, M. Levancini, M. Villarroel, E. Pizarro, P. Vanhauwaert, S. Velasco, M. Fernadez, M. Gambacciani, N. Biglia, A. Cagnacci, S. Caruso, E. Cicinelli, V. De Leo, C. Di Carlo, M. Farris, A. Gambera, S. Guaschino, A. Lanzone, A. M. Paoletti, N. Russo, F. Vicariotto, P. Villa, A. Volpe, P. Lavin, M. Lopez, C. Campusano, P. Barriga, and S. Brantes. 2018. "Menopause and Menopause Hormone Therapy Recommendations 2018 of the Gynecological Endocrinology Unit of the German Clinic of Santiago -Italian Society of Menopause and the Chilean Society of Gynecological Endocrinology." *Chilean Journal of Obstetrics and Gynecology* 83:527-550. doi: https://dx.doi.org/10.4067/s0717-75262018000500527.

Pinkerton, Joann V., Lucy Abraham, Andrew G. Bushmakin, Joseph C. Cappelleri, and Barry Komm. 2016. "Relationship between changes in vasomotor symptoms and changes in menopause-specific quality of life and sleep parameters." *Menopause* 23(10):1060–1066. doi: 10.1097/GME.0000000000000678.

Polesel, Daniel N., Camila Hirotsu, Karen T. Nozoe, Andre C. Boin, Lia Bittencourt, Sergio Tufik, Monica L. Andersen, and Helena Hachul. 2015. "Waist circumference and postmenopause stages as the main associated factors for sleep apnea in women: a cross-sectional population-based study." *Menopause* 22(8):835–844. doi:10.1097/GME.0000000000000406.

Rance, Naomi E., Penny A. Dacks, Melinda A. Mittelman-Smith, Andrei A. Romanovsky, and Sally J. Krajewski-Hall. 2013. "Modulation of body temperature and LH secretion by hypothalamic KNDy (kisspeptin, neurokinin B and dynorphin) neurons: a novel hypothesis on the mechanism of hot flushes." *Frontiers in Neuroendocrinology* 4(3): 211–227. doi: 10.1016/j.yfrne.2013.07.003.

Randolph, J. F. Jr., H. Zheng, M. R. Sowers, C. Crandall, S. Crawford, E. B. Gold, and M. Vuga. 2011. "Change in follicle-stimulating hormone and estradiol across the menopausal transition: effect of age at the final menstrual period." *The Journal of Clinical Endocrinology and Metabolism* 96:746–754. doi: 10.1210/jc.2010-1746.

Riemann, Dieter, Christoph Nissen, Laura Palagini, Andreas Otte, Michael L. Perlis, and Kai Spiegelhalder. 2015. "The neurobiology, investigation, and treatment of chronic insomnia." *Lancet Neurology* 14(5):547–558. doi: 10.1016/S1474-4422(15)00021-6.

Rigol, O. 2014. "Gynaecological Semiology." In *Obstetrics and Gynecology,* edited by S. Rigol, 55-62. Havana: Medical Sciences.

Roche Diagnostics FSH, Follicle Stimulating Hormone. 2006–11. *Package Insert: Elecsys and cobas analyzers*. V 13 Roche Diagnostics, Indianapolis, IN.

Romans, Sarah, David Kreindler, Gillian Einstein, Sheila Laredo, Michele Petrovic, and James Stanley. 2015. "Sleep quality and the menstrual cycle." *Sleep Medicine* 16:489-495. doi: 10.1016/j.sleep.2014.12.001

Rosolowich, V., E. Saettler, and B. Szuck. 2006. "Mastalgia: clinical practice guideline" *Journal of Obstetrics and Gynaecology Canada* 28:49-57. doi: https://doi.org/10.1016/S1701-2163(16)32027-8.

Rueda-Beltz, Camilo, María Cecilia Arturo-Rojas, Jorge Cantor-Guarnizo, and Ana María Arias-Galeano. 2016. "Neuroendocrinology: regulation and central action of steroid hormones (hypothalamic-pituitary-ovarian axis, regulation of the synthesis and release of gonadotropins by hypothalamic peptides)." In *Climacteric and Menopause*. Edited by Latin American Federation of Obstetrics and Gynecology Societies (FLASOG), 181-206. Mexico City: Nieto.

Santoro, N., and K. Sutton-Tyrrell. 2011. "The SWAN song: Study of Women's Health Across the Nation's recurring themes." *Obstetrics and Gynecology Clinics of North America* 38:417-423. doi: 10.1016/j.ogc.2011.05.001.

Santoro, N., C. N. Epperson, and S. B. Mathews. 2015. "Menopausal Symptoms and Their Management." *Endocrinology and metabolism clinics of North America* 44:497–515. doi: 10.1016/j.ecl.2015.05.00.

Sateia, Michael. 2014. "International Classification of Sleep Disorders-Third Edition." *Chest* 146(5):1387–1394. doi:10.1378/chest.14-0970.

Scharf, Martin B., Michael D. McDannold, Robin Stover, Nadia Zaretsky, and David V. Berkowitz. 1997. "Effects of estrogen replacement therapy on rates of cyclic alternating patterns and hot-flush events during sleep in postmenopausal women: a pilot study." *Clinical Therapeutics* 19(2):304-311. doi.org/10.1016/S0149-2918(97)80118-X.

Scheyer, O., A. Rahman, H. Hristov, C. Berkowitz, R.S. Isaacson, R. Diaz Brinton, and L. Mosconi. 2018. "Female Sex and Alzheimer's Risk: The Menopause Connection." *The journal of prevention of Alzheimer's disease* 5:225–230. doi: 10.14283/jpad.2018.34.

Schneider, H.P., L.A. Heinemann, H.P. Rosemeier, P. Potthoff, and H.M. Behre. 2000 "The Menopause Rating Scale (MRS): reliability of scores of menopausal complaints." *Climacteric* 3:59-64. doi: 10.3109/13697130009167600.

Schweizer, Edward, George Case, Felipe Garcia-Espana, David J. Greenbladt and Karl Rickels. 1995. "Progesterone co-administration in patients discontinuing long-term benzodiazepine therapy: effects on withdrawal severity and taper outcome." *Psychopharmacology* 117:424-429. doi.org/10.1007/BF02246214.

Shaver, Joan L., Sandra K. Johnston, Martha J. Lentz, and Carol A. Landis. 2002. "Stress exposure, psychological distress, and physiological stress activation in midlife women with insomnia." *Psychosomatic Medicine* 64(5):793–802. doi: http://dx.doi.org/10.1097/01.PSY.0000024235.11538.9A.

Shin, Chol, Sangyeol Lee, Taewook Lee, Kyungrim Shin, Hyeryeon Yi, Kuchan Kimm, and Namhan Cho. 2005. "Prevalence of insomnia and its relationship to menopausal status in middle-aged Korean women." *Psychiatry and Clinical Neurosciences* 59:395–402. doi: https://doi.org/10.1111/j.1440-1819.2005.01391.x.

Shneerson, John M. 2000. *Handbook of sleep medicine*. Oxford: Blackwell Science.

Solomon, Eldra, Linda Berg, and Diana W. Martin. 2008. "Reproduction." In *Biology*, edited by Eldra Solomon, Linda Berg and Diana W. Martin 1051-1079. Mexico City: Mc Graw - Hill.

Sowers, Mary Fran, Huiyong Zheng, Howard Kravitz, Karen Matthews, Joyce Bromberger, Ellen Gold, Jane Owens, Flavia Consens, and Martica Hall. 2008. "Sex steroid hormone profiles are related to sleep measures from polysomnography and the Pittsburgh Sleep Quality Index." *Sleep* 31(10):1339-1349. doi: 10.5665/sleep/31.10.1339.

Stricker, R., R. Eberhart, M.C. Chevailler, F.A. Quinn, and P. Bischof. 2006. "Establishment of detailed reference values for luteinizing hormone, follicle stimulating hormone, estradiol, and progesterone during different phases of the menstrual cycle on the Abbott ARCHITECT analyzer." *Clinical Chemistry and Laboratory Medicine* 44:883–887. doi: 10.1515/CCLM.2006.160.

Studd, J. W. 2011. "A guide to the treatment of depression in women by estrogens." *Climacteric* 14:637–642. doi: 10.3109/13697137.2011.609285.

Tendler, A., and S. Wagner. 2015. "Different types of theta rhythmicity are induced by social and fearful stimuli in a network associated with social memory." *eLife* 4: e03614. doi: 10.7554/eLife.03614.

Thurston, Rebecca C., Nanette Santoro, and Karen A. Matthews. 2012. "Are vasomotor symptoms associated with sleep characteristics among symptomatic midlife women? Comparisons of self-report and objective measures." *Menopause* 19(7):742–748. doi: 10.1097/gme.0b013e3182422973.

Toffol, Elena, Nea Kalleinen, Anna Sofia Urrila, Sari-Leena Himanen, Tarja Porkka-Heiskanen, Timo Partonen, and Päivi Polo-Kantola. 2014. "The relationship between mood and sleep in different female reproductive states." *BMC Psychiatry* 14. doi: 10.1186/1471-244X-14-177.

Tom, S. E., D. Kuh, J. M. Guralnik, and G. D. Mishra. 2010. "Self-reported sleep difficulty during the menopausal transition: results from a prospective cohort study." *Menopause* 17:1128–1135. doi: 10.1097/gme.0b013e3181dd55b0.

Torres-Jiménez, Ana Paola, and José María Torres-Rincón. 2018. "Climacteric and Menopause." *Journal of Faculty of Medicine,* UNAM 61:51-58. doi: http://www.medigraphic.com/pdfs/facmed/un-2018/un182j.pdf.

Valenzuela, M. P., E. Arteaga, R. Pou, and P. Villaseca. 2016. "Reproductive stages in adult women: transition to menopause." In *Climacteric and Menopause*. Edited by Latin American Federation of Obstetrics and Gynecology Societies (FLASOG), 1-14. Mexico City: Nieto.

Vaughn, Bradley, and Peterson Giallanza. 2008. "Technical Review of Polysomnography." *Chest* 134(6):1310–1319. doi:10.1378/chest.08-0812.

Vousoura, Eleni, Areti Spyropoulou, Kalliopi Koundi, Chara Tzavara, Helen Verdeli, Thomas Paparrigopoulos, Areti Augoulea, Irene Lambrinoudaki, and Iannis Zervas. 2015. "Vasomotor and depression symptoms may be associated with different sleep disturbance patterns in postmenopausal women." *Menopause* 22:1053–1057. doi: 10.1097/GME.0000000000000442.

Wildman, R. P., A. B. Colvin, L. H. Powell, K. A. Matthews, S. A. Everson-Rose, S. Hollenberg, J.M. Johnston, and K. Sutton-Tyrrell K. 2008. "Associations of endogenous sex hormones with vasculature in menopausal women: the Study of Women's Health Across the Nation (SWAN)." *Menopause* 15:414-421. doi: 10.1097/gme.0b013e318154b6f5.

Woods, N. F., and E. S. Mitchell. 2010. "Sleep symptoms during the menopausal transition and early postmenopause: observations from the Seattle Midlife Women's Health Study." *Sleep* 33:539–549. doi: 10.1093/sleep/33.4.539.

Woodward, Suzanne, and Robert R. Freedman. 1994. "The thermoregulatory effects of menopausal hot flashes on sleep." *Sleep* 7(6):497–501. doi: https://doi.org/10.1093/sleep/17.6.497.

Xu, Min, Lynda Bélanger, Hans Ivers, Bernard Guay, Junjian Zhang, and Charles M. Morin. 2011. "Comparison of subjective and objective sleep quality in menopausal and non-menopausal women with insomnia." *Sleep Medicine* 12(1):65-69. doi: doi.org/10.1016/j.sleep.2010.09.003.

Xu, Qunyan, and Cathryne Lang. 2014. "Examining the relationship between subjective sleep disturbance and menopause: a systematic review and meta-analysis." *Menopause* 21:1301–1318. doi: 10.1097/GME.0000000000000240.

Xua, Hanna, Rebecca C. Thurstonb, Karen A. Matthews, Cindy L. Bryce, Ron D. Hays, Wishwa N. Kapoor, Roberta B. Ness, and Rachel Hess. 2012. "Are hot flashes associated with sleep disturbance during midlife? Results from the STRIDE cohort study." *Maturitas* 71(1):34–38. doi: 10.1016/j.maturitas.2011.10.003.

Young, T., D. Rabago, A. Zgierska, D. Austin, and F. Laurel. 2003. "Objective and subjective sleep quality in premenopausal, perimenopausal, and postmenopausal women in the Wisconsin Sleep Cohort Study." *Sleep* 26:667–672. doi: 10.1093/sleep/26.6.667.

In: Sleep, Cognition and Emotion
Editors: I. Ramírez Salado et al. ISBN: 978-1-53617-561-5

Chapter 4

AMYGDALA OSCILLATIONS AND PLASTICITY DURING SLEEP

Zeidy Muñoz-Torres[1,2,*], PhD
and Corinne J. Montes-Rodríguez[3,†], PhD
[1]Psicobiología & Neurociencias, Facultad de Psicología,
Universidad Nacional Autónoma de México, Mexico City, Mexico
[2]Grupo Dinámica Neural, Centro de Ciencias de la Complejidad (C3),
Universidad Nacional Autónoma de México, Mexico City, Mexico
[3]Grupo de Plasticidad Sináptica y Ensambles Neuronales,
Centro de Ciencias de la Complejidad (C3),
Universidad Nacional Autónoma de México, Mexico City, Mexico

ABSTRACT

This chapter reviews the role of amygdala oscillations in emotional processing during sleep. Sleep is a dynamic global state during which organisms disconnect from the outside world, though intrinsic mechanisms continue to generate specific oscillations. It has been suggested that these patterns of neural activity may reorganize brain circuits to sharpen the meaning of experiences, consolidate memories and release unnecessary information from the system. A bidirectional relationship between emotion and sleep has been revealed by studies of neuropsychiatric disorders and sleep deprivation, while oscillations during Rapid Eye Movement sleep (REM) have been related to the consolidation and regulation of emotional information. Oscillations during non-REM sleep, in contrast, are associated with declarative and contextual memory consolidation. In both stages, amygdala oscillations may play a central role in consolidation fear memories through coherent activity with other brain structures involved in memory, like

[*] Corresponding Author's: Email: zeimt@comunidad.unam.mx.
[†] Corresponding Author's: Email: montes@unam.mx.

the hippocampus. It is well-known that sleep-dependent synaptic plasticity in the amygdala and hippocampus correlates with the correct acquisition, consolidation and retrieval of fear memories. While many studies show the hippocampus' central role in memory processes, the amygdala's influence on emotional memory processing in relation to sleep is little-explored to date; hence our understanding of underlying mechanisms is insufficient. This review analyzes amygdala oscillations during sleep as the brain mechanism that generates the plasticity associated with emotional memory processing.

Keywords: amygdala, oscillations, sleep, synaptic plasticity, memory, emotion

INTRODUCTION

Oscillations, Plasticity and Sleep

Neural oscillations are periodic fluctuations of neuronal excitability driven by intrinsic mechanisms that place brain information in pertinent time and space domains. Synchronized rhythmic oscillations are considered key elements of brain functioning and have been associated with information processing, memory processes and the expression of various global brain states; for example, sleep vs. wakefulness.

Several brain rhythms have been described. Slow rhythms are involved in long-range, global neural communication, while fast oscillations are related to local interactions (Buzsáki 2011). Delta rhythms fall in the 0.1-4 Hz frequency range and are observed initially during sleep and under anesthesia. Two main cellular sources of delta activity have been described: the thalamus and cortex. Thalamic neurons from distinct sensory, motor, associational and intralaminar nuclei that project into the cortex display delta rhythms spontaneously or by hyperpolarizing current stimuli. In cats, this activity was maintained even after decortication (Steriade et al. 1991). Some delta oscillations survive in the cortex after thalamectomy (Villablanca 1974), and they are present in cortical slices (Sanchez-Vives and McCormick 2000), suggesting that this rhythm also is of cortical generation.

Theta rhythms are observed in the 4-7 Hz band, especially during locomotion, Rapid Eye Movement (REM) sleep and freezing after fear induction. They appear conspicuously in rodents and have been recorded in several limbic system structures: the amygdala, hippocampal formation, entorhinal and perirhinal cortices, and cingulate cortex (Leung and Borst 1987). However, these areas do not generate theta rhythms in isolation; rather, they facilitate their expression by receiving inputs from other brain circuits, primarily fibers from the medial septum and entorhinal cortex (Buzsáki 2011; Vanderwolf and Leung 1983; Hummos and Nair 2015).

Alpha rhythms present frequencies in the 8-13 Hz range and are best detected over posterior head regions during relaxed wakefulness with eyes closed, though they can be reduced or blocked by mental effort and during sleep onset. They may overlap other EEG events, such as sleep spindles, thus accounting for the sigma rhythm. Modules of cortical alpha rhythm generators have been described in visual areas as well as spread across the brain's surface (Cooper et al. 1965).

Beta and gamma rhythms are fast activities in the 20-150 Hz band observed during wakefulness and REM sleep in such areas as the amygdala, hippocampus, striatum, thalamus, hypothalamus and cortex (Buzsáki and Wang 2012). A generation of gamma (30-70 Hz) activity in somatosensory cortical layers II/III has been seen *in vitro*, while beta activity (20-30 Hz) is generated in layer V, and both co-exist in layer IV (Roopun et al. 2006).

Coordinated neuronal activity during specific oscillations opens temporal windows that promote plasticity mechanisms in the brain. Synaptic plasticity is the process that underlies long-term changes in synaptic connections by strengthening or weakening them; changes that correlate with memory formation, consolidation and retrieval (Citri and Malenka 2008). Thus, synaptic plasticity may be the mechanism that re-organizes and shapes neuronal ensembles. At the cellular level, the expression of Immediate Early Genes (IEGs) is one of the most important mediators of activity-dependent synaptic plasticity, while at the network level, the synchronization (coherent activity) of brain structures during specific oscillations is the main mechanism that communicates and transfers information between neurons.

Sleep is a dynamic process with transient activations and changes at all levels: biochemical, metabolic, electrical and magnetic. It has been divided into two stages according to the activity observed (Rechtschaffen and Kales 1968): sleep with rapid eye movements or REM sleep, and sleep without these movements (non-REM). The facts that sleep is conserved along several species, the amount of time spent in this state, and that prolonged sleep deprivation can kill organisms (Rechtschaffen and Bergmann 2002), indicate that this is a critically-important process. Though sleep has long been studied, its functions remain open to debate. One proposal holds that emotional processing and the regulation/consolidation of emotional memories all occur during REM sleep (Hutchison and Rathore 2015; Corsi-Cabrera and Poe, 2014; Walker and van der Helm 2009), while non-REM sleep is related to declarative and contextual memory consolidation (Puentes-Mestril et al. 2019; Corsi-Cabrera and Poe 2014; Wilson and McNaughton 1994). Some authors argue that the sharpening of emotional information may also occur during REM sleep (Sterpenich et al. 2014), accompanied by a decrease in the emotional tone of intense experiences (Walker and van der Helm 2009); in contrast, memory reactivation has been reported during non-REM sleep (Tononi and Cirelli 2019; Wilson and McNaughton 1994). It is widely-accepted that the sharp wave-ripples characteristic of slow-wave sleep promote information transfers between the hippocampus and cortex (McNaughton 2010; Girardeua et al. 2009). During slow-wave sleep, the cerebral cortex displays slow frequencies and sleep spindles characterized by slow oscillations between transient periods of neuronal silence (down states) and periods of intense neuronal firing (up states) (Steriade et al. 1993). This slow oscillatory activity permits the recruitment of neurons over a large neuronal space and is thought to facilitate the transfer of information between structures, thereby mediating the consolidation of memories acquired during wakefulness (Sirota and Buzsáki 2005).

Interestingly, it has been proposed that sleep serves to promote forgetting, cleaning redundant information (Poe 2017; Crick and Mitchison 1983); and maintaining the correct amount of synaptic connections by elimination of unnecessary formed during waking (Cirelli and Tononi, 2019). In fact, REM sleep does prune recent dendritic spines and facilitate subsequent spine formation during development or new learning, but it also strengthens and maintains newly-formed spines (Li et al. 2017; Yang et al. 2014). All this evidence suggests that REM sleep plays a crucial role in balancing and maintaining newly-selected dendritic spines.

Though several theories of sleep function have been proposed, they can be summarized as follows: sleep may function by re-processing information acquired during waking in order to re-organize brain connections. Thus, specific oscillations during sleep and the associated synaptic plasticity processes could shape neuronal ensembles (Puentes-Mestril et al. 2019; Montes-Rodríguez et al. 2006).

In light of this, the aim of this chapter is to review literature on the role of oscillatory activity in the amygdala in sleep-induced plasticity and emotional processing during sleep. The following sections address the relationship between emotions and sleep supported by mood disorders and sleep deprivation studies, discuss neural activity during sleep, and review the neuroanatomy of the amygdala, its functional role and oscillatory activity. It then examines experimental studies that describe the effects of amygdala sleep regulation on emotional processing and plasticity. Finally, it approaches the question of how neuronal ensembles are sculpted by amygdala oscillations during sleep.

Relationship between Emotions and Sleep

Mood Disorders and Sleep

Clinical evidence supports the hypothesis of interaction between emotions and sleep. In fact, almost all psychiatric and neurological disorders that manifest sleep disturbances display symptoms of an affective imbalance (Benca et al. 1992). In major depressive patients, for instance, sleep is characterized by intrusions of periods of wakefulness that can produce insomnia and alter the proportions and patterns of sleep stages. Specifically, slow-wave sleep is diminished but the amount of REM sleep is normal, though its latency is shortened. The first REM episode is longer and accompanied by an increase in transient events, such as eye movements (Borbely and Wirz-Justice 1982). We also know that anxiety-reducing drugs like diazepam are sleep inducers that impact the control of impulsive responses associated with emotional stimuli, decision-making and attention (Muñoz-Torres et al. 2011a; 2011b). This evidence supports an interaction or overlapping of the neurochemical mechanisms of emotions and sleep.

In another vein, a relation among sleep, brain oscillations during sleep and emotional memories has been suggested due to evidence that traumatic events affect sleep. It has been proposed that sleep disturbances (particularly during REM sleep) are a hallmark of post-traumatic stress disorder (PTSD) because, for example, of the negative correlation between relative beta EEG power and PTSD severity, and discontinuous REM episodes duration in PTSD patients (Mellman et al. 2007).

Observations of a rat model of PTSD showed a correlation between the reduction in theta and sigma (10-15 Hz) band power during transitions to REM sleep and impaired fear-associated memory processing (Vanderheyden et al. 2015). In addition, some reports indicate that during REM sleep the theta band (4-8 Hz) is higher in humans who were resistant to developing PTSD than in subjects who met the criteria for PTSD diagnoses (Cowdin et al. 2014). Theta power was higher in the electrode in the right prefrontal cortex in resilient subjects; a finding that concurs with previous results that showed a positive correlation between the amount of REM sleep and right prefrontal-dominant theta power during REM with emotional memory facilitation when subjects were napping vs. non-napping (Nishida et al. 2009).

Taken together, these studies suggest that oscillations during REM sleep may underlie the memory processing of emotional experiences. Specifically, theta activity seems to facilitate emotional memory organization and balance. As mentioned, some researchers suggest that REM-related consolidation may reduce the emotional charge of content but preserve salient components, thus serving an adaptive or therapeutic function (Walker and van der Helm 2009). Although these studies reveal a clear relationship between REM sleep oscillations and memory processing, the underlying brain structures and related synaptic plasticity have been less explored. We discuss this issue in the final part of this chapter.

Sleep Deprivation and Emotional Processing

Measurements of the effects of a reduced or temporary elimination of sleep provide information on the contribution of sleep to various aspects of emotional processing. In fact, studies of total or partial sleep deprivation show a deterioration in the evaluation of emotional stimuli, as in one study one night of total sleep deprivation increased both the negative valence scores assigned to neutral images (Tempesta et al. 2010), and impulsiveness towards negative stimuli on an emotional Go/NoGo task where automatic responses had to be withheld. This suggests a decrease in the ability to inhibit responses (Anderson and Platten 2011).

Recognition of facial emotions is also altered after sleep deprivation as deprived subjects show worse performance in identifying sad faces and slower reaction times in response to all emotional expressions compared to subjects who slept (see Cote et al. 2014). Studies that compared faces to words with positive, negative or neutral valence have found that sleep deprivation deteriorated the recognition of the emotional valence of words more significantly than faces, accompanied by more severe problems in identifying emotionally-negative (sad) than emotionally-positive (happy) faces (Maccari et al. 2014).

An increase in emotional disturbances has also been identified in sleep-deprived subjects based on increases in subjective reports of episodes of irritability, emotional volatility (Horne 1985), and stress, anxiety and anger (Minkel et al. 2014a), with more numerous emotional difficulties during the day, as measured by scales and diaries of emotions (Dinges et al. 1997). People with high levels of anxiety traits are more prone to suffering these emotional disorders when sleep-deprived (Goldstein et al. 2013).

Other work shows that lack of sleep affects responses to emotional stimuli measured in controlled environments. In Franzen and Buysse's study (2008), one night of sleep deprivation increased pupil dilation (an autonomous sympathetic system response) as subjects passively viewed pictures with negative emotional valence. Excessive responses by the sympathetic system are associated with a lack of flexibility, an impaired ability to respond to emotional challenges (Sauvet et al. 2010), and psychopathologies (Appelhans and Luecken 2006).

Similarly, partial sleep deprivation reduced the speed of generating emotional facial expressions when subjects in a study were asked to respond with a movement of facial muscles compatible, or incompatible, with the emotional stimulus presented (Schwarz et al. 2013). Likewise, sleep-deprived participants showed less emotional expressiveness than controls, especially in response to videos with positive content, according to judges' observations (Minkel et al. 2014b).

Several neuroimaging studies have confirmed abnormal activity in brain structures related to emotional processing under conditions of sleep deprivation. After 35 hours of total deprivation, subjects in the study by Yoo et al. (2007) presented activity in the amygdala of greater intensity and extension in response to negative stimuli than controls who slept. This approach also revealed an increase in brain activity to positive stimuli after 32 hours of deprivation, specifically in regions of the reward system (ventral tegmental area and striatum) and the amygdala. Upon analyzing the influence of sleep stages on fear consolidation and extinction using a split-night protocol, Menz et al. (2016) found that sleep episodes rich in REM periods (second half of the night) improved discrimination between neutral and fear stimuli during recall. Slow-wave sleep did not improve recall but, rather, triggered an increase of activity in the amygdala and ventromedial prefrontal cortex. In this vein, after selective REM sleep deprivation, subjects in Rosales-Lagarde et al.'s experiment (2012) showed an increase of impulsive defensive responses to threatening stimuli. In addition, a reduction of functional connectivity between the amygdala and frontal regions associated with the executive functions (Shao et al. 2014), medial prefrontal cortex (Yoo et al. 2007), orbitofrontal cortex, and fusiform gyrus was observed (Gujar et al. 2011), resulting in the emission of inappropriate emotional responses to the context (Sotres-Bayon et al. 2004). These results support the role of REM sleep in re-establishing control of the amygdala through the higher cortical regions and in promoting an optimal emotional state to face the demands of the following day.

Sleep Deprivation in Animals

Several studies have effectively demonstrated that sleep manipulations like REM (Rossi et al. 2014; Alvarenga et al. 2008; Silva et al. 2004a; 2004b) or total sleep deprivation (Fernandes-Santos et al. 2012; Hagewoud et al. 2011; Patti et al. 2010; Graves et al. 2003) can impair the acquisition, consolidation and retrieval of contextual and fear memories.

Other research sustains that IEGs like c-Fos and Arc act as mediators of memory consolidation processes (Datta et al. 2008; Fleischmann et al. 2003; Guzowski 2002; Guzowski et al. 2001) and are reactivated during contextual fear memory recall (Besnard et al. 2014). Moreover, IEGs expression in the basolateral amygdala (BLA) seems to be essential for the retrieval and persistence of reactivated contextual fear memories (Nakayama et al. 2016; Hall et al. 2001). In this regard, IEGs could be the cellular mediators of fear memory consolidation during sleep. In this case, sleep deprivation would impair the fear memory evocation-induced IEGs expression in the amygdala. Though few studies exist in this field, Hagewoud et al. (2011) found a decrease in pCREB expression in CA1 and the amygdala associated with impairment of contextual fear memory consolidation by total sleep deprivation. There, sleep deprivation was applied after fear acquisition, so the question remains of whether this affected the consolidation or retrieval of contextual fear memory while plasticity in the hippocampus and amygdala was impaired. A recent study by Montes-Rodríguez et al. (2019) showed that 24 h of total sleep deprivation impairs the acquisition, consolidation and retrieval of contextual fear memory, and the retrieval-induced increase of c-Fos in the hippocampus and BLA, although sleep deprivation before retrieval only impaired c-Fos expression in the BLA (Figure 1), suggesting a crucial role of sleep in fear memory retrieval and the associated increase of c-Fos in the BLA.

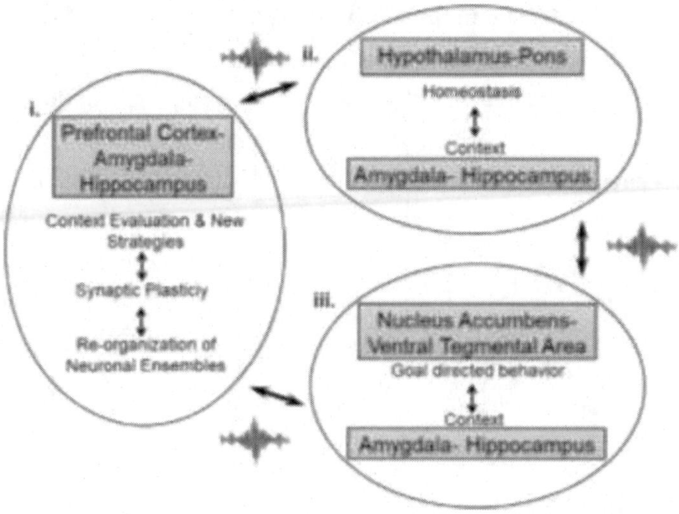

Figure 1. Total sleep deprivation impairs the retrieval-induced increase of cFos in the basolateral amygdala. Almost no c-Fos signal (arrowhead) is visible in the basolateral amygdala of the home cage animals (Home Cage); c-Fos expression in the basolateral amygdala increased after retrieval of the contextual fear memory (Control); 24 h of total sleep deprivation impaired the retrieval-induced increase of cFos in the basolateral amygdala when sleep deprivation was performed before acquisition (TSD-Acquisition), during consolidation (TSD-Consolidation), or before retrieval (TSD-Retrieval) on a contextual fear task.

The influences between sleep and emotional processing run in both directions. On the one hand, emotional experiences during the day impact architecture and brain activity patterns during sleep; on the other, the quantity and quality of sleep modulates the emotional perception of stimuli –by increasing or decreasing their valence– and the behavioral responses directed to facial stimuli that can be negative/threatening or positive/reinforcing. Sleep thus seems to be important for accurately evaluating and discriminating emotional stimuli, acquiring new information, and memory consolidation and retrieval. At the cellular level, sleep-induced synaptic plasticity in the amygdala appears to be significant for fear memory consolidation and retrieval. Given this evidence, the mechanism proposed for achieving emotional processing during sleep is the oscillatory activity of limbic structures, specifically the amygdala.

Neuroimaging Sleep Activity

Studies of metabolic activity during sleep in healthy subjects have been conducted with neuroimaging techniques that demonstrate an overlapping of brain networks involved in emotional processes and sleep with the activity of the limbic and paralimbic systems, two regions involved in emotional processes. Positron emission tomography (PET) studies that measure the distribution of marked radioisotopes involved in neural communication have explored activity changes associated with different alert states, revealing a global reduction from wakefulness to sleep estimated at around 40% of brain metabolism (Maquet et al. 1990), mainly in the frontal, temporal and occipital association cortices, the primary visual cortex, and sub-cortical regions like the anterior and dorsomedial thalamic nuclei, brainstem and forebrain (Dang-Vu et al. 2010; Nofzinger et al. 2002). However, when the absolute metabolism is controlled, relative metabolism increases from wakefulness to sleep in regions linked to emotional processing, such as the amygdala, anterior cingulate gyrus, ventral striatum, hypothalamus, forebrain and pontine tegmentum (Nofzinger et al. 2002).

Regarding REM sleep compared to wakefulness and non-REM sleep, studies have identified an increase in activity in regions related to emotion, including the amygdala, medial prefrontal cortex, and anterior cingulate gyrus (Nofzinger et al. 1997; Maquet et al. 1996). But REM sleep is also marked by a deactivation of the regions involved in cognition and executive processes; for example, the dorsolateral prefrontal cortex and inferior parietal cortex (Maquet et al. 2005; 1996; Nofzinger et al. 1997). This result was confirmed by measuring functional connectivity using quantitative electroencephalography (EEG) (Perez-Garci et al. 2001), which showed a disconnection between prefrontal/cognitive and posterior/sensory brain regions during REM sleep. It has been suggested that this configuration of brain activity during REM sleep might explain the typically illogical, bizarre content of dreams (Muzur 2005). Though dreams also occur in non-REM sleep, their content is simpler and less emotionally intense than during REM sleep (Oudiette et al. 2012), a result consistent with the decreased activity in structures related to the emotions observed in non-REM compared to REM sleep.

The Amygdala

Structural Organization

The amygdala, or amygdaloid complex, is formed by a group of nuclei in the dorsomedial portion of the temporal lobe, in front of and above the inferior horn of the lateral ventricle, located rostral to the hippocampus and caudally at the tail of the caudate nucleus (Parent 1996). It is divided into three main nuclear groups called corticomedial, central and basolateral. The corticomedial nuclear group corresponds to the dorsomedial region of the amygdala in humans, and is sub-divided into the anterior amygdala area, the nucleus of the lateral olfactory tract, the medial amygdaloid nucleus, and the cortical amygdaloid nucleus, with the dorsal part pertaining to the periamygdaloid cortex (Parent 1996). The basolateral (BLA) nuclear group is the largest and best-differentiated part of the amygdala in humans. It is divided into the lateral amygdaloid nucleus, the basal amygdaloid nucleus, and the accessory basal amygdaloid nucleus. The central amygdaloid nucleus, finally, in conjunction with the bed nucleus of the stria terminalis and the subventricular portion of the substantia innominata forms the extended amygdala (De Olmos 1990).

The amygdala has broad connections with numerous brain regions (Aggleton et al. 1980). It contains two main sub-cortical fiber systems: the ventral amygdalofugal pathway and the stria terminalis. The former emerges from the dorsomedial surface of the amygdala and extends rostromedially below the lenticular nucleus. Its fibers penetrate the lateral pre-optic area, Broca's diagonal band and the septal region. After passing through the hypothalamic and preoptic regions, the fibers of these systems are incorporated into the lower thalamic peduncle before terminating in the magnocellular part of the dorsomedial nucleus of the thalamus (Krettek and Price 1977). The stria terminalis sends fibers that pass into, and terminate in, the bed nuclei of the stria terminalis, located lateral to the columns of the fornix (Parent 1996). Another group of sub-cortical axons reaches the anterior hypothalamic area, some of which join the medial forebrain bundle.

The amygdala receives sensory information from the thalamus and sensory cortices and sends fibers to establish connections with the hippocampus, basal nuclei, forebrain, hypothalamus, peri- and entorhinal cortices, and prefrontal cortex, among other areas (LeDoux 2007; Amaral et al. 2003). This anatomical configuration has led to the hypothesis –later confirmed empirically– that the amygdala plays a central role in integrating emotional processes with other cognitive, social and biological functions (Pitkänen et al. 1997).

Functions of the Amygdala

The amygdala is involved in many processes, including appetitive behaviors like affiliation, sex and drug abuse, but its role in various properties of fear behavior has been widely explored. The central nucleus regulates fear responses, including the signal to release cortisol through the paraventricular nucleus of the hypothalamus that, in communication with the midbrain, intensifies the startle response and modulates the autonomic nervous system through the lateral hypothalamus (Davis 1992). The lateral and basal amygdala nuclei are involved in associative learning by receiving projections from sensory areas (auditory, visual), and establishing an association between previously neutral and aversive stimuli (e.g., electric shock) that results in the acquisition of conditioned fear. Studies have found that the amygdala also modulates the fear response in humans, as brain imaging techniques reveal amygdala activity associated with fearful stimuli like fear-conditioned cues, threatening faces and fear-inducing images, as well as positive stimuli (Sergei et al. 2008; Phan et al. 2002).

Network oscillations in the amygdala's circuits have been related to the storage, consolidation and retrieval of emotional and appetitive experiences. During fear memory retrieval, for example, theta synchrony between the lateral amygdala and CA1 in the hippocampus is observed, but when a threat is perceived, theta power increases, activity among the BLA, mPFC and CA1 becomes coherent, and the fast gamma band is phase-locked to theta oscillation. In contrast, when a safe situation is perceived, synchronization of fast gamma power in the mPFC, BLA and CA1 increases. Also, when a conditioned stimulus predicts a reward, slow gamma increases and coherence among the BLA, ventral striatum and entorhinal cortex is observed (for a review, see Bocchio et al. 2017). These interactions show that synchronization between specific brain structures and the amygdala underlies various emotional neuronal processes. As we have seen, the amygdala is considered the core of emotional processing, so if amygdala oscillations during sleep play a role in emotional regulation, then alterations in mood and sleep mechanisms will arise.

Amygdala Oscillations during Sleep

In humans, intracranial recordings offer outstanding techniques that allow direct access to the electrical activity of neuronal responses with high temporal and spatial resolution. Analyses of simultaneous polysomnography EEG and amygdala activity in patients undergoing pre-operative evaluation during an entire night revealed significantly higher delta, theta and sigma oscillations during slow-wave sleep than wakefulness and REM sleep (Figure 2), indicating that the amygdala is subject to similar influences as other brain structures during non-REM sleep (Muñoz-Torres et al. 2018). Unexpectedly, the activity patterns observed in the amygdala during wakefulness and REM sleep were similar and show that the amygdala is not tonically more activated during REM sleep than wakefulness, but receives transient phasic activation likely arising from pontine activation associated with ponto-geniculo-occipital activity (PGO) or P-waves, which are oscillations seen during REM sleep that have been related to sleep-induced synaptic plasticity (Datta et al. 2013; 2008; Datta 2000). In fact, analyses of these patients' eye movements during REM sleep revealed an associated increase in gamma absolute power in the amygdala (Corsi-Cabrera et al. 2016), a change that did not occur during the eye movements typical of wakefulness. Magnetoencephalographic (MEG) studies also report an increased current density in the amygdala several milliseconds before the onset of REM sleep eye movements (Ioannides et al. 2004).

Andrillon et al. (2015) designed a study to determine the relation between eye movements during REM sleep and activity in visual-mnemonic brain regions, and to compare those activity patterns with the ones that appear in scanning images during wakefulness. By performing intracranial EEG recordings of epileptic patients, they found robust evoked potentials in medial temporal lobe regions, including the amygdala, that suggest a similar modulation of neuronal activity in relation to eye movements during sleep and wakefulness.

Other transient changes in electrical activity recorded in the human amygdala during non-REM sleep are sleep spindles; that is, brief oscillations that emerge, increase and decrease in amplitude with an approximate duration of 1 s at typical frequencies of 11-16 Hz. Sleep spindles have been linked to memory and learning facilitation (Ulrich 2016), greater brain plasticity, and enhanced functioning after brain injuries (Urakami 2012). In this regard, Giaquinto (1973) showed that these spindles appear with a caudo-rostral gradient that was more frequent in posterior than anterior regions of limbic structures, with none in the uncus, and very few in the amygdala, though they were frequent in the hippocampal gyrus and mid-portions of the hippocampus. This spindle gradient was confirmed by Andrillon et al. (2011), who observed more frequent sleep spindles in neocortical regions, fewer in the hippocampus and parahippocampal gyrus, and only rarely in the amygdala and entorhinal cortex.

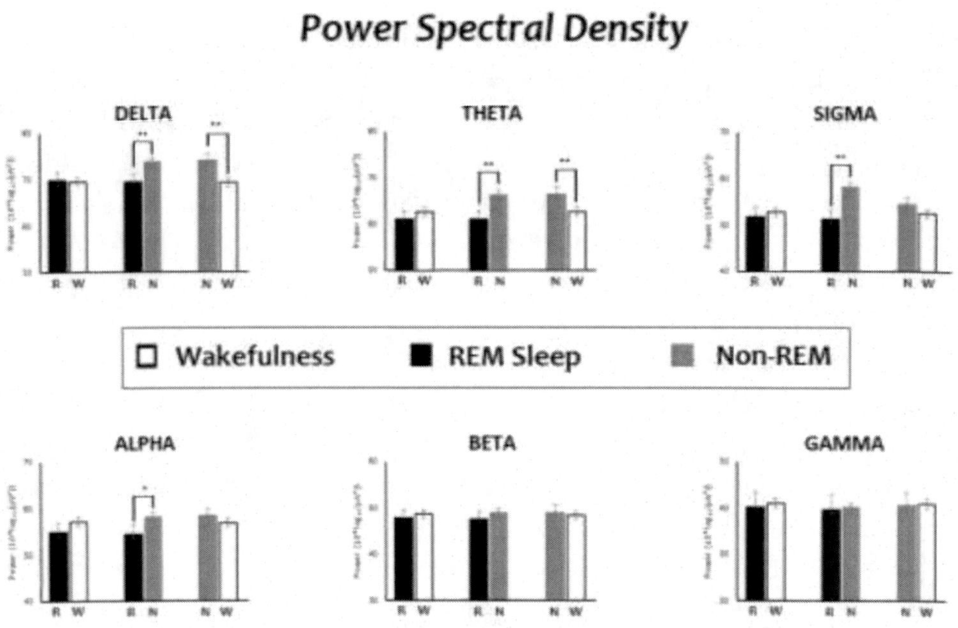

Figure 2. Mean power spectral density and standard errors of the amygdala activity for each broad band among brain states: wakefulness (W) white; REM sleep (R) black; and non-REM sleep (N) gray. Brackets and asterisks indicate statistical differences ** $p < 0.01$, * $p < 0.05$.

Few studies, however, have explored the relationship between sleep spindles and emotional information processing. Ward et al. (2014) found that the density of sleep spindles on the night in which participants classified images according to their valence correlated positively with the recognition of emotional images 6 days later. Also, an increase in the density of sleep spindles after administration of zolpidem and sodium oxybate improved the discrimination of stimuli classified as emotionally intense and negative (Kaestner et al. 2013). That research indicates the probable participation of elements of non-REM sleep in the processes that underlie emotional memory, possibly as a way to facilitate consolidation. Finally, a high percentage (87% ± 11%) of neurons in the amygdala showed significant phase-locking spike activity to slow waves during sleep (< 1Hz) by means of simultaneous scalp EEG, intracerebral EEG, and unit firing recordings of medicated epilepsy patients (Nir et al. 2011).

In animals, Paré and Gaudreau (1996) described that BLA cells are virtually silent during waking but increase their firing rate during slow-wave sleep. They also found a clear modulation of spiking activity in the amygdala due to cortical delta and theta oscillations during sleep. In slow-wave sleep, many types of neurons in the lateral nucleus displayed an increased probability of firing during the depth-negative component of delta, while in the basolateral nucleus bursting neurons fired preferentially during the depth-positive phase of delta, while the fast-firing ones did so in the negative phase. During REM sleep, bursting cells discharged mainly during the negative phase of entorhinal theta waves, whereas fast-firing cells fired preferentially during the depth-positive phase.

Amygdala Oscillations and Plasticity during Sleep

Amygdala, Sleep and Emotional Memory

The role of the amygdala in sleep-regulated emotional memory has been documented by BLA inactivation studies. Wellman et al. (2014) showed that REM sleep decreased in a rat model of PTSD because inactivation of the BLA before aversive training (i.e., when fear was learned) blocked the post-training reduction of REM sleep, and inactivation after context re-exposure also returned REM sleep to normal levels with attenuated freezing. Other studies report individual differences in REM disturbances after shock-training and re-exposure to an aversive context. There, some animals exhibited decreased REM sleep (vulnerable), but others showed no such alteration of REM sleep (resilient). REM and stress alterations were later restored by BLA inactivation in the vulnerable animals (Wellman et al. 2017; 2016), indicating that BLA activity is crucial to sleep alterations induced by experiences of fear.

Regarding the role of oscillations in the amygdala involved in fear-induced sleep alterations, research shows that chronic immobilization stress induces a biphasic change in REM sleep, as some animals exhibited enhanced REM sleep, but others showed a decrease. In the experimental animals with enhanced REM sleep, synchronization between the hippocampus and lateral amygdala in the theta frequency decreased 21 days after chronic immobilization stress (Hedge et al. 2011); specifically, the coherence of theta activity between the lateral amygdala and hippocampus (CA1 and CA3) and CA3-CA1 (Hedge et al. 2011). These findings suggest that a decrease of synchronized theta oscillations in the hippocampus and amygdala during REM sleep may be associated with anxiety disorders.

Sampath et al. (2014) showed that maternal separation and isolation stress during the hyporesponsive stress period produced generalized fear with increased fear memory retention, and increased both REM sleep and total sleep time with changes in neural synchronization in hippocampal-amygdala-cortical loops during theta, beta and gamma oscillations. Maternal separation increased coherent activity in theta (4-12 Hz) but decreased it in the beta (12-20 Hz) and gamma (20-40 Hz) frequency bands. Directional coupling was enhanced from CA1 to the infralimbic cortex in theta and beta oscillations, from the lateral amygdala to the infralimbic cortex in theta and beta band oscillations, and from CA1 to the lateral amygdala in theta band oscillations alone. Those authors sustained that synchronization during enhanced REM sleep was related to a stressful environment that may increase the unidirectional flow of information from the amygdala and hippocampus to the infralimbic cortex. This suggests that changes in behavioral fear responses can be explained by alterations of coherent oscillations during REM sleep among the amygdala, hippocampus and prefrontal cortex, induced by fear/stressful experiences.

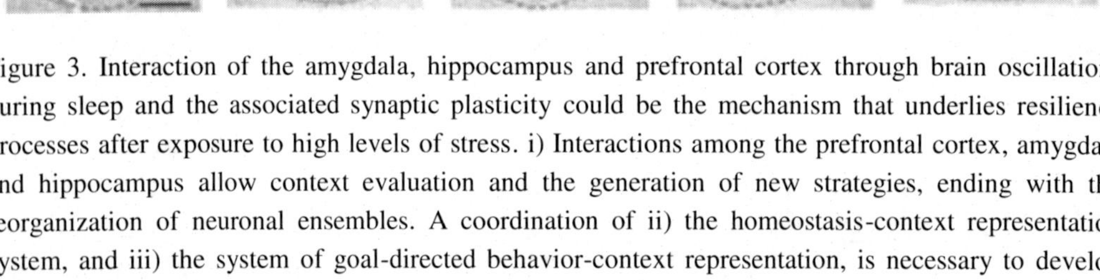

Figure 3. Interaction of the amygdala, hippocampus and prefrontal cortex through brain oscillations during sleep and the associated synaptic plasticity could be the mechanism that underlies resilience processes after exposure to high levels of stress. i) Interactions among the prefrontal cortex, amygdala and hippocampus allow context evaluation and the generation of new strategies, ending with the reorganization of neuronal ensembles. A coordination of ii) the homeostasis-context representation system, and iii) the system of goal-directed behavior-context representation, is necessary to develop resilience. These interactions may be mediated by oscillatory activity during sleep (Figure based on Montes-Rodríguez and Urteaga-Urías 2018).

All these data indicate that coherent activity among the amygdala, hippocampus and prefrontal cortex during sleep may constitute the mechanism through which vulnerability and resilience processes develop. This interaction was suggested previously by Montes-Rodríguez and Urteaga-Urías (2018) but during wakefulness. It may be, then, that the occurrence of this synchronized activity during sleep presents a cue to become resilient or vulnerable after a highly-stressful experience (Figure 3).

Sleep Oscillations, Plasticity and Emotional Memory

The presence of PGO in the amygdala was first suggested by Calvo and Fernández-Guardiola (1984), and Datta (2000) described that in rats fear memory consolidation and retention increase the amount of REM sleep, transitions of non-REM to REM sleep, and P-waves. It should be noted that PGO/P-waves density correlated with improvements on fear task performance; suggesting that P-waves are oscillations that facilitate sleep-dependent behavioral plasticity. An increased synchronization between hippocampal and amygdala theta waves has been associated with an increase of P-waves. Also, a frequency increase of theta waves in the amygdala and hippocampus positively correlates with an increase of P-waves density (Karashima et al. 2010). These findings suggest that synaptic plasticity associated with P-waves during REM sleep and the coordination of the hippocampus and amygdala in the theta frequency may open temporal windows for emotional memory processes.

Additional evidence of sleep oscillation-induced plasticity and emotional memory is the correlation of the consolidation of fear extinction with increased REM sleep and P-waves in rats (Datta and O'Malley 2013). This correlation included animals that were not naturally able to extinguish a fear memory and others without the extinction protocol, leading to the suggestion that the extinction of an emotional memory is strongly-related to specific oscillations during sleep that, presumably, will induce the plasticity changes that underlie emotional memory consolidation; that is, P-waves. In fact, learning a fear task (e.g., two-way active avoidance) can activate the P-waves generator and increase the expression of the plasticity-related genes pCREB, ARC, BDNF and Egr-1 in the dorsal hippocampus and amygdala (Datta et al. 2008). Meanwhile, the elimination of the P-waves generator using ibotenic acid abolished P-waves and decreased the fear learning-induced expression of plasticity genes in the hippocampus and amygdala. That study suggests that P-waves during REM sleep do not affect the normal acquisition process, but may be important in memory consolidation and the related expression of plasticity-related genes in the hippocampus and amygdala.

Sleep oscillation-induced plasticity in the amygdala was explored elegantly by Popa et al. (2010), who showed that coherent theta activity occurs intermittently during REM sleep in the BLA, medial prefrontal cortex and hippocampus, though individual variations in fear memory consolidation (high or low freezing time during evocation) correlated strongly with changes in theta coherence from pre-to-post-training REM sleep episodes for the BLA-medial prefrontal cortex and BLA-hippocampus, but not for the hippocampus-medial prefrontal cortex. These correlations were not found in other frequency bands or vigilance states, indicating that coherent activity associated with a fear memory consolidation is attributable to REM sleep. Moreover, those authors evaluated the directionality of these interactions using Granger causality analysis and theta entrainment of the BLA and medial prefrontal cortex by the hippocampus before and after conditioning was found. The directionality during theta oscillations among the three structures after training was related to the efficacy of the consolidation process. During post-training REM sleep episodes, the rats with stronger influences from the hippocampus over BLA theta, or of the BLA over the medial prefrontal cortex theta, had higher efficacy after overnight consolidation. In contrast, the directionality of the hippocampus-medial prefrontal cortex theta interactions was not predictive of consolidation efficacy; suggesting a crucial role for amygdala theta oscillations and their synchronization with the medial prefrontal cortex and hippocampus during REM sleep to promote fear memory consolidation. Here, the conclusion is that memory consolidation and sleep oscillation-induced amygdala plasticity involve more than one brain structure and, therefore, should be thought of as a process that requires the coordination of several regions (here, at least the amygdala, medial prefrontal cortex and hippocampus) where oscillations along with amygdala plasticity play a crucial role.

Amygdala Oscillations and Plasticity during Sleep

Most studies that have explored sleep oscillation-induced plasticity relate REM sleep to emotional memory consolidation, but work on memory reactivation indicates that reactivation occurs mainly during non-REM sleep, specifically when sharp wave-ripples occur. The coordinated reactivation of the hippocampus and amygdala during non-REM sleep after contextual fear training was demonstrated recently by Buzsáki's group (Girardeau et al. 2017). In their work, rats were trained on a contextual fear memory task where they learned to associate a place with an aversive stimulus. This type of task includes emotional and spatial components that depend on amygdala and hippocampus activity, respectively. That study included multiple unit recordings in the hippocampus, BLA and piriform cortex to capture neuronal ensembles. Data showed that the hippocampus-BLA ensembles were reactivated during non-REM sleep, but that the hippocampus-piriform cortex and hippocampus-central amygdala did not present reactivation of the contextual fear memory. Furthermore, a small sub-sample of pyramidal ripple-modulated BLA cells exhibited the largest reactivation. Interestingly, hippocampus-BLA interaction during sharp-wave ripples was able to reactivate the aversive trajectory but not the safe one. These data show that non-REM sharp wave-ripples are specific time windows where the place-threat association is replayed during sleep. Girardeau et al.'s (2017) study indicates that the reactivation of hippocampus-BLA interaction during sleep allows a union of the representation of localization and threat to stabilize the new contextual fear memory.

Tripathi et al. (2018) observed an increase of slow-wave activity during non-REM sleep and in theta waves during REM sleep in the central amygdala after appetitive-conditioning training and testing. Moreover, the increase of oscillatory waves (slow wave activity and theta) in the central amygdala and parietal cortex correlated positively with performance; indicating a crucial role for the amygdala and cortical oscillations during sleep on the consolidation of an appetitive-conditioned task. Although memory evocation correlated positively with the amount of REM sleep, the amount of non-REM sleep was not affected, and only changes in specific oscillatory activity were observed. These results suggest that the amount of sleep does not necessarily reflect learning and memory-induced plasticity processes during sleep. Indeed, it seems more likely that the acquisition and consolidation of new information depend on communication among specific brain structures during precise global neuronal states, such as sleep, that, in fact, has been considered as the function of brain oscillations.

Finally, Totty et al. (2017) recorded local field potentials from the pre-limbic and infra-limbic prefrontal cortex, ventral hippocampus and lateral amygdala to evaluate synchronization and phase differences among these structures during sleep and after fear conditioning and extinction. Strong synchrony was seldom observed during wakefulness, but during non-REM and REM sleep synchrony appeared in the delta and theta bands, respectively. Synchronization during non-REM sleep in the delta band was observed between the lateral amygdala and limbic cortices (infra- and prelimbic), the dorsal and ventral hippocampus and prelimbic cortex, and the prelimbic and infralimbic cortices; whereas during REM sleep –in addition to those structures– synchronization was observed among the lateral amygdala, infralimbic cortex and ventral and dorsal hippocampus, and within the hippocampus (ventral and dorsal hippocampus). Moreover, delta rhythms were more synchronized within the cortex (infralimbic cortex-prelimbic cortex) and between specific cortical and sub-cortical regions (infralimbic cortex with the lateral amygdala, prelimbic cortex with the lateral amygdala and dorsal and ventral hippocampus) during non-REM sleep than REM sleep and wakefulness, while theta rhythms showed higher synchronization among sub-cortical structures during REM sleep. It is important to note that the mean phase difference between the lateral amygdala and ventral hippocampus predicted individual changes in freezing behavior following fear extinction; that is, animals with reduced freezing after extinction exhibited $< 180°$ phase difference (out-of-phase), whereas the ones that displayed higher freezing had a 0/360° phase lag difference (in-phase). Since it has been suggested that in-phase synchronization facilitates neuronal plasticity, and that the phase of neural input (peak vs. through) determines synapse strengthening (long-term potentiation) or weakening (long-term depression), those authors proposed that theta rhythms during REM sleep modulate synaptic plasticity by dynamically modulating temporal coding and spike-timing-dependent plasticity.

Conclusion

We have reviewed experimental evidence suggesting that amygdala oscillations during sleep and related synaptic plasticity may be the mechanism through which sleep promotes learning and the consolidation and retrieval of emotional experiences. Since oscillations open spatiotemporal windows to establish functional connectivity between specific neural circuits, synchrony among regions forms dynamic ensembles that facilitate a contextual/declarative or emotional memory, depending on the specific regions that become synchronized (e.g., the dorsal and ventral hippocampus with the amygdala).

The characteristic oscillations observed during non-REM sleep, such as sharp wave-ripples, facilitate a replay in the hippocampus and prefrontal cortex. Acting through thalamo-cortical circuits, sleep spindles enable the consolidation independently of the hippocampus, while the slow activity during slow-wave sleep participates in the down-scaling of redundant connections. Activities observed during REM sleep, such as PGO and theta waves, promote a flexible amygdala-hippocampus-medial prefrontal cortex network that may contribute to the integration of new information, its reorganization and generalization, together with an emotional re-evaluation and emotional adjustment.

The orchestration of all these oscillatory elements with amygdala activity may allow the replay of relevant memories that enable a sharpening of the different memories to build a meaning that remains integrated in cortical circuits. The temporal coupling of activity in the amygdala, hippocampus and prefrontal cortex during sleep may, therefore, be a primary mechanism that binds contextual and emotional memories into brain ensembles that contribute to survival.

Acknowledgments

We thank Paul Kersey for correcting the English version of the manuscript. This work was financed partially by Universidad Nacional Autónoma de México, PAPIIT: IA208018.

References

Aggleton, J. P., Burton M. J., & Passingham, R. E. (1980). "Cortical and Subcortical Afferents to the Amygdala of the Rhesus Monkey (Macaca Mulatta)." *Brain Research* 190: 347–68. https://doi.org/10.1016/0006-8993(80)90279-6.

Alvarenga, Tathiana A., Camilla L. Patti, Monica L. Andersen, Regina H. Silva, Mariana B. Calzavara, Giorgia B. Lopez, Roberto, Frussa-Filho, & Sergio Tufik. (2008). "Paradoxical sleep deprivation impairs acquisition, consolidation, and retrieval of a discriminative avoidance task in rats." *Neurobiology of Learning and Memory 90*: 624-32. https://doi.org/10.1016/j.nlm.2008.07.013.

Amaral, David G., Hamraz Behniea, & Jennifer L. Kelly. (2003). "Topographic Organization of Projections from the Amygdala to the Visual Cortex in the Macaque Monkey." *Neuroscience* 118 (4): 1099–120. http://www.ncbi.nlm.nih.gov/pubmed/12732254.

Anderson, Clare, & Charlotte R. Platten. (2011). "Sleep Deprivation Lowers Inhibition and Enhances Impulsivity to Negative Stimuli." *Behavioural Brain Research* 217 (2): 463–66. https://doi.org/10.1016/j.bbr.2010.09.020.

Andrillon, Thomas, Yuval Nir, Chiara Cirelli, Guilio Tononi, & Itzhak Fried. (2015). "Single-neuron activity and eye movements during human REM sleep and awake vision." *Nat Commun.* 6: 7884. https://doi.org/10.1038/ ncomms8884.

Andrillon, Thomas, Yuval Nir, Richard J. Staba, Fabio Ferrarelli, Chiara Cirelli, Giulio Tononi, G., et al. (2011). "Sleep spindles in humans: insights from intracranial EEG and unit recordings." *J Neurosci.* 31: 17821–34. https://doi.org/10.1523/JNEUROSCI.2604-11.2011.

Appelhans, Bradley M., & Linda J. Luecken. (2006). "Heart Rate Variability as an Index of Regulated Emotional Responding." *Review of General Psychology* 10 (3): 229–40. https://doi.org/10.1037/1089-2680.10.3.229.

Benca, R. M., Obermeyer, W. H., Thisted, R. A., & Gillin, J. C. (1992). "Sleep and Psychiatric Disorders. A Meta-Analysis." *Archives of General Psychiatry 49* (8): 651–68; discussion 669-70. http:// www.ncbi.nlm.nih.gov/pubmed/1386215.

Besnard, Antoine, Serge Laroche, & Jocelyne Caboche. (2014). "Comparative dynamics of MAPK/ERK signalling components and immediate early genes in the hippocampus and amygdala following contextual fear conditioning and retrieval." *Brain Structure and Function* 219 (1): 415-30. https://doi.org/10.1007/s00429-013-0505-y.

Bocchio, Marco, Sadegh, Nabavi, & Marco Capogna. (2017). "Synaptic Plasticity, Engrams, and Network Oscillations in Amygdala Circuits for Storage and Retrieval of Emotional Memories." *Neuron* 94: 731-43. doi: 10.1016/j.neuron.2017.03.022.

Borbely, Alexander A., & Anna Wirz-Justice. (1982). "Sleep, Sleep Deprivation and Depression. A Hypothesis Derived from a Model of Sleep Regulation." *Human Neurobiology*, 1: 205–10.

Buzsáki, György, & Xiao-Jing Wang. (2012). "Mechanisms of gamma oscillations." *Annu. Rev. Neurosci.* 35: 203–25.

Buzsáki, György. (2011). Rhythms of the Brain. New York: Oxford University Press.

Calvo, Jose María, & Augusto Fernández-Guardiola. (1984). "Phasic activity of the basolateral amygdala, cingulate gyrus, and hippocampus during REM sleep in the cat." *Sleep* 7 (3): 202-10.

Cirelli, Chiara., & Giulio Tononi. (2019). "Linking the need to sleep with synaptic function." *Science* 366 (6462): 189-90. doi: 10.1126/science.aay5304.

Citri, Ami, & Robert C. Malenka. (2008). "Synaptic plasticity: multiple forms, functions, and mechanisms." *Neuropsychopharmacology 33*: 18-41. https://doi.org/10.1038/sj.npp.1301559.

Cooper, Ray, A. L. Winter, H. J. Crow, & W. Grey Walter. (1965). "Comparison of subcortical, cortical and scalp activity using chronically indwelling electrodes in man." *Electroencephalogr Clin Neurophysiol.* 18: 217-28.

Corsi-Cabrera, María, & Gina R. Poe. (2014). "The role of sleep in processing emotional and contextual information: from mechanism to impact on everyday life and emotional health." *Experimental Brain Research* 232: 1399. https://doi.org/10.1007/s00221-014-3940-y.

Corsi-Cabrera, María., Francisco Velasco, Yolanda del Río-Portilla, Jorge L. Armony, David Trejo-Martínez, Miguel A. Guevara, & Ana L. Velasco. (2016). "Human Amygdala Activation during Rapid Eye Movements of Rapid Eye Movement Sleep: An Intracranial Study." *Journal of Sleep Research* 25 (5): 576–82. https://doi.org/10.1111/jsr.12415.

Cote, K. A., C. J. Mondloch, V. Sergeeva, M. Taylor, & T. Semplonius. (2014). "Impact of total sleep deprivation on behavioural neural processing of emotionally expressive faces." *Experimental Brain Research* 232 (5): 1429–42. https://doi.org/10.1007/s00221-013-3780-1.

Cowdin, Nancy, Ihori Kobayashi, & Thomas A. Mellman. (2014). "Theta frequency activity during rapid eye movement (REM) sleep is greater in people with resilience versus PTSD." *Experimental Brain Research* 232 (5): 1479-85. doi: 10.1007/s00221-014-3857-5.

Crick, Francis, & Graeme Mitchison. (1983). "The function of dream sleep." *Nature* 304 (5922): 111-4. doi: 10.1038/304111a0.

Dang-Vu, Thien Thanh, Manuel Schabus, Martin Desseilles, Virginie Sterpenich, Maxime Bonjean, & Pierre Maquet. (2010). "Functional Neuroimaging Insights into the Physiology of Human Sleep." *Sleep* 33 (12): 1589–603.

Datta, Subimal, & Matthew W. O'Malley. (2013). "Fear extinction memory consolidation requires potentiation of pontine-wave activity during REM sleep." *Journal of Neuroscience* 33 (10): 4561-9. doi: 10.1523/JNEUROSCI.5525-12.2013.

Datta, Subimal. (2000). "Avoidance task training potentiates phasic pontine-wake density in the rat: A mechanism for sleep-dependent plasticity." *Journal of Neuroscience* 20: 8607-13.

Datta, Subimal, Guangmu Li, & Sanford Auerbach. (2008). "Activation of phasic pontine-wave generator in the rat: mechanism for expression of plasticity-related genes and proteins in the dorsal hippocampus and amygdala." *European Journal of Neuroscience* 27: 1876-92. https:// doi.org/10.1111/j.1460-9568.2008.06166.x.

Davis, Michael (1992). "The role of the amygdala in conditioned fear." In *The Amygdala: Neurobiological Aspects of Emotion, Memory and Mental Dysfunction*, edited by Aggleton J, New York: Wiley-Liss.

De Olmos, J. S. (1990). "Amygdala." In *The Human Nervous System*, edited by Paxinos, G, 583-710. New York: Academic Press.

Dinges, David F., Frances Pack, Katherine Williams, Kely A. Gillen, John W. Powell, Geoffrey E. Ott, Caitlin Aptowicz, & Allan I. Pack. (1997). "Cumulative Sleepiness, Mood Disturbance, and Psychomotor Vigilance Performance Decrements during a Week of Sleep Restricted to 4-5 Hours per Night." *Sleep* 20 (4): 267–77. http://www.ncbi.nlm.nih.gov/pubmed/9231952.

Fernandes-Santos, Luciano, Camilla L. Patti, Karina A. Zanin, Helaine A. Fernandes, Sergio Tufik, Monica L. Andersen, & Roberto Frussa-Filho. (2012). "Sleep deprivation impairs emotional memory retrieval in mice influence of sex." *Progress in Neuro-psychopharmacology and Biological Psychiatry* 38 (2): 216-22. https://doi.org/10.1016/j.pnpbp.2012.03.014.

Fleischmann, Alexander, Oivind Hvalby, Vidar Jensen, Tatyana Strekalova, Christiane Zacher, Liliana E. Layer, Ane Kvello, Markus Reschke, Rainer Spanagel, Rolf Sprengel, Erwin F. Wagner, & Peter Gass. (2003). "Impaired long-term memory and NR2A-type NMDA receptor-dependent synaptic plasticity in mice lacking c-Fos in the CNS." *Journal of Neuroscience* 23: 116-22.

Franzen, Peter L., & Daniel J. Buysse. (2008). "Sleep Disturbances and Depression: Risk Relationships for Subsequent Depression and Therapeutic Implications." *Dialogues in Clinical Neuroscience* 10 (4): 473–81.

Giaquinto, Salvatore. (1973). "Sleep recordings from limbic structures in man." *Confinia neurologica* 35: 285–303.

Girardeau, Gabrielle, Ingrid Inema, & György Buzsáki. (2017). "Reactivations of emotional memory in the hippocampus-amygdala system during sleep." *Nature Neuroscience* 20: 1634–42. doi: 10.1038/nn.4637.

Girardeau, Gabrielle., Kim Benchenane, Sydney I. Wiener, György Buzsáki, & Michaël B. Zugaro. (2009). "Selective suppression of hippocampal ripples impairs spatial memory." *Nature Neuroscience* 12: 1222–3. doi:10.1038/nn.2384.

Goldstein, Andrea N., Stephanie M. Greer, Jared M. Saletin, Allison G. Harvey, Jack B. Nitschke, & Matthew P. Walker. (2013). "Tired and Apprehensive: Anxiety Amplifies the Impact of Sleep Loss on Aversive Brain Anticipation." *Journal of Neuroscience* 33 (26): 10607–15. https://doi.org/10.1523/JNEUROSCI.5578-12.2013.

Graves, Laurel A., Elizabeth A. Heller, Allan I. Pack, & Ted Abel. (2003). "Sleep deprivation selectively impairs memory consolidation for contextual fear conditioning." *Learning and Memory* 10 (3): 18-76. doi: 10.1101/lm.48803.

Gujar, Ninad, Seung-Schik Yoo, Peter Hu, & Matthew P. Walker. (2011). "Sleep Deprivation Amplifies Reactivity of Brain Reward Networks, Biasing the Appraisal of Positive Emotional Experiences." *Journal of Neuroscience* 31 (12): 4466–74. https://doi.org/10.1523/JNEUROSCI.3220-10.2011.

Guzowski, John F. (2002). "Insights into immediate-early gene function in hippocampal memory consolidation using antisense oligonucleotide and fluorescent imaging approaches." *Hippocampus* 12 (1): 86-104. https://doi.org/10.1002/hipo.10010.

Guzowski, John F., Barry Setlow, Edward K. Wagner, & James L. McGaugh. (2001). "Experience-dependent gene expression in the rat hippocampus after spatial learning: a comparison of the immediate-early genes Arc, c-fos, and zif268." *Journal of Neuroscience* 21 (14): 5089-98. https://doi.org/10.1523/JNEUROSCI.21-14-05089.2001.

Hagewoud, Roelina, Lillina J. Bultsma, Pauline Barf R., Jaap M. Koolhaas, & Peter Meerlo. (2011). "Sleep deprivation impairs contextual fear conditioning and attenuates subsequent behavioural, endocrine and neuronal responses." *Journal of Sleep Research* 20 (2): 259-66. https://doi.org/10.1111/j.1365-2869.2010.00895.x.

Hall, Jeremy, Kerrie L. Thomas, & Barry J. Everitt. (2001). "Fear memory retrieval induces CREB phosphorylation and Fos expression within the amygdala." *European Journal of Neuroscience* 13 (7): 1453-8.

Hegde, Preethi, H.R. Jayakrishnan, Sumantra Chattarji, Bindu M. Kutty, & T. Rao Laxmi. (2011). "Chronic stress-induced changes in REM sleep on theta oscillations in the rat hippocampus and amygdala." *Brain Reserach* 1382: 155-64. doi: 10.1016/j.brainres.2011.01.055.

Horne, J. A. (1985). "Sleep function, with particular reference to sleep deprivation." *Annals of Clinical Research*, *17* (5), 199–208. http://www.ncbi.nlm.nih.gov/pubmed/3909914.

Hummos, Ali, & Satish S. Nair. (2017). "An integrative model of the intrinsic hippocampal theta rhythm." *PLoS One* 12 (8): e0182648. doi: 10.1371/journal.pone.0182648.

Hutchison, Isabel, & Shailendra Rathore. (2015). "The role of REM theta activity in emotional memory." *Frontiers Psychology 6*: 1439. https:// doi.org/10.3389/fpsyg.2015.01439.

Ioannides, Aandreas A. Corsi-Cabrera María, Fenwick Peter B. C., Yolanda Del Rio Portilla, Nikos A. Laskaris, Ara Khurshudyan, et al. (2004). "MEG tomography of human cortex and brainstem activity in waking and REM sleep saccades." *Cereb Cortex* 14: 56–72. https://doi.org/10.1093/cercor/ bhg091.

Kaestner, Erik J., John T. Wixted, & Sara C. Mednick. (2013). "Pharmacologically Increasing Sleep Spindles Enhances Recognition for Negative and High-Arousal Memories." *J Cogn Neurosci*. 25 (10): 1597–610.

Karashima, Akihiro., Norihiro Katayama, & Mitsuyuki Nakao. (2010). "Enhancement of Synchronization Between Hippocampal and Amygdala Theta Waves Associated With Pontine Wave Density." *Journal of Neurophysiology* 103: 2318–25. https://doi.org/10.1152/ jn.00551.2009.

Krettek, J. E., & Joseph L. Price. (1977). "Projections from the amygdala complex to the cerebral cortex and thalamus in the rat and cat." *J Comput Neurol.* 172: 723-52.

LeDoux, Joseph. (2007). "The Amygdala." *Current Biology*, *17* (20), 868–74. https://doi.org/10.1016/j.cub.2007.08.005.

Leung, L. W. S., & J. G. G. Borst. (1987). "Electrical activity of the cingulate cortex. I. Generating mechanisms and relations to behavior." *Brain Research*, *407*, 68-80.

Li, Wei, Lei Ma, Guang Yang, & Wen-Biao Gan. (2017). REM sleep selectively prunes and maintains new synapses in development and learning." *Nature Neuroscience* 20: 427–37. doi: 10.1038/nn.4479.

Maccari, Lisa., Diana Martella, Andrea Marotta, Mara Sebastiani, Nerisa Banaj, Luis J. Fuentes, & Maria Casagrande. (2014). "Effects of Sleep Loss on Emotion Recognition: A Dissociation between Face and Word Stimuli." *Experimental Brain Research* 232 (10): 3147–57. https://doi.org/10.1007/s00221-014-3995-9.

Maquet, Pierre, Dominique Dive, Eric Salmon, Bernard Sadzot, Gianni Franco, Robert Poirrier, R., et al. (1990). "Cerebral Glucose Utilization during Sleep-Wake Cycle in Man Determined by Positron Emission Tomography and [18F]2-Fluoro-2-Deoxy-D-Glucose Method." *Brain Research* 513 (1): 136–43. http://www.ncbi.nlm.nih.gov/ pubmed/ 2350676.

Maquet, Pierre, Jean-Marie Péters, Joël Aerts, Guy Delfiore, Christian Degueldre, André Luxen, & Georges Franck. (1996). "Functional Neuroanatomy of Human Rapid-Eye-Movement Sleep and Dreaming." *Nature* 383 (6596): 163–6. https:// doi.org/10.1038/383163a0.

Maquet, Pierre, Perrine Ruby, Audrey Maudoux, Genevieve Albouy, Virginie Sterpenich, Thanh Dang-Vu, Martin Desseilles, et al. (2005). "Human Cognition during REM Sleep and the Activity Profile within Frontal and Parietal Cortices: A Reappraisal of Functional Neuroimaging Data." *Progress in Brain Research* 150: 219–27. https://doi.org/10.1016/S0079-6123(05)50016-5.

McNaughton, Bruce L. (2010). "Cortical hierarchies, sleep, and the extraction of knowledge from memory." *Artificial Intelligence* 174: 206-14. https://doi.org/10.1016/j.artint.2009.11.013.

Mellman, Thomas A., Wilfred R. Pigeon, Peter D. Nowell, & Bruce Nolan. (2007). "Relationships between REM sleep findings and PTSD symptoms during the early aftermath of trauma." *Journal of Traumatic Stress* 20 (5): 893-901. https://doi.org/10.1002/jts.20246.

Menz, Mareike M., Julia S. Rihm, & Christian Bcüche. (2016). "Rem Sleep Is Causal to Successful Consolidation of Dangerous and Safety Stimuli and Reduces Return of Fear after Extinction." *Journal of Neuroscience* 36 (7): 2148–60. https://doi.org/10.1523/ JNEUROSCI.3083-15.2016.

Minkel, Jared D., Siobhan Banks, Marisa C. Moreta, Christopher W. Jones, Norah S. Simpson, & David F. Dinges. (2014a). "Sleep Deprivation and Stressors: Evidence from Elevated Negative Affect in Response to Mild Stressors When Sleep Deprived." *Emotion* 12 (5): 1015–20. https://doi.org/10.1037/a0026871.Sleep.

Minkel, Jared D., Oo Htaik, Siobhan Banks, & David F. Dinges. (2014b). "Emotional Expressiveness in Sleep-Deprived Healthy Adults." *Behavioral Sleep Medicine* 9 (1): 5–14. https://doi.org/10.1038/ nature13314.A.

Montes-Rodríguez, Corinne J., & Emiliano Urteaga-Urías. (2018). "Synaptic Plasticity as a substrate of resilience." *Revista de Neurología* 67: 453-60. https://doi.org/10.33588/rn.6711.2018258.

Montes-Rodríguez, Corinne J., Pavel E. Rueda-Orozco, & Oscar Prospéro-García. (2019). "Total Sleep Deprivation impairs fear memory retrieval by decreasing the basolateral amygdala activity." *Brain Research* 1719: 17-23. doi: 10.1016/j.brainres.2019.05.030.

Montes-Rodríguez, Corinne J., Pavel E. Rueda-Orozco, Emiliano Urteaga-Urías, Raúl Aguilar-Roblero, & Oscar Prospéro-García. (2006). "From neuronal recovery to the reorganization of neuronal circuits: a review of the functions of sleep." *Revista de Neurología* 43, 409-15. https://doi.org/10.33588/rn.4307.2005285.

Muñoz-Torres, Zeidy, Francisco Velasco, Ana L. Velasco, Yolanda Del Río-Portilla, & María Corsi-Cabrera. (2018). "Electrical Activity of the Human Amygdala during All-Night Sleep and Wakefulness." *Clinical Neurophysiology* 129 (10): 2118–26. https://doi.org/10.1016/ j.clinph.2018.07.010.

Muñoz-Torres, Zeidy, Jorge L. Armony, David Trejo-Martínez, Rubén Conde, & María Corsi-Cabrera. (2011a). "Behavioural and Neural Effects of Diazepam on a Rule-Guided Response Selection Task." *Neuroscience Research* 70 (3): 260–8. https://doi.org/10.1016/j.neures.2011.03.009.

Muñoz-Torres, Zeidy, Yolanda del Río-Portilla, & María Corsi-Cabrera. (2011b). "Diazepam-Induced Changes in EEG Oscillations During Performance of a Sustained Attention Task." *Journal of Clinical Neurophysiology 28* (4): 394–9.

Muzur, Amir. (2005). "Toward an Integrative Theory of Sleep and Dreaming." *Journal of Theoretical Biology 233* (1): 103–18. https:// doi.org/10.1016/j.jtbi.2004.09.011.

Nakayama, Daisuke, Yoshiko Hashikawa-Yamasaki, Yuji Ikegaya, Norio Matsuki, & Hiroshi Nomura. (2016). "Late Arc/Arg3.1 expression in the basolateral amygdala is essential for persistence of newly-acquired and reactivated contextual fear memories." *Scientific Reports* 6: 21007. doi: 10.1038/srep21007.

Nir, Yuval, Richard J. Staba, Thomas Andrillon, Vladyslav V. Vyazovskiy, Chiara Cirelli, Itzhak Fried, & Giulio Tononi. (2011)."Regional slow waves and spindles in human sleep." *Neuron*, 70 (1): 153-69. doi: 10.1016/j.neuron.2011.02.043.

Nishida, Masaki, Jori Pearsall, Randy L. Buckner, & Matthew P. Walker. (2009). "REM sleep, prefrontal theta, and the consolidation of human emotional memory." *Cerebral Cortex* 19 (5): 1158-66. doi: 10.1093/cercor/bhn155.

Nofzinger, Eric A., Mark A. Mintun, MaryBeth Wiseman, David J. Kupfer, & Robert Y. Moore. (1997). "Forebrain Activation in REM Sleep: An FDG PET Study." *Brain Research* 770 (1–2): 192–201. http://www.ncbi.nlm.nih.gov/ pubmed/9372219.

Nofzinger, Eric A., Daniel J. Buysse, Jean M. Miewald, Carolyn C. Meltzer, Julie C. Price, Robert C. Sembrat, Hernando Ombao, et al. (2002). "Human Regional Cerebral Glucose Metabolism during Non-Rapid Eye Movement Sleep in Relation to Waking." *Brain* 125, 1105–15. https://doi.org/ 10.1093/brain/awf103.

Oudiette, Delphine, Marie José Dealberto, Ginevra Uguccioni, Jean Louis Golmard, Milagros Merino-Andreu, Mehdi Tafti, Lucile Garma, Sophie Schwartz, & Isabelle Arnulf. (2012). "Dreaming without REM Sleep." *Consciousness and Cognition* 21 (3): 1129–40. https://doi.org/10.1016/j.concog.2012.04.010.

Paré, Denis, & Hélène Gaudreau. (1996). "Projection cells and interneurons of the lateral and basolateral amygdala: distinct firing patterns and differential relation to theta and delta rhythms in conscious cats." *Journal of Neuroscience* 16: 3334-50. DOI: https://doi.org/10.1523/ JNEUROSCI.16-10-03334.1996.

Parent, Andre. (1996). Carpenter's Human Neuroanatomy. Baltimore: Williams & Wilkins.

Patti, Camilla L., Karina A. Zanin, Leandro Sanday, Sonia R. Kameda, Luciano Fernandes-Santos, Helaine A. Fernandes, Monica L. Andersen, Sergio Tufik, & Roberto Frussa-Filho. (2010). "Effects of sleep deprivation on memory in mice: role of state-dependent learning." *Sleep* 33 (12): 1669-79.

Pérez-Garci, Enrique, Yolanda Del-Río-Portilla, Miguel A. Guevara, Consuelo Arce, & María Corsi-Cabrera. (2001). "Paradoxical Sleep Is Characterized by Uncoupled Gamma Activity between Frontal and Perceptual Cortical Regions." *Sleep* 24 (1): 118–26.

Phan, K. Luan, Tor Wager, Stephan F. Taylor, & Israel Liberzon. (2002). "Functional neuroanatomy of emotion: A meta-analysis of emotion activation studies in PET and fMRI." *Neuroimage* 16: 331–48.

Pitkänen, Asia, Vesa Savander, & Joseph E. LeDoux. (1997). "Organization of Intra-Amygdaloid Circuitries in the Rat: An Emerging Framework for Understanding Functions of the Amygdala." *Trends in Neurosciences* 20 (11): 517–23. https://doi.org/10.1016/ S0166-2236(97)01125-9.

Poe, Gina R. (2017). "Sleep is for forgetting." *Journal of Neuroscience* 37: 464-73. doi: 10.1523/JNEUROSCI.0820-16.

Popa, Daniela, Sevil Duvarci, Andrei T. Popescu, Clément Léna, & Denis Paré. (2010). "Coherent amygdalocortical theta promotes fear memory consolidation during paradoxical sleep." *Proceedings of the National Academy of Science* 107: 6516-9.

Puentes-Mestril, Carlos, James Roach, Niels Niethard, Michal Zochowski, & Sara J. Aton. (2019). "How rhythms of the sleeping brain tune memory and synaptic plasticity." *Sleep* 42, pii: zsz095. https://doi.org/10.1093/sleep/zsz095.

Rechtschaffen, A. & Kales, A. (1968). A Manual of Standardized Terminology, Techniques and Scoring System for Sleep Stages of Human Subjects. Edited by A Rechtschaffen and A Kales. University of California, Los Angeles: Bethesda, Maryland, USA: U.S. Department of Health.

Rechtschaffen, Allan. & Bernanrd M. Bergmann. (2002). "Sleep Deprivation in the Rat: An update of the 1989 Paper." *Sleep* 25, 18-24. https://doi.org/10.1093/sleep/25.1.18.

Roopun, Anita K., Steven J. Middleton, Mark O. Cunningham, Fiona E. N. LeBeau, Andrea Bibbig, Miles A. Whittington, et al. (2006). "A beta2-frequency (20-30 Hz) oscillation in nonsynaptic networks of somatosensory cortex." *Proc Natl Acad Sci USA* 103: 15646-50.

Rosales-Lagarde, Alejandra, Jorge L. Armony, Yolanda del Río-Portilla, David Trejo-Martínez, Rubén Conde, & María Corsi-Cabrera. (2012). "Enhanced emotional reactivity after selective REM sleep deprivation in humans: an fMRI study." *Frontiers in behavioral neuroscience* 6: 1–13. https://doi.org/ 10.3389/ fnbeh.2012.00025.

Rossi, Vanessa Contatto, Paula Ayako Tiba, Karin Di Monteiro Moreira, Tatiana Lima Ferreira, Maria Gabriela Menezes Oliveira, & Deborah Suchecki. (2014). "Effects of sleep deprivation on different phases of memory in the rat: dissociation between contextual and tone fear." *Frontiers in Behavioral Neuroscience* 8: 389. doi: 10.3389/ fnbeh.2014.00389.

Sampath, Dayalan, K.R. Sabitha, Preethi Hegde, H. R. Jayakrishnan, Bindu M. Kutty, Sumantra Chattarji, Govindan Rangarajan, & T. R. Laxmi. (2014). "A study on fear memory retrieval and REM sleep in maternal separation and isolation stressed rats." *Behavioural Brain Research* 273: 144-54. doi: 10.1016/j.bbr.2014.07.034.

Sanchez-Vives, Maria V. & David A. McCormick. (2000). "Cellular and network mechanisms of rhythmic recurrent activity in neocortex." *Nature Neuroscience* 3: 1027-34.

Sauvet, Fabien., Georges Leftheriotis, Danièlle Gomez-Merino, Christophe Langrume, Catherine Drogou, Pascal Van Beers, Cyprien Bourrilhon, Geneviève Florence, & Mounir Chennaoui. (2010). "Effect of Acute Sleep Deprivation on Vascular Function in Healthy Subjects." *Journal of Applied Physiology* 108 (1): 68–75. https://doi.org/10.1152/ japplphysiol.00851.2009.

Schwarz, Johanna F. A., Roland Popp, Jessica Haas, Jürgen Zulley, Peter Geisler, Georg W. Alpers, Michael Osterheider, & Hedwig Eisenbarth. (2013). "Shortened Night Sleep Impairs Facial Responsiveness to Emotional Stimuli." *Biological Psychology* 93 (1): 41–4. https://doi.org/10.1016/j.biopsycho.2013.01.008.

Sergerie, Karine, Chochol Caroline. & Jorge L. Armony. (2008). "The role of the amygdala in emotional processing: a quantitative meta-analysis of functional neuroimaging studies." *Neuroscience and biobehavioral reviews* 32 (4): 811-30.

Shao, Yongcong, Yu Lei, Lubin Wang, Tianye Zhai, Xiao Jin, Wei Ni, Yue Yang, et al. (2014). "Altered Resting-State Amygdala Functional Connectivity after 36 Hours of Total Sleep Deprivation." *PLoS ONE* 9 (11). https://doi.org/10.1371/journal.pone.0112222.

Silva, R. H., A. B. Chehin, S. R. Kameda, A. L.Takatsu-Coleman, V. C. Abílio, S. Tufik, & R. Frussa-Filho. (2004a). "Effects of pre- or post-training paradoxical sleep deprivation on two animal models of learning and memory in mice." *Neurobiology of Learning and Memory* 82: 90-8. https://doi.org/10.1016/j.nlm.2004.04.005.

Silva, R. H., V. C. Abílio, A. L. Takatsu, S. R. Kameda, C. Grassl, A. B. Chehin, W. A. Medrano, M. B. Calzavara, S. Registro, M. L. Andersen, R. B. Machado, R. C. Carvalho, A. Ribeiro, Sergio Tufik, & R. Frussa-Filho. (2004b). "Role of hippocampal oxidative stress in memory deficits induced by sleep deprivation in mice." *Neuropharmacology* 46: 895-903. https://doi.org/10.1016/j.neuropharm.2003.11.032.

Sirota, Anton, & György Buzsáki. (2005). "Interaction between neocortical and hippocampal networks via slow oscillations." *Thalamus Relat Syst.* 3 (4): 245-59.

Sotres-Bayon, Francisco, David E. A. Bush, & Joseph E. LeDoux. (2004). "Emotional Perseveration: An Update on Prefrontal-Amygala Interactions in Fear Extinction." *Learning & Memory 11*: 525–35. https://doi.org/10.1101/lm.79504.

Steriade Mircea, Diego Contreras, R. Curró Dossi, & Angel Nuñez. (1993). "The slow (<1 Hz) oscillation in reticular thalamic and thalamocortical neurons: scenario of sleep rhythm generation in interacting thalamic and neocortical networks." *Journal of Neuroscience* 13: 3284–99.

Steriade, Mircea, R. Curró Dossi, & Angel Nuñez. (1991). "Network modulation of a slow intrinsic oscillation of cat thalamocortical neurons implicated in sleep delta waves: cortically induced synchronization and brainstem cholinergic suppression." *Journal of Neuroscience* 11: 3200-17.

Sterpenich, Virginie., Camille Piguet, Martin Desseilles, Leonardo Ceravolo, Markus Gschwind, Dimitri Van De Ville, Patrik Vuilleumier, & Sophie Schwartz. (2014). "Sleep sharpens sensory stimulus coding in human visual cortex after fear conditioning." *NeuroImage* 100: 608-18. https://doi.org/10.1016/j.neuroimage.2014.06.003.

Tempesta, Daniela, Alessandro Couyoumdjian, Giuseppe Curcio, Fabio, Moroni, Cristina Marzano, Luigi De Gennaro, & Michele Ferrara. (2010). "Lack of Sleep Affects the Evaluation of Emotional Stimuli." *Brain Research Bulletin 82*, (1–2): 104–8. https://doi.org/10.1016/ j.brainresbull.2010.01.014.

Tononi, Giulio, & Chiara Cirelli. (2019). "Sleep and synaptic down-selection." *European Journal of Neuroscience* 1-9. https://doi.org/ 10.1111/ejn.14335.

Totty, Michael S., Chesney Logan A., Phillip A. Geist, & Subimal Datta. (2017). "Sleep-Dependent Oscillatory Synchronization: A Role in Fear Memory Consolidation." *Frontiers in Neural Circuits* 11: 49. doi:10.3389/fncir.2017.00049.

Tripathi, Shweta, Pankaj Taneja, & Sushil K. Jha. (2018). "Training on an Appetitive (Delay)-Conditioning Task Enhances Oscillatory Waves During Sleep in the Cortical and Amygdalar Network." *Frontiers in Behavioral Neuroscience* 12: 260. doi:10.3389/fnbeh.2018.00260.

Ulrich, Daniel. (2016). "Sleep spindles as facilitators of memory formation and learning." *Neural Plasticity* 1–7. http://dx.doi.org/10.1155/2016/1796715

Urakami, Yuko. (2012). "Relationship between, sleep spindles and clinical recovery in patients with traumatic brain injury: A simultaneous EEG and MEG study." *Clinical EEG and Neuroscience 43* (1): 39–47.

Vanderheyden, William M., Sophie A. George, Lea Urpa, Michaela Kehoe, Israel Liberzon, & Gina R. Poe. (2015). "Sleep alterations following exposure to stress predict fear-associated memory impairments in a rodent model of PTSD." *Experimental Brain Research* 233: 2335. https://doi.org/10.1007/s00221-015-4302-0.

Vanderwolf, C. H. & Leung, L. S. (1983). "Hippocampal rhythmical slow activity: a brief history and effects entorhinal lesions and phencyclidine." In: The Neurobiology of the Hippocampus, edited by Seifert W, 275-302. London: Academic Press.

Villablanca, Jaime. (1974). "Role of the thalamus in sleep control: Sleep-wakefulness studies in chronic diencephalic and athalamic cats." In Basic Sleep Mechanisms, edited by Petre-Quadens O, and J, Schlag, 51-81. New York: Academic.

Walker, Matthew P. & Elms van der Helm. (2009). "Overnight therapy? The role of sleep in emotional brain processing." *Psychological Bulletin* 135(5), 731-48. http://dx.doi.org/10.1037/a0016570.

Ward, Marcus P., Kevin R. Peters, & Carlyle T. Smith. (2014). "Effect of Emotional and Neutral Declarative Memory Consolidation on Sleep Architecture." *Experimental Brain Research* 232 (5): 1525–34.

Wellman, Laurie L., Mairen E. Fitzpatrick, Olga Y. Hallum, Amy M. Sutton, Brook L. Williams, & Larry D. Sanford. (2016). "Individual Differences in Animal Stress Models: Considering Resilience, Vulnerability, and the Amygdala in Mediating the Effects of Stress and Conditioned Fear on Sleep" *Sleep* 39: 1293–303. https://doi.org/ 10.5665/sleep.5856.

Wellman, Laurie L., Mairen E. Fitzpatrick, Olga Y. Hallum, Amy M. Sutton, Brook L. Williams, & Larry D. Sanford. (2017). "The basolateral amygdala can mediate the effects of fear memory on sleep independently of fear behavior and the peripheral stress response." *Neurobiology of Learning and Memory* 137: 27-35. doi: 10.1016/j.nlm.2016.11.004.

Wellman, Laurie L., Mairen E. Fitzpatrick, Mayumi Machida, & Larry D. Sanford. (2014). "The basolateral amygdala determines the effects of fear memory on sleep in an animal model of PTSD." *Experimental Brain Research* 232 (5): 1555-65. doi: 10.1007/s00221-014-3850-z.

Wilson, Matthew A., & Bruce L. McNaughton. (1994). "Reactivation of hippocampal ensemble memories during sleep." *Science 265*, 676-9. doi: 10.1126/science.8036517.

Yang, Guang, Cora Sau Wan Lai, Joseph, Cichon, Lei Ma, Wei Li, & Wen-Biao Gan. (2014). "Sleep promotes branch-specific formation of dendritic spines after learning." *Science* 344 (6188): 1173-8. doi: 10.1126/science.

Yoo, Seung Schik, Ninad Gujar, Peter Hu, Ferenc A. Jolesz, & Matthew P. Walker. (2007). "The Human Emotional Brain without Sleep - a Prefrontal Amygdala Disconnect." *Current Biology* 17 (20): 877–8. https://doi.org/10.1016/j.cub.2007.08.007.

In: Sleep, Cognition and Emotion
Editors: I. Ramírez Salado et al.
ISBN: 978-1-53617-561-5

Chapter 5

ALTERED BEHAVIOR AND CORTICAL ACTIVITY DURING SLEEP IN NEURODEGENERATIVE DISORDERS AND NON-INVASIVE NEUROMODULATORY TECHNIQUES AS A POSSIBLE TREATMENT

Ruth Alcalá-Lozano*
Laboratorio de Neuromodulación, Subdirección de Investigaciones Clínicas,
Instituto Nacional de Psiquiatría Ramón de la Fuente Muñiz,
Ciudad de México, México

ABSTRACT

In order to understand the changes in behavior and cortical activity related to neurodegenerative diseases (ND) it is important to know the changes that take place in the sleep cycle and structure, as well as the electrophysiological changes that occur due to aging, since they can serve as a guide for the creation of simple and easy to implement treatments. Repetitive Transcranial Magnetic Stimulation (rTMS) and Transcranial Direct Current Stimulation (tDCS) are non-invasive methods of brain stimulation used to explore neurophysiological mechanisms such as cortical inhibition, excitability and plasticity. Electrical modulation of cortical activity can benefit patients with sleep disorders and, as is the case with other neurophysiological methods, it is considered safe as well as effective.

Keywords: cortical activity, sleep, Neurodegenerative Diseases (ND), Transcranial Magnetic Stimulation (rTMS), Transcranial Direct Current Stimulation (tDCS)

* Corresponding Author's Email: ruthalcala_@hotmail.com; ruthalcalalozano@gmail.com.

In order to understand the changes in behavior and cortical activity that take place in neurodegenerative diseases (ND), also known as dementias, it is important to recognize the changes in the sleep cycle and structure, as well as the electrophysiological changes that occur in non-patient older adults, which can serve as a guideline for the identification and treatment of these alterations. Older people are at high risk for sleep disorders because of a variety of factors, including multiple psychological and neurological conditions such as ND as well as general systemic diseases that indirectly affect the central nervous system (CNS) such as metabolic, cardiovascular, infectious, hematological, skeletal and muscular alterations, as well as a history of prolonged alcoholism. These conditions have been associated with nocturnal hypoxemia, which in turn is related to sleep apnea, and it has been suggested that these disease states may interact with sleep disorders (Anand, Gill, and Mahdi 2014; Ronald C. Petersen 2016; Sanford 2017).

Aging is considered a risk factor for neurodegenerative diseases, and their diagnosis represents a constant challenge for health professionals. Among the most commonly used and revised classifications is the American Psychiatric Association's Diagnostic and Statistical Manual of Mental Illnesses (DSM). In previous versions, neurodegenerative diseases were included in dementias; in their current version (DSM-5) they are classified as major and minor neurocognitive disorders that differ depending on the intensity of the symptoms and their effect on the functionality of the patient, and include various etiological subtypes mentioned in Figure 1 (Ronald C. Petersen 2016; López-Álvarez and Agüera-Ortiz 2015).

- Neurocognitive Disorder Due to Alzheimer's Disease
- Frontotemporal Neurocognitive Disorder
- Neurocognitive Disorder With Lewy Bodies
- Vascular Neurocognitive Disorder
- Neurocognitive Disorder Due to Traumatic Brain Injury
- Substance/Medication-Induced Neurocognitive Disorder
- Neurocognitive Disorder Due to HIV Infection
- Neurocognitive Disorder Due to Prion Disease
- Neurocognitive Disorder Due to Parkinson's Disease
- Neurocognitive Disorder Due to Huntington's Disease
- Neurocognitive Disorder Due to Another Medical Condition
- Neurocognitive Disorder Due to Multiple Etiologies
- Unspecified Neurocognitive Disorder

Figure 1. Subtypes of neurocognitive disorders according to DSM-5.

Aging has a profound effect on brain structure and function, causing physiological changes in the CNS that influence cognition and behavior. According to the DSM-5, Mild Cognitive Impairment (MCI) or Minor Neurocognitive Disorder is a diagnostic entity that designates individuals with a decrease in performance in one or more cognitive functions, compared to people of the same age and schooling, without significantly affecting their daily function (Ronald C. Petersen 2016; Albert et al. 2011). MCI is considered to be an intermediate condition between normal aging and dementia. However, not all people with this diagnosis will progress to present a ND, some will remain stable with this impairment and others will regain their cognitive function. Research topics related to the variables that predict this process, and the potential strategies to prevent it, are extremely relevant today (Ronald C. Petersen 2016).

This chapter focuses mainly on MCI and Alzheimer's disease (AD) patients due to its insidious and gradual progression, but also on those who develop other forms of dementia which cause changes in thinking and behavior, such as Frontotemporal neurodegenerative disorder (FTD), Vascular neurodegenerative disorder (VND) and Neurodegenerative disorder induced by lewy bodies (LBD). Some other causes of neurocognitive disorders will also be mentioned, as well as some changes due to aging considered normal, and those that accompany ND. These include some cortical changes that can influence behavior, considering the limitations in the differentiation of what might be part of normal aging and those defined as pathological. The intention of this chapter is to contribute to the identification of neurophysiological and sleep disorders, which enable us to not only suggest pharmacological interventions, but also to use neuromodulatory techniques which represent non-invasive forms of treatment, such as repetitive transcranial magnetic stimulation (rTMS) or transcranial direct current stimulation (tDCS). It is known that these neuromodulatory techniques can reduce the severity of the symptoms and slow the progression of NDs by preserving normal function of patients, as well as help prevent side effects. There is ample scientific evidence regarding their safety when used under standard guidelines (Rossi et al. 2009). Alongside its clinical efficacy, a growing body of research is currently focusing on pinpointing the neurophysiological changes that can be used as diagnostic tools, and on therapeutic response biomarkers. In view of the above, previous approaches centered on the identification of such markers and slowing down their evolution towards dementia, taking as a starting point those initial stages where the cognitive reserve and physiological resources preserved by the patient can be used as a means to study and develop therapeutic approaches.

The relationship between changes in behavior and cortical activity associated with aging and sleep is not fully understood. Physiological and clinical changes in the aging brain are highly variable, making it difficult to predict a trajectory in cognitive, behavioral and neurophysiological symptoms to monitor the transition from non-pathological states to ND (Gomes-Osman et al. 2018). Its study is of great and growing importance because ND is one of the main causes of disability during old age, which is a growing large-scale public health issue. Currently, around 46.8 million people worldwide are living with dementia, with a projection of 131.5 million by 2050 (Prince et al. 2015). When considering only adults over the age of 60, up to 18% live with MCI, 8 to 15% of whom will develop dementia over the course of the next year, and 60% during the next 10 years (Ronald C. Petersen 2016). Sleep alterations can be a direct result of brain damage caused by ND, if the condition affects the brain centers responsible for the process of sleep-wake. Alternatively, sleep alterations can arise from drug treatments, or from other comorbid diseases. The ND-sleep disorders binomial is usually complex due to the variety of their etiology, which makes their interdisciplinary study a necessity.

The behavioral and psychological symptoms of dementia (BPSD) can occur in people with NDs in variable patterns and at any stage of the disease, and constitute multiple specialized segments which can be evaluated and treated. They are of potential interest to the prodromal phases of ND and their clinical management is essential ("An Introduction to BPSD: The IPA Complete Guides To" 2012). They are also extremely frequent in the course of ND disease, and it is estimated that they are present in 70 to 90% of patients throughout the disease. Sleep disturbances, such as the loss of the physiological sleep-wake cycle, hypersomnia, insomnia, cycle inversion, fragmented sleep, etc. belong to this set of BPSDs and are highly prevalent (Steinberg et al. 2010). Of all sleep disorders, the ones that affect the elderly the most are restless legs syndrome, obstructive sleep apnea syndrome, periodic limb movements during sleep, insomnia, advanced sleep phase syndrome and REM sleep behavior disorders. Normal aging and NDs are associated with sleep disorders due to physiological modifications, drug use and associated pathological processes. Recognizing these alterations in early and preclinical stages as possible manifestations of ND is essential for the development of timely interventions since the currently available treatments seem insufficient. It may be essential to be able to relate them to cortical alterations in general, and/or cortical alterations during sleep, as sleep quality is crucial for cognitive and daily function, and may also precipitate hospitalization or institutionalization (Zanto et al. 2011).

Although age is a determining factor to the physiology of sleep, changes in its structure can be partially determined by a decline in the ability of the elderly to remain asleep, thereby decreasing slow wave sleep (SWS) and increasing light sleep. In sleep disorders of the elderly, the first episode of rapid eye movement sleep (REM sleep) has a longer latency, therefore the time spent in REM sleep is reduced; there is an increase in light sleep (non REM sleep, phase 2, or N2), and a decrease in the amplitude of delta waves in SWS. Total sleep time and sleep efficiency index are also reduced. A shift in sleep stages, particularly repeated regressions to lighter sleep stages or failure to achieve deeper sleep, as well as desynchronization of circadian rhythms have also been extensively reported. It is common for this population group to take longer to fall asleep, wake up more easily and experience frequent awakenings at night and during the early morning, which generates the need to take naps during the day. The aforementioned changes also take place in ND disease, and even more frequently and/or more severely, especially in the more advanced stages of the degenerative process.

In Alzheimer's disease (AD), which is the most frequent ND, a *sunset* can develop which is characterized by a set of symptoms that occur during sunset and can continue until night, and may include agitation, delusions, confusion, disorganized thinking, lack of attention, restlessness, wandering, hallucinations, anger and anxiety. For patients with LBD, it may be difficult to differentiate between dreams and reality, sometimes accompanied by visoperceptive dysfunction, although these changes are not exclusive to LBD or AD. However, REM sleep behavior disorder, when accompanied by visual hallucinations, cognitive impairment and parkinsonism can be pathognomonic of LBD and the symptoms can be interpreted as a marker for the beginning of the disease (Chan et al. 2018).

REM sleep behavior disorder is associated with Parkinson's disease, DBL, and multiple system atrophy (synucleinopathies) and may appear many years before ND disease. This disorder is characterized by simple or complex movements of the limbs that can harm the bed partner, often accompanied by aggressive vocalizations.

Polysomnography shows REM sleep with increased muscle tone in the extremities and a decrease of electroencephalogram (EEG) activity during this same phase. In Parkinson's disease (PD), abnormalities of dopaminergic neurons in the ventral tegmental area (VTA) alter the sleep cycle and, as a result, the EEG is not synchronized (Boeve et al. 2001).

Cortical activity is maintained during sleep, so it has been proposed that the alterations seen during sleep and related to memory consolidation in patients with MCI and ND may reflect the same alterations that occur during wakefulness, and/or be a result of altered perception during the waking state.

This is because the dream could facilitate the consolidation of memory through the repetition of dreams of important events, as well as the maintenance of synaptic competence by activating poorly stimulated neuronal systems during waking, which would maintain their integrity and effectiveness thanks to the nocturnal compensatory process. It is known that important metabolic processes are also carried out during sleep, such as the restoration of neuronal glycogen stores that are consumed during wakefulness. Cyclic phenomena, such as sleep, depend on complex interactions involving multiple neurotransmitters, peptides and neuronal pathways, and there is talk of an asynchronous or erratic oscillation of these processes during sleep, that is, wakefulness and sleep can vary erratically in MCI and NDs and some processes which should occur during wakefulness may take place during sleep, and *vice versa*. These alterations, however, may not affect the brain as a whole in a homogeneous and synchronous way (Chokroverty 2017).

It has been proposed that age-related changes in cognition and behavior may be caused by subtle disruption of the microvasculature of cerebral white matter, as found in studies of brain imaging (Salat 2014; Chokroverty 2017). A general decrease in cortical thickness has also been found which presents in an anteroposterior gradient, that is, the frontal and parietal regions show higher rates of decline than the temporal and occipital areas. At the cellular level, a marked reduction in the activation of prefrontal cortex neurons (PFC) has been documented with advanced age. It has been shown that the decrease in the age-related response speed can be partially reverted by addressing the molecular aspects of PFC circuits (for example, inhibiting cyclic signaling of AMP or blocking potassium channels HCN or KCNQ) (Chokroverty 2017). We believe that taking these findings into account could allow us to take advantage of the time before the onset of disease by stimulating the PFC with non-invasive neuromodulatory strategies that interfere with response speed.

EEG can be used as a diagnostic tool in ND by recording the synchronization of the brain's electrical activity between different areas of the brain, which are responsible for different cognitive tasks. EEG studies are also effective, non-invasive and much less expensive than brain imaging studies. Studies with functional magnetic resonance imaging (fMRI) have been successfully correlated to electrophysiological oscillations identified in EEG studies (Marino et al. 2019). Quantitative analysis of the EEG is also a useful approach to assess the neurophysiological changes that occur during aging. In recent years, excellent studies have been designed to find predictive biomarkers of ND pathologies, with iAD and MCI being the most thoroughly studied, through structural, functional and molecular imaging studies, as well as cerebrospinal fluid, beta-amyloid (Aβ) protein and tau protein studies (Hampel et al. 2008; Rossum et al. 2010; Drago et al. 2011). Few studies have focused on the electrophysiological biomarkers in MCI (Trebbastoni et al. 2016) even though it is considered that these are suitable to directly detect rapid oscillations associated with neuronal processes, and that they can provide a measure of electrophysiological activity. This has already been evidenced in EEG measurements and fMRI where the same brain activity is reflected (Kwong et al. 1992; Scheeringa et al. 2012; Marino et al. 2019). For a better understanding of the importance of proper sleep and functioning brain activity related to aging, the relationship between fluctuations in neuronal communication signals and blood oxygenation dependent levels (BOLD), or in other words, the electrical activity of the brain in the form of oscillatory patterns and evoked potentials, has been studied. EEG quantitative analysis is a useful approach for evaluating the neurophysiological changes due to aging while considering the role of oscillatory brain activity in the transfer of information through neuroplasticity, which has effects in cognition. However, not all changes recorded by EEG are pathological, as some could be interpreted as maturational changes.

On its own, the EEG is of limited value for the evaluation of ND because of its low sensitivity and specificity. For example, in AD, the findings of EEG are the slowing of the subsequent dominant rhythm, increased diffuse slow activity (theta or delta) and generalized slow-activity discharges, but none of these cortical function alterations are specific to AD. However, it is important to perform more careful EEG analyses to determine whether significant changes can in fact be detected in early or premorbid stages. Although some work has been done, it may be worth to replicate some of the findings with equipment that uses better technology and allows for more specific analysis with greater sensitivity. Of the NDs studied with EEG, only Creutzfeldt-Jakob disease has been found to present a characteristic pattern of electrical activity, which consists of progressive disorganization of the basic rhythms and an increase in the amount of generalized slow activity (delta-theta) and, as the disease progresses, the EEG is characterized by biphasic and triphasic acute, periodic, bilaterally synchronous waves, which appear at irregular intervals of one or two per second.

Quantitative electroencephalography and long-latency cortical potentials, such as P300, do not appear to increase the specificity of clinical diagnosis in diseases that evolve to ND, although due to their reliability and objectivity they may continue to be a topic of clinical research. Their importance lies in the fact that they express neuroelectric sensory activity in a perceptual and cognitive context. For example, the 300ms potential (P300) represents the electrophysiological substrate of cognitive functions, so it is very suitable for analyzing the age-related changes in the human brain and, therefore, could indirectly relate to behavior and the architecture of the sleep. P300 can be obtained through various stimulus modalities. The temporal and parietal lobes are P300 generators, as is the hippocampus, a brain structure related to the functioning of memory. Other subcortical structures, such as the primary auditory cortex, also produce P300 potentials. In addition, this wave of smaller amplitude has been recorded in the frontal lobe and in the thalamus (Coyle et al. 1991). However, the frontal lobe and thalamus registered this wave in smaller amplitudes (Barón-Herrera et al. 2011).

Clinical applications of P300 are varied. Since its latency is related to processing time, its application could be important for the diagnosis of ND. Attention is a complex cognitive function that involves conditions of awareness and wakefulness. It is part of a set of domains with which it works in a coordinated manner, and is affected by sleep cycle alterations. The function of attention is to select from the environment those stimuli, either internal or external, that are relevant to the current cognitive state of the subject and that serve to carry out an action and achieve one or more objectives. These domains manage to connect with each other through cortical and subcortical functional networks (Knott et al. 2003; Coyle et al. 1991) and it has been proposed that studying the amplitude and latency of the P300 and relating it to cognitive functions, based on fundamentals such as brain reserve could help guide rehabilitation to damage. It could be useful to explore the observed variations in the amplitude of these late potentials, particularly if there is some relationship to other potentials, such as the potential for disparity. The P300 is also mediated by attention, that is, if the potential for disparity is considered as a preattenional test that is related to the automatic firing of neurons, and p300 as an attentional test, both tests have a cognitive correlation and have repercussions on aging, since a shortening in the amplitudes of the disparity potentials and an elongation in the latencies of P300 can be observed, and from this perspective, it is considered that they may have the same support, since they are both related to functions that express plastic modifications which manifest on EEG recordings, and could be taken as a study parameter and be then correlated to other measurable variables, such as sleep.

Distinguishing between pathological EEG abnormalities and those considered product of old age is critical because these disorders can lead to metabolic encephalopathy. It has been reported that electrophysiological oscillations in the alpha frequency range (8-13 Hz) are the most prominent during the resting phase (Vecchio et al. 2011) and aging has been correlated with an increase in spectral power in the slowest frequency bands delta (1-3 Hz) and theta (4-7 Hz) (Finnigan and Robertson 2011). This was similarly reported in an old, but no less important study by Katz and Horowitz (1982) conducted in healthy older adults, where they took subjects that were evaluated with a neurological, psychiatric and neuropsychological examination and found that normative data were presented with an average alpha frequency of 9.8 Hz. In other words, it was similar to young adults. (Tatti et al. 2016) mention in their review that the alpha frequency band below 10.5 Hz could be related to the overall level of attention preparation and long-range connectivity, and higher alpha waves could be involved in the transmission of information from sensorimotor centers and in the recovery of semantic memory (Vecchio et al. 2011). So, the alpha slowdown seems to be related to a decrease in mental function, which may be an early stage of ND (Barón-Herrera et al. 2011). In contrast, alpha intensity was negatively associated with hippocampal atrophy in MCI and AD patients, indicating a continuum from normal aging to the initiation of AD, characterizing alpha amplitude and reliability as a biomarker (Bian et al. 2014). The possibilities of quick operation in older adults as a representation of retained brain function have been researched for a long time, but results remain unclear.

The delta and theta powers have also been studied, but with more inconsistent results (Castro Chavira et al. 2014). Changes in delta waves have been described in the temporal region of adults over 60, notably in the temporal region of the left hemisphere in its anterior part, the voltage is usually below 70mv and these waves do not interrupt background activity, but they change with the state of mental alertness and the opening of the eyes, increase with drowsiness and hyperventilation and occur as individual waves. The interesting thing about these findings is that they have helped identify changing patterns in oscillatory activity, and could be thought of as a compensatory attempt to control information processing to meet cognitive demands and maintain behavioral stability. However, it is also unclear how they interfere with the dream architecture. Thus, if there is a cognitive decline that results in a dysfunction that can be related to EEG oscillations (Gildner et al. 2014) in sleep, this could be interpreted as difficulty in achieving its restorative function, not only in the cognitive domain, but in general, added to the fact that we know that sleep affects other conditions, such as metabolic and cardiovascular diseases, which have been identified as having a negative impact on cognition and behavior when they are not well regulated. Proper sleep could lead to better cognitive performance by minimizing problems related to the accumulation of bad sleep nights (Gildner et al. 2014; Low, Wu, and Spira 2019) especially with regards to alpha, as it is known to be the dominant inactive electroencephalographic rhythm during the resting state (Gonçalves et al. 2006). Taking these improvements into account may lead to the steps currently known to induce neuroplasticity, such as non-invasive stimulation techniques.

Sleep is a necessary state in humans, and is associated with a characteristic pattern of changes in the brain's electrical activity during the night. Among the main changes during aging, K-complexes (KCs) share a common generation mechanism with the continuous delta activity observed in SWS. With a decline in the incidence of spontaneous KCs, aging has been shown to affect delta generation during sleep. Nevertheless, the specific meaning of the frequency and amplitude (N550) of evoked KCs remains unclear. Slow wave sleep activity (SWA) is assumed to indicate the need for sleep, increasing in proportion to waking time, and decreasing during sleep. Nevertheless, the SWA control mechanism is not understood (Massimini et al. 2007).

It is also mentioned that sleep has been reliably associated with decreases in the delta power of NREM sleep, which is reflected in the amplitude and incidence of delta waves. It has been proposed that the measurements of KC evoked as an index of delta production that allows the separation of the incidence and amplitude variables that are confused in the more traditional delta and SWS power measurements. For example, Colrain et al. (2010) hypothesized that the delta frequency response evoked as indicated by the amplitude of the motor potential evoked from N550, would show a linear decrease with increasing age throughout adulthood and that it could thus provide a functional index of age-related macrostructural and microstructural changes of the brain. They found that the amplitude of N550 in the average of evoked KCs could be a very sensitive marker of aging and that the evoked KCs seemed to reflect a functional measure of the ability of the aged brain to produce large scale synchronized waveforms, and could act as an indirect measure of how aging affects the functional integrity of brain networks, especially in frontal regions. In their conclusions they propose that these means being derived from EEG performed during sleep could provide measures free of confusing factors associated with cognition (ability to remember and follow instructions).

In the EEG, sleep spindles are waves characteristic of the second phase of sleep which presents without rapid eye movements. They are generated in the thalamus and the cerebral cortex. These are bursts of 10 to 16 Hz, with a duration of between 0.5 and two seconds. It is believed that this phenomenon is related to various functions, such as memory consolidation, and can show a variety of changes due to the aging process (Chokroverty 2017). On the other hand, it is considered that these spindles can help modulate the influence of external stimuli during sleep. In this way we are interested in whether they can be a way to indirectly, but objectively measure the potential for altered behavior in adults with ND.

It is common to hear the complaint that older adults do not sleep like when they were young, usually regarding the duration of their sleep. However, their total sleep time during a circadian period is very similar due to the fact that they usually take naps during the day. According to the cyclic pattern of REM to NRM, the overall sleep time remains unchanged, but the first cycle is decreased and interrupted due to frequent interruptions during dense periods of REM. In addition, there are more intermittent arousals. This could be a reason to wake up in both ill or vulnerable brains, and in healthy older adults (Chokroverty 2017), and a reason to produce daytime sleepiness to compensate for the lost quality of night sleep. However, this cycle quickly becomes erratic. The possible brain damage results not only from ND, but also from the state of delirium which may affect cortical activity, as well as the conditions that may be affected by cerebral nocturnal hypoxemia and be associated with obstructive sleep apnea, such as diabetes, obesity and hypertension. They affect or are reflected in the alteration of the sleep-wake cycle, and might become a challenge for both the doctor and the family, since the use of drugs as treatment can complicate these sleep alterations even further.

When electrical brain activity is altered during wakefulness, we can usually predict that there will be alterations during sleep. The finding of the decrease in delta waves in the frontal cortex in older adults during the early stages of sleep is significant, and reflects the level of caution that must be taken to differentiate this occurrence from certain sleep disorders, particularly as these are so common in this age group, but can also be found in ND patients. Slow and diffuse waves in both delta and theta frequencies have been identified in several types of ND. It is possible to correlate prefrontal atrophy, slow waves in disrupted non-REM and memory changes (Bryce A Mander, Vikram Rao, Brandon Lu, Jared M Saletin, John R Lindquist, Sonia Ancoli-Israel 2013). Another group of researchers report that the linear decrease in the amplitude of delta waves evoked during sleep could be related to the reduction of neuronal synchronization in the neocortex, changes in subarachnoid spaces, reductions in very specific neuronal subpopulations, changes in receptor function and neurochemical alterations and changes in synaptic communication, which leads to a decrease in synaptic density resulting in a decrease in brain connectivity. At this point it would be worth mentioning the polysomnographic characteristics of dementia: decreased sleep efficiency, increased awake time after onset of sleep, increased NREM stage 1 sleep, decreased NREM sleep stage 3 and decreased REM sleep (Chokroverty 2017).

Neuroplasticity is known as the brain's default reaction to adapt to new circumstances, and to regain equilibrium after a damaging accident. In response to different sensory experiences, the brain has a remarkable ability to reorganize itself, and the processes related to neuroplasticity are extremely variable between individuals, and throughout life (Pascual-Leone et al. 2012). Interpersonal variability is attributable to several factors, which include variability in functioning of inhibitory networks and neuromodulatory systems, age and sex, diseases that affect the brain, and psychological traits. We currently know that neuroplasticity can be manipulated in both healthy and diseased brains. Therefore, recent proposals to identify which mechanisms alter neuroplasticity could take advantage of the ability to establish personalized strategies for recovery of adequate function after aging or brain disease (Voss et al. 2017; Boggio et al. 2011).

As such, plasticity is an intrinsic property of the nervous system that allows its function and structure to be changed in response to environmental demands by strengthening, weakening, pruning or adding synaptic connections and fostering neurogenesis. Neuroplasticity includes modifications to existing synapses and the formation and loss of synaptic connections that modify brain structure and function throughout life and in response to experience (Voss et al. 2017; Pascual-Leone et al. 2012). This means that the brain is not static, but continues to change as the mandatory consequence of each sensory input, motor act, association, reward signal, plan, action, and even of awareness (Pascual-Leone et al. 2012).

The precise changes that take place in healthy adults are not yet clear, and even less is known about patient populations. Disorders that affect plasticity regulators increase the likelihood of maladaptive plastic changes in the brain. On the other hand, if the brain is too plastic, the structural connections may become unstable and the functional systems necessary for cognition and behavior may be compromised and result in unwanted alterations such as sleep disturbances. Plasticity is strictly regulated by a variety of cellular and molecular processes. Functional and structural elements that both promote and limit plasticity include the inhibitory activity of GABAergic interneurons such as positive parvalbumin cells (PV +), extracellular matrix components, including perineuronal networks (PNN), and myelin-associated proteins. These components can form structural barriers to limit plasticity and stabilize cortical representations. In aging, changes in their functioning could lead to a state of cortical instability, and it has been suggested that inhibitory regulation of plasticity could be reduced in the elderly brain, which contradicts the traditional view that aging is a period of plasticity. It is thus proposed to carry out interventions aimed at plasticity, with enriched or facilitating environments. Added to these neuroplasticity proposals, we consider the possibility of understanding and integrating how sleep is influenced by these changes and how commonly used treatments, such as drugs, do not have the expected effect and alter neuroplasticity (Voss et al. 2017) which is why neuromodulatory strategies and treatments could have great benefits.

The therapeutic potential of neuromodulatory systems to promote prevention of or recovery from age-related deficits and pathological aging suggests therapeutic interventions directed at several neuromodulatory systems which tend to be deregulated, not primarily to treat cognition, but to influence BPSD. Studies that concentrate on this aim will be critical in developing paradigms with procedures that allow addressing functional disruption, and of particular relevance in determining causality, taking into account that neuromodulatory systems show great variation among individuals, creating great potential for personalized treatments. This increases the likelihood of individualized therapies by harnessing the innate ability of the brain to change (Lee et al. 2014; Devi et al. 2014).

As is well known, the brain is highly interconnected, so plasticity develops throughout multiple levels at the nervous system, from cells through microcircuits to large circuits and networks. Lower-level activity can influence higher-level activity, and vice versa. Changes in local plasticity can be compensated by circuit and network adaptations in such a way that behavior does not deteriorate (in fact, some behaviors can paradoxically improve). Conversely, maladaptive circuits and network dynamics that lead to symptoms of illness and disease may aggravate local changes. Changes in local plasticity may therefore be the first in a chain of events that culminate in "pathological circuits," in which the symptoms arise from neural circuit and network dysfunctions. Furthermore, cortical plasticity tests can provide very early local and network biological markers of neuropsychiatric diseases, such as ND. Evidence has indicated that interventions promoting neural plasticity may cause substantial cognitive changes, especially in at-risk subjects, which means that we might be able to improve the BPSDs.

Sleep has been proposed to restore "synaptic homeostasis," aided by thc brain's extraordinary capacity for plasticity. The plasticity processes that occur during wakefulness result in a net increase in synaptic strength in many brain circuits. The increase in synaptic strength has several costs at the cellular and system levels, including increased energy consumption, increased need for synaptic cellular resources triggering cell pressure, and the associated changes in glial cells. The increase in synaptic strength also reduces neuronal response selectivity and saturates the ability to learn, as the electrophysiological evidence in neocortex and hippocampus suggests. Sleep can restore the brain to a reference condition by renormalizing the synaptic force (Reato et al. 2013; Stenberg 2007).

AD, which is the most studied ND, is known to be a progressive and degenerative process of the basal forebrain, in which there are a large number of cholinergic neurons, resulting in a decrease in acetylcholine. Patients with AD experience a gradual decline in the most basic and instrumental activities of daily life, and some cognitive domains such as memory, attention, perception, and language. Neuronal pathway damage that initiates and maintains sleep is the factor that can explain sleep disorders in AD, fundamentally cholinergic decrease, given the role of acetylcholine and its precursors in REM sleep induction. The ability to perform daily activities becomes impaired as the disease progresses (Hansen et al. 2008; Schwartz and Roth 2008). Understanding the areas affected by this disorder is helpful for designing strategies aimed at selective regulation of cortical function, which could also be expressed in behavioral changes, and sleep could also be approached in this same manner (Gomes-Osman et al. 2018).

Recent evidence suggests that interventions that promote neural plasticity can induce significant cognitive gains, especially in subjects at risk of or with mild AD. In view of this, the possibility has been highlighted of designing research models with innovative treatments that have few side effects and can alter the course of the disease. More specifically we refer to non-invasive stimulation techniques, the most studied of which are rTMS and tDCS. Although they have already shown some positive effects, they have not been compared to each other, and it is still necessary to determine which parameters have stronger clinical effects (cognitive, behavioral and functional) and to examine neurophysiological and neurogenesis interventions that can be regarded as biomarkers of neuroplasticity in both the acute and maintenance phases (Lozano et al. 2017; Nardone et al. 2012).

TMS is used as a tool for research, therapeutic and diagnostic purposes, since phenomena such as inhibition, excitation, and cortical plasticity can be evaluated. It has been shown that a single pulse applied to the primary motor cortex (M1) can stimulate the cortical tissues associated with the muscles of the hands or legs and thus cause motor evoked potentials (MEP) which is the general reaction of a peripheral muscle measured and reported with electromyography (EMG). TMS allows us to study the neurobiology of cognitive function and behavior (Lefaucheur et al. 2014). In rTMS, several successive pulses are applied in the same sequence in a rhythmic manner, which allows the modulation of cortical excitability. Based on this fact it is postulated that cellular plasticity is induced by repetitive use of stimulation when generating a prolonged depolarizing response, which in turn facilitates communications that depend on the activity, thus strengthening the circuits (Miniussi and Ruzzoli 2013).

rTMS can be applied at high or low frequencies producing different effects on cortical activity. The prevailing hypothesis is that low stimulation frequencies ($\leq$ 1 Hz) reduce or inhibit cortical excitability, while high frequency rTMS ($\geq$ 5 Hz) promotes increased cortical excitability (Jiang et al. 2013; Pascual-Leone et al. 2012). It is important to emphasize that these effects have been shown to last longer than the duration of the stimulation itself, and that their therapeutic efficacy has also been explored by analyzing their applications and effects on the memory, behavior and pathophysiology of various neurological and psychiatric disorders. The positive effects of these treatments on NDs are known and it is understood that the potential cumulative effects of regular sessions create the basis for therapeutic applications for diseases with neurocognitive manifestations, and some studies have shown that cognitive skills are modulated. Such findings show that rTMS is useful in restoring brain function, which could reflect its ability to recruit compensatory networks that underlie memory coding and other cognitive processes. It was therefore proposed that stimulation could modulate properties in a neural network by amplifying or reducing its operation, thus producing long-term effects (Guerra et al. 2011).

TMS-induced MEPs amplitude, area, latency and duration can be used to investigate the integrity of the corticospinal pathways as well as the motor cortex activation threshold. This discovery, which manages to combine TMS and EMG, allows for the study of processes such as excitability, plasticity, corticocortical connectivity and the interacting relationship between excitatory and inhibitory cortical processes, which can give us information about different neurotransmission systems. The combination of these tools is a valuable aid for ND and its effects on sleep. When administered in healthy subjects with 5 Hz (5 Hz-rTMS) and with a power greater than the threshold for M1, rTMS enhances cortical excitability and improves the MEPs amplitude during the train. This facilitation survives the train for about 1 second, which demonstrates plasticity shifts in the cortex. These effects appear to be related to the synaptic mechanisms of long-term potentiation (LTP) and are mainly related to the glutamatergic system through the activation of postsynaptic N-methyl-d-aspartate (NMDA) acid receptors (NMDAr). This neurophysiological evidence suggests that rTMS may be an appropriate instrument for researching changes in synaptic plasticity and excitability within the human cortex. Many researchers have introduced TMS methods to examine patterns of neuroplasticity changes in Alzheimer's disease, corroborating results suggesting that cortical physiology is affected by an underlying neurodegenerative mechanism which influences neurotransmission of acetylcholine and glutamate (Pilato et al. 2012; Nardone et al. 2012).

MCI has been addressed by few studies as a predictive factor, and even less as a mechanism of change after neuromodulatory strategies have been applied. 5 Hz-rTMS in particular provided useful data in patients with AD on various aspects of impaired glutamatergic neurotransmission. The aim of Trebbastoni and its collaborators was to investigate glutamatergic function in patients with MCI by evaluating cortical excitability and synaptic plasticity through the use of rTMS (5Hz) applied on M1 and correlating it with cognitive measurements. He proposes using this as a biomarker test. Note that changes in MEP and motor threshold (MT) amplitudes reflect plasticity indices in the motor cortex, as it is believed that increases and decreases in MEP amplitudes indicate LTP-related synaptic plasticity, and changes in MT may be considered an indirect measure of intrinsic plasticity (Trebbastoni et al. 2016).

Theta Bursts (TBS) is a form of rTMS which continuously stimulates the cerebral cortex, and involves the application of three bursts of 50 Hz rTMS repeated every 200ms, either continuously for a total of 40 s, or intermittently (every 8 s) for a total of 3 min. It has been shown to reduce cortical excitability for up to 60 minutes when applied continuously and for at least 15 minutes if applied intermittently. Because of its shorter stimulation period and lower intensity compared to traditional rTMS, it is believed to be a more efficient way to alter brain activity, but further evidence is required (Kraus et al. 2019).

tDCS is another non-invasive method commonly used to examine cerebral cortex disease, which requires applying a constant electrical current at 1-2 A, resulting in a current strength of about 0.03-0.06 mA/cm2. It is applied by electrodes placed on the scalp pointing to the brain's surface areas ("Transcranial Direct Current Stimulation (TDCS) for Depression" 2015).

In order to induce relevant changes to excitability and modify physiological, perceptual and cognitive processes in the prefrontal, parietal, temporal and occipital cortices, modulating stimulus intensity is essential. Although tDCS does not trigger the potential for action, it modulates the neuronal transmembrane potential toward either hyperpolarization or depolarization during stimulation to induce neuroplasticity. Anodic stimulation is hypothesized to cause depolarization and increase neural excitability, while cathodic stimulation would cause hyperpolarization and decrease cortical excitability (Hsu et al. 2015).

The general idea of brain stimulation is to induce changes in cortical excitability that lead to the possibility of restoring or reorganizing the functional networks responsible for impaired cognitive function or behavioral changes, such as sleep, provided that the function can be restored or compensated by mechanisms involving structural and functional changes in the brain circuits involved. The growing interest in applying rTMS and tDCS to ND patients is due to the potential physiological effects, based on the evidence known to date and what it implies: a) alteration of synaptic plasticity through the potentiation mechanism and long-term inhibition. Proof from research dealing with changes in cortical excitability showed that early intervention with rTMS could favor the restoration of partially lost connections. The reconstruction of compensatory networks can be used to restore damaged areas or contralateral homologous cortical regions that may have a similar anatomical structure and may therefore perform impaired functions; b) improved cerebral blood flow when stimulation is applied at high frequencies when using rTMS; c) it has been proposed that repeated exposure to TMS may have positive effects due to the modification of cortical oscillatory activity. rTMS can provide a reliable means of characterizing important neurophysiological and physiopathological aspects of NDs, and a decrease in plasticity and connectivity/cortical reactivity have been identified by comparing them to healthy people and even with patients suffering dementia, which is why rTMS, as well as tDCS, represent promising therapies for these conditions (Bentwich et al. 2011; Brem et al. 2013; Freitas, Mondragón-Llorca, and Pascual-Leone 2011; Manenti, Cotelli, and Miniussi 2011; Sole et al. 2006).

The neuromodulatory effects of cortical stimulation have a significant impact on glutamatergic plasticity because these treatments induce the release of gamma-aminobutyric acid (GABA) and activate NMDA receptors which in turn affect calcium channels; the enzyme cascades are activated depending on the quantity of intraneuronal calcium channels which, if sufficient, modify α-amino-3-hydroxy-5-methyl-4-isoxazolepropionic acid (AMPA) receptors, which is the main basis for LTP and long term depression (LTD). While GABA is the main neurotransmitter involved in excitability and neuroplasticity, these neuroplastic processes are mediated by other neurotransmitters such as dopamine, acetylcholine, serotonin, adrenaline and norepinephrine, and support the mechanism of action of these non-invasive techniques. GABAergic interneurons, located in the upper layers of the cortex, form synaptic networks that control low excitability and generate neuroplasticity as their behavior is coordinated through a combination of inhibitory and excitatory postsynaptic potentials (IPSP and EPSP) which they generate through specific receptors, and have an effect on glutamatergic receptors and ion channels. This is how the inhibition of GABA becomes imperative in order to achieve the synchronization of cortical circuits and, through these synchronized IPSP waves, they may spread to different networks and can be detected through oscillating rhythms if their intensity is high enough. Monoamines and acetylcholine have an impact on TDS and TMS plasticity, which increases the activation of dopamine receptors by the non-selective L-dopa precursor, thereby achieving excitability (Nardone et al. 2012).

The combination/combined application of TMS and EEG with different paradigms can be used as a plasticity index in the dorsolateral prefrontal cortex (DLPFC) and can provide crucial advantages when attempting to understand the primary brain processes underlying cognition and behavior. Future studies may also be performed to investigate possible regional pharmacological effects that may be of particular importance to neurodegenerative diseases. Considering that its manifestations are clinically different, they could prove to be a valuable asset in differentiating between diagnoses. Many studies have tried to do so, however, the need still exists for more accurate, more homogeneous samples, and greater samples of patients need to be included, so that measurements of TMS and MEP or TMS and EEG can be used as a diagnostic or response indicator of treatment by an accurate analysis of oscillations triggered by TMS.

On the other hand, the efficacy of the aforementioned non-invasive neurostimulation methods, rTMS, TBS and tDCS seemingly promise to improve the quality of sleep, and TMS is likely to provide data on ND's neurobiological foundations, as its measurements can be used to identify and control changes in cortical reactivity, integrity of inhibitory and excitatory intracortical circuits, LTP/LTD-like plasticity mechanisms and central cholinergic function. Repetitive TMS and tDCS can be used to modulate neuronal excitability and improve cortical function and, therefore, offer a potential means to curb or reverse neurodegenerative processes and influence sleep disorders (Gomes-Osman et al. 2018).

TMS has been used in combination with EEG to assess the effects of electromagnetic induction on cortical oscillations, a combination of methods that generated important neurophysiological ways to obtain information and opened several new avenues of research, for both healthy and disease states (Chokroverty 2017; Marshall et al. 2006). Simultaneous EEG recording during TMS stimulation was previously unachievable due to the technical shortcomings of EEG amplifiers that would saturate for an extended period due to the large artifacts produced by magnetic stimulation.

Non-invasive neuromodulatory strategies such as tDCS and rTMS have been developed by achieving stimulation with electric currents and magnetic fields respectively. A study published in nature in 2006, applied tDCS at 0.75 Hz for 5 minutes with 1 minute rest intervals after starting SWS, measured with EEG at rest intervals and showed that the induction of potential oscillation-like fields of slow waves during early nocturnal non-rapid eye movement (NREM) sleep improves retention of hippocampus-dependent declarative memories. In the frontal cortex, an increase in SWS and slow spindle activity can be induced using tDCS. They conducted an experiment under the same conditions but with stimulating at 5 Hz which, although a frequency that predominates during REM sleep, yielded no improvements in declarative memory. The researchers concluded that endogenous slow-potential oscillations play an important role in consolidating sleep-related memory (Reato et al. 2013). They emphasize the relative lack of studies dealing with non-invasive techniques, which leads to the need for research comparing neurophysiological and clinical findings to determine whether the use of low-frequency stimulation with non-invasive neuromodulatory techniques can not only alter cortical activity but also improve patient's quality of sleep, which is of highest concern to older adults living with NDs.

Under the same hypothesis of nocturnal brain activity characterized by slow wave activity which is modified during the night as a result of homeostatic synaptic reduction, Reato et al. (2013) used a paradigm similar to that of Marshall et al. (2006) with the intention of promoting synaptic homeostasis with 25 minutes of tDCS. They showed that an acceleration of the homeostatic decomposition of SWA in subjects stimulated by tDCS at the beginning of SWS, reduced the decomposition of slow wave oscillations (SWO) during the rest of the night.

In much the same way, but using TMS, Massimini et al. (2007) reported that, in sleeping subjects, slow waves and sleep spindles can be activated non-invasively and reliably by TMS using frequencies below 1Hz, which evokes an individual slow wave of high amplitude that originates below the coil and extends to the cortex. The research group suggests that the activation of slow-wave TMS stabilizes thalamic-cortical networks during NREM sleep (thalamo-cortical networks during non-rapid eye movement sleep). They explain that the evoked slow waves allow sleep to be deeper, and at the same time manage to increase the activity of slow waves in the EEG (0.5-4-5 Hz), which leads to the hypothesis that this allows to restore the brain and consolidate the memory (Massimini et al. 2007).

Huber et al. (2007), based on the precedent that there are changes in SWA and that it is believed to be linked to the need for sleep and occur in the waking state after being subjected to a learning task, which explains that SWA implies a change in the neuroplasticity of the brain, applied rTMS to show that the need for sleep may be related to synaptic plasticity. Their objective was to find out whether the induction of synaptic potentiation in cortical regions increases SWA. They found that the application of a protocol with pre and post phase (200 and 1500 pulses respectively) of rTMS at 5 Hz on the motor cortex at 90% of the MT induces localized potentiation of cortical EEG responses, and that SWA increases significantly, which provides evidence for a direct link between plastic changes and sleep regulation (Huber et al. 2007).

Cheng-gang Jiang et al. (2013) studied the effectiveness of rTMS in the treatment of patients with chronic primary insomnia, applied to adults of 48.31 ± 8.45 years and found that rTMS, applied to DLPFC at 1 Hz (low frequencies) continuously for two weeks at 80 percent of MT enhanced both SWS and REM sleep. He demonstrated greater efficacy than pharmacological and psychotherapeutic care for patients of the same age group. Low-frequency rTMS on DLPFC increases melatonin secretion in the pineal gland as well as brain serotonin and noradrenaline concentration. Such compounds play a major role in sustaining the sleep-wake cycle and other physiological functions of the body. Low-frequency rTMS facilitates hippocampal neurogenesis, reducing the symptoms of chronic insomnia and reducing damage to the hippocampal nerves (Jiang et al. 2013). Also quantified was the Pittsburgh Sleep Quality Index (PSQI), which was lower in the rTMS category, possibly due to a more rational sleep pattern and a more pronounced decrease in the level of excitement. Another study conducted in China also evaluated the effectiveness of the low frequency rTMS (1Hz) at 90% of resting MT, applied on the right parietal lobe in patients with comorbid Anxiety disorders and insomnia. Stimulation consisted of 3 trains of 500 pulses with an inter-train interval of 10 minutes, administered daily for 10 consecutive days. This protocol showed improvement in the Pittsburgh Sleep Quality Index (PSQI) scores (Huang et al. 2018). Notwithstanding, the long-term impact of repeated tDCS and TMS remains unclear. Other research has therefore focused on the possibility of using sensory stimulation to induce slow waves in a more natural physiological way.

The neuronal systems of the cortex, thalamus, anterior brain and brainstem, including dopaminergic, cholinergic, serotonergic and glutamatergic systems of the ascending reticular system, are affected due to the neurodegenerative processes characteristic of Alzheimer's Disease (AD). This is reflected in an increase in the amplitude of delta (2-4 Hz) and theta (4-8 Hz) waves and a decrease in the amplitude of alpha waves. Because TMS showed changes on the EEG waves, a therapeutic effect could be established in which the TMS functioned as a modulating measure of altered waves in AD: 1 Hz and 5 Hz trains were associated with alpha (8-12 Hz) and beta (14-30 Hz) wave changes. Other studies have also found that cognitive impairment caused by sleep deprivation in rodents can be treated with TMS so that work and spatial memory improvements have been found.

In summary, neuromodulatory approaches provide a new, safe and effective way to study and treat cortical changes influencing sleep in ND patients. Specifically, rTMS can replace sedative and hypnotic medication, thereby reducing drug abuse and dependence on sedative and hypnotic treatment. The clinical trials, and in particular, research based on sleep patterns, were carried out in the general population, and few reported in the ND. Although this is encouraging for this category of diseases, the etiology of both NDs and sleep mechanisms is not fully understood. We believe that the best approach should focus on different treatment protocols, compare all therapies and set stimulation criteria to develop it as an ND therapy and advance in the medical area, preventing polypharmacy and the adverse effects of inadequate medicine in the population suffering from NDs who also present sleep disorders.

References

Albert, Marilyn S., Steven T. DeKosky, Dennis Dickson, Bruno Dubois, Howard H. Feldman, Nick C. Fox, Anthony Gamst, et al. 2011. "The Diagnosis of Mild Cognitive Impairment Due to Alzheimer's Disease: Recommendations from the National Institute on Aging-Alzheimer's Association Workgroups On." *Alzheimer's & Dementia* 7 3): 270–79. https://doi.org/10.1016/j.jalz.2011.03.008.The.

An Introduction to BPSD: The IPA Complete Guides To. 2012. The International Psychogeriatric Association (IPA) Owns. 2012.

Anand, R., Kiran Dip Gill, and Abbas Ali Mahdi. 2014. "Therapeutics of Alzheimer's Disease: Past, Present and Future." *Neuropharmacology* 76 (PART A): 27–50. https://doi.org/10.1016/j.neuropharm.2013.07.004.

Barón-Herrera, Rosa Angélica, Yolanda Rebeca Peñaloza-López, Teodoro Bernardo Flores-Rodríguez, Blanca Gracicla Flores-Ávalos, Felipe García-Pedroza, and Aline B. Herrera-Rangel. 2011. "Potenciales Auditivos de Latencia Larga: Potencial de Disparidad y P300 En Dos Grupos de Adultos Mayores." *Revista Mexicana de Neurociencia* 12 (5): 242–49.

Bentwich, Jonathan, Evgenia Dobronevsky, Sergio Aichenbaum, Ran Shorer, Ruth Peretz, Michael Khaigrekht, Revital Gandelman Marton, and Jose M. Rabey. 2011. "Beneficial Effect of Repetitive Transcranial Magnetic Stimulation Combined with Cognitive Training for the Treatment of Alzheimer's Disease: A Proof of Concept Study." *Journal of Neural Transmission (Vienna, Austria : 1996)* 118 (3): 463–71. https://doi.org/10.1007/s00702-010-0578-1.

Bian, Zhijie, Qiuli Li, Lei Wang, Chengbiao Lu, Shimin Yin, and Xiaoli Li. 2014. "Relative Power and Coherence of EEG Series Are Related to Amnestic Mild Cognitive Impairment in Diabetes." *Frontiers in Aging Neuroscience* 6 (FEB): 1–9. https://doi.org/10.3389/fnagi.2014.00011.

Boeve, Bradley F., Michael H. Silber, Tanis J. Ferman, John A. Lucas, and Joseph E. Parisi. 2001. "Association of REM Sleep Behavior Disorder and Neurodegenerative

Disease May Reflect an Underlying Synucleinopathy." *Movement Disorders* 16 (4): 622–30. https://doi.org/10.1002/mds.1120.

Boggio, Paulo Sérgio, Claudia Aparecida Valasek, Camila Campanhã, Ana Carolina Alem Giglio, Nathalia Ishikawa Baptista, Olivia Morgan Lapenta, and Felipe Fregni. 2011. "Non-Invasive Brain Stimulation to Assess and Modulate Neuroplasticity in Alzheimer's Disease." *Neuropsychological Rehabilitation* 21 (5): 703–16. https://doi.org/10.1080/09602011.2011.617943.

Brem, Anna-Katharine, Lukas Schilberg, Catarina Freitas, Natasha Atkinson, Erica Seligson, and Alvaro Pascual-Leone. 2013. "Effects of Cognitive Training and RTMS in Alzheimer's Disease." *Alzheimer's & Dementia* 9 (4, Supplement): P664. https://doi.org/http://dx.doi.org/10.1016/j.jalz.2013.05.1367.

Bryce A Mander, Vikram Rao, Brandon Lu, Jared M Saletin, John R Lindquist, Sonia Ancoli-Israel, William Jagust & Matthew P Walker. 2013. "Prefrontal Atrophy, Disrupted NREM Slow Waves and Impaired Hippocampal-Dependent Memory in Aging." *Nat Neurosci* 16: 357–364.

Castro Chavira, Susana Angélica, Thalía Fernández, Catalina Alatorre, Sergio Sánchez-Moguel, Marisol Rincón, Thalía Harmony, Jimena Sandoval, and Marbella Espino. 2014. "Relationship between the Excess of Eeg Theta Activity and Cognitive Performance in Healthy Elderly Subjects." *Alzheimer's & Dementia* 10 (4): P903–4. https://doi.org/10.1016/j.jalz.2014.07.092.

Chan, Po Chi, Hsun Hua Lee, Chien Tai Hong, Chaur Jong Hu, and Dean Wu. 2018. "REM Sleep Behavior Disorder (RBD) in Dementia with Lewy Bodies (DLB)." *Behavioral Neurology* 2018. https://doi.org/10.1155/2018/9421098.

Chokroverty, Sudhansu. 2017. "Sleep Disorders in the Elderly." *Sleep Disorders Medicine: Basic Science, Technical Considerations and Clinical Aspects: Fourth Edition*, 1243. https://doi.org/10.1007/978-1-4939-6578-6.

Colrain, Ian M., Kate E. Crowley, Christian L. Nicholas, Lania Afifi, Fiona C. Baker, Mayra Padilla, Sharon R. Turlington, and John Trinder. 2010. "Sleep Evoked Delta Frequency Responses Show a Linear Decline in Amplitude Across the Adult Lifespan." *Neurobiology of Aging* 31 (5): 874–83. https://doi.org/10.1016/j.neurobiolaging.2008.06.003.SLEEP.

Coyle, Stephen, Evian Gordon, Alan Howson, and Russell Meares. 1991. "The Effects of Age on Auditory Event-Related Potentials." *Experimental Aging Research* 17 (2): 103–11. https://doi.org/10.1080/03610739108253889.

Devi, G, H U Voss, D Levine, D Abrassart, L Heier, J Halper, L Martin, and S Lowe. 2014. "Open-Label, Short-Term, Repetitive Transcranial Magnetic Stimulation in Patients with Alzheimer's Disease with Functional Imaging Correlates and Literature Review." *American Journal of Alzheimer's Disease & Other Dementias* 29 (3): 248–55. https://doi.org/http://dx.doi.org/10.1177/1533317513517047.

Drago, Valeria, Claudio Babiloni, David Bartrés-Faz, Anna Caroli, Beatriz Bosch, Tilman Hensch, Mira Didic, et al. 2011. "Disease Tracking Markers for Alzheimers Disease at the Prodromal (MCI) Stage." *Journal of Alzheimer's Disease* 26 (SUPPL. 3): 159–99. https://doi.org/10.3233/JAD-2011-0043.

Finnigan, Simon, and Ian H. Robertson. 2011. "Resting EEG Theta Power Correlates with Cognitive Performance in Healthy Older Adults." *Psychophysiology* 48 (8): 1083–87. https://doi.org/10.1111/j.1469-8986.2010.01173.x.

Freitas, Catarina, Helena Mondragón-Llorca, and Alvaro Pascual-Leone. 2011. "Noninvasive Brain Stimulation in Alzheimer's Disease: Systematic Review and Perspectives for the Future." *Experimental Gerontology* 46 (8): 611–27. https://doi.org/10.1016/j.exger.2011.04.001.

Gildner, Theresa E., Melissa A. Liebert, Paul Kowal, Somnath Chatterji, and J. Josh Snodgrass. 2014. "Associations between Sleep Duration, Sleep Quality, and Cognitive Test Performance among Older Adults from Six Middle Income Countries: Results from the Study on Global Ageing and Adult Health (SAGE)." *Journal of Clinical Sleep Medicine* 10 (6): 613–21. https://doi.org/10.5664/jcsm.3782.

Gomes-Osman, Joyce, Aprinda Indahlastari, Peter J. Fried, Danylo L.F. Cabral, Jordyn Rice, Nicole R. Nissim, Serkan Aksu, Molly E. McLaren, and Adam J. Woods. 2018. "Non-Invasive Brain Stimulation: Probing Intracortical Circuits and Improving Cognition in the Aging Brain." *Frontiers in Aging Neuroscience* 10 (JUN). https://doi.org/10.3389/fnagi.2018.00177.

Gonçalves, S. I., J. C. De Munck, P. J.W. Pouwels, R. Schoonhoven, J. P.A. Kuijer, N. M. Maurits, J. M. Hoogduin, E. J.W. Van Someren, R. M. Heethaar, and F. H. Lopes Da Silva. 2006. "Correlating the Alpha Rhythm to BOLD Using Simultaneous EEG/FMRI: Inter-Subject Variability." *NeuroImage* 30 (1): 203–13. https://doi.org/10.1016/j.neuroimage.2005.09.062.

Guerra, Andrea, Federica Assenza, Federica Bressi, Federica Scrascia, Marco Del Duca, Francesca Ursini, Stefano Vollaro, et al. 2011. "Transcranial Magnetic Stimulation Studies in Alzheimer's Disease." *International Journal of Alzheimer's Disease* 2011: 263817. https://doi.org/10.4061/2011/263817.

Hampel, Harald, Katharina Bürger, Stefan J. Teipel, Arun L.W. Bokde, Henrik Zetterberg, and Kaj Blennow. 2008. "Core Candidate Neurochemical and Imaging Biomarkers of Alzheimer's Disease**This Paper Was Presented in Part by the 1st Author at the 10th International Conference of Alzheimer's Disease and Related Disorders (ICAD), Madrid, Spain, July 2006, as an Invited Plenary Lecture." *Alzheimer's and Dementia* 4 (1): 38–48. https://doi.org/10.1016/j.jalz.2007.08.006.

Hansen, Richard A., Gerald Gartlehner, Aaron P. Webb, Laura C. Morgan, Charity G. Moore, and Daniel E. Jonas. 2008. "Efficacy and Safety of Donepezil, Galantamine, and Rivastigmine for the Treatment of Alzheimer's Disease: A Systematic Review and Meta-Analysis." *Clinical Interventions in Aging* 3 (2): 211–25. https://doi.org/10.1002/gps.

Hsu, Wan-Yu, Yixuan Ku, Theodore P Zanto, and Adam Gazzaley. 2015. "Effects of Noninvasive Brain Stimulation on Cognitive Function in Healthy Aging and Alzheimer's Disease: A Systematic Review and Meta-Analysis." *Neurobiology of Aging* 36 (8): 2348–59. https://doi.org/10.1016/j.neurobiolaging.2015.04.016.

Huang, Zhaoyang, Yue Li, Matt T. Bianchi, Shuqin Zhan, Fengying Jiang, Ning Li, Yan Ding, et al. 2018. "Repetitive Transcranial Magnetic Stimulation of the Right Parietal Cortex for Comorbid Generalized Anxiety Disorder and Insomnia: A Randomized, Double-Blind, Sham-Controlled Pilot Study." *Brain Stimulation* 11 (5): 1103–9. https://doi.org/10.1016/j.brs.2018.05.016.

Huber, Reto, Steve K. Esser, Fabio Ferrarelli, Marcello Massimini, Michael J. Peterson, and Giulio Tononi. 2007. "TMS-Induced Cortical Potentiation during Wakefulness Locally Increases Slow Wave Activity during Sleep." *PLoS ONE* 2 (3): 1–7. https://doi.org/10.1371/journal.pone.0000276.

Jiang, Cheng gang, Ting Zhang, Fa guo Yue, Ming ling Yi, and Dong Gao. 2013. "Efficacy of Repetitive Transcranial Magnetic Stimulation in the Treatment of Patients with Chronic Primary Insomnia." *Cell Biochemistry and Biophysics* 67 (1): 169–73. https://doi.org/10.1007/s12013-013-9529-4.

Katz, Richard, and Gary Horowitz. 1982. "Electroencephalogram in the Septuagenarian: Studies in a Normal Geriatric Population." *Journal of the American Geriatrics Society* 30: 273–75. https://doi.org/10.1111/j.1532-5415.1982.tb07101.x.

Knott, Verner, Lisa Bradford, Louise Dulude, Anne Millar, Fahad Alwahabi, Tim Lau, Catherine Shea, and Andrew Wiens. 2003. "Effects of Stimulus Modality and Response Mode on the P300 Event-Related Potential Differentiation of Young and Elderly Adults." *Clinical EEG and Neuroscience* 34 (4): 182–90. https://doi.org/10.1177/155005940303400404.

Kraus, Christoph, Bashkim Kadriu, Rupert Lanzenberger, Carlos A. Zarate, and Siegfried Kasper. 2019. "Prognosis and Improved Outcomes in Major Depression: A Review." *Translational Psychiatry* 9 (1). https://doi.org/10.1038/s41398-019-0460-3.

Kwong, K. K., J. W. Belliveau, D. A. Chesler, I. E. Goldberg, R. M. Weisskoff, B. P. Poncelet, D. N. Kennedy, B. E. Hoppel, M. S. Cohen, and R. Turner. 1992. "Dynamic Magnetic Resonance Imaging of Human Brain Activity during Primary Sensory Stimulation." *Proceedings of the National Academy of Sciences* 89 (12): 5675–79. https://doi.org/10.1073/pnas.89.12.5675.

Lee, Nancy Raitano, Armin Raznahan, Gregory L. Wallace, Aaron Alexander-Bloch, Liv S. Clasen, Jason P. Lerch, and Jay N. Giedd. 2014. "Anatomical Coupling among Distributed Cortical Regions in Youth Varies as a Function of Individual Differences in Vocabulary Abilities." *Human Brain Mapping* 35 (5): 1885–95. https://doi.org/10.1002/hbm.22299.

Lefaucheur, Jean-Pascal, Nathalie Andre-Obadia, Andrea Antal, Samar S Ayache, Chris Baeken, David H Benninger, Roberto M Cantello, et al. 2014. "Evidence-Based Guidelines on the Therapeutic Use of Repetitive Transcranial Magnetic Stimulation (RTMS)." *Clinical Neurophysiology: Official Journal of the International Federation of Clinical Neurophysiology* 125 (11): 2150–2206. https://doi.org/10.1016/j.clinph.2014.05.021.

López-Álvarez, Jorge, and Luis F. Agüera-Ortiz. 2015. "Nuevos Criterios Diagnósticos de La Demencia y La Enfermedad de Alzheimer: Una Visión Desde La Psicogeriatría." *Psicogeriatria* 5 (1): 3–14.

Low, Dominique V., Mark N. Wu, and Adam P. Spira. 2019. "Sleep Duration and Cognition in a Nationally Representative Sample of U.S. Older Adults." *The American Journal of Geriatric Psychiatry*, 1–11. https://doi.org/10.1016/j.jagp.2019.07.001.

Lozano, Ruth Alcalá, Erik Daniel Morelos Santana, Orely Osorio Rojas, and Jorge Julio González Olvera. 2017. "An Overview of Repetitive Transcranial Magnetic Stimulation (RTMS) in Alzheimer's Disease." *Salud Mental*. Instituto Nacional de Psiquiatria Ramon de la Fuente. https://doi.org/10.17711/SM.0185-3325.2017.022.

Manenti, Rosa, Maria Cotelli, and Carlo Miniussi. 2011. "Successful Physiological Aging and Episodic Memory: A Brain Stimulation Study." *Behavioral Brain Research* 216 (1): 153–58. https://doi.org/10.1016/j.bbr.2010.07.027.

Marino, Marco, Giorgio Arcara, Camillo Porcaro, and Dante Mantini. 2019. "Hemodynamic Correlates of Electrophysiological Activity in the Default Mode Network." *Frontiers in Neuroscience* 13 (October): 1–12. https://doi.org/10.3389/fnins.2019.01060.

Marshall, Lisa, Halla Helgadóttir, Matthias Mölle, and Jan Born. 2006. "Boosting Slow Oscillations during Sleep Potentiates Memory." *Nature* 444 (7119): 610–13. https://doi.org/10.1038/nature05278.

Massimini, Marcello, Fabio Ferrarelli, Steve K. Esser, Brady A. Riedner, Reto Huber, Michael Murphy, Michael J. Peterson, and Giulio Tononi. 2007. "Triggering Sleep Slow Waves by Transcranial Magnetic Stimulation." *Proceedings of the National Academy of Sciences of the United States of America* 104 (20): 8496–8501. https://doi.org/10.1073/pnas.0702495104.

Miniussi, Carlo, and Manuela Ruzzoli. 2013. "Transcranial Stimulation and Cognition." *Handbook of Clinical Neurology* 116: 739–50. https://doi.org/10.1016/B978-0-444-53497-2.00056-5.

Nardone, Raffaele, Jürgen Bergmann, Monica Christova, Francesca Caleri, Frediano Tezzon, Gunther Ladurner, Eugen Trinka, and Stefan Golaszewski. 2012. "Effect of Transcranial Brain Stimulation for the Treatment of Alzheimer Disease: A Review." *International Journal of Alzheimer's Disease* 2012. https://doi.org/10.1155/2012/687909.

Pascual-Leone, Alvaro, Catarina Freitas, Lindsay Oberman, Jared Horvath, Mark Halko, Mark Eldaief, Shahid Bashir, et al. 2012. "Characterizing Brain Cortical Plasticity and Network Dynamics Across the Age-Span in Health and Disease with TMS-EEG and TMS-FMRI." *Brain Topography* 24: 302–15. https://doi.org/10.1007/s10548-011-0196-8.Characterizing.

Pilato, F., P. Profice, F. Ranieri, F. Capone, R. Di Iorio, L. Florio, and V. Di Lazzaro. 2012. "Synaptic Plasticity in Neurodegenerative Diseases Evaluated and Modulated by in Vivo Neurophysiological Techniques." *Molecular Neurobiology* 46 (3): 563–71. https://doi.org/10.1007/s12035-012-8302-9.

Prince, Martin, Anders Wimo, Maëlenn Guerchet, Ali Gemma-Claire, Yu-Tzu Wu, and Matthew Prina. 2015. "World Alzheimer Report 2015: The Global Impact of Dementia - An Analysis of Prevalence, Incidence, Cost and Trends." *Alzheimer's Disease International*, 84. https://doi.org/10.1111/j.0963-7214.2004.00293.x.

Reato, Davide, Fernando Gasca, Abhishek Datta, Marom Bikson, Lisa Marshall, and Lucas C. Parra. 2013. "Transcranial Electrical Stimulation Accelerates Human Sleep Homeostasis." *PLoS Computational Biology* 9 (2). https://doi.org/10.1371/journal.pcbi.1002898.

Ronald C. Petersen 2016. "'Mild Cognitive Impairment.'" *Continuum Review Article*, 404–18. https://doi.org/10.1007/s00117-010-2094-6.

Rossi, Simone, Mark Hallett, Paolo M. Rossini, Alvaro Pascual-Leone, Giuliano Avanzini, Sven Bestmann, Alfredo Berardelli, et al. 2009. "Safety, Ethical Considerations, and Application Guidelines for the Use of Transcranial Magnetic Stimulation in Clinical Practice and Research." *Clinical Neurophysiology* 120 (12): 2008–39. https://doi.org/10.1016/j.clinph.2009.08.016.

Rossum, Ineke A. Van, Stephanie Vos, Ron Handels, and Pieter Jelle Visser. 2010. "Biomarkers as Predictors for Conversion from Mild Cognitive Impairment to Alzheimer-Type Dementia: Implications for Trial Design." *Journal of Alzheimer's Disease* 20 (3): 881–91. https://doi.org/10.3233/JAD-2010-091606.

Salat, D. H. 2014. "Imaging Small Vessel-Associated White Matter Changes in Aging." *Neuroscience* 276 (December): 174–86. https://doi.org/10.1016/j.neuroscience.2013.11.041.

Sanford, Angela M. 2017. "Mild Cognitive Impairment." *Clinics in Geriatric Medicine*. https://doi.org/10.1016/j.cger.2017.02.005.

Scheeringa, René, Karl Magnus Petersson, Andreas Kleinschmidt, Ole Jensen, and Marcel C.m. Bastiaansen. 2012. "EEG Alpha Power Modulation of FMRI Resting-State Connectivity." *Brain Connectivity* 2 (5): 254–64. https://doi.org/10.1089/brain.2012.0088.

Schwartz, Jonathan R L, and Thomas Roth. 2008. "Neurophysiology of Sleep and Wakefulness: Basic Science and Clinical Implications." *Current Neuropharmacology* 6 (4): 367–78. https://doi.org/10.2174/157015908787386050.

Sole, Cristina, Carme Junque, Imma C Clemente, and Luis Molinuevo. 2006. "Repetitive Transcranial Magnetic Stimulation Effects on Brain Function and Cognition among Elders with Memory Dysfunction. A Randomized Sham-Controlled Study," no. October. https://doi.org/10.1093/cercor/bhj083.

Steinberg, Martin, Huibo Shao, Peter Zandi, Constantine G Lyketsos, A Kathleen, Maria C Norton, John C S Breitner, David C Steffens, and T Joann. 2010. "Point and 5-Year Period Prevalence of Neuropsychiatric Symptoms in Dementia : The Cache County Study." *International Journal* 23 (2): 170–77. https://doi.org/ 10.1002/gps.1858.Point.

Stenberg, D. 2007. "Neuroanatomy and Neurochemistry of Sleep." *Cellular and Molecular Life Sciences : CMLS* 64 (10): 1187–1204. https://doi.org/ 10.1007/ s00018-007-6530-3.

Tatti, Elisa, Simone Rossi, Iglis Innocenti, Alessandro Rossi, and Emiliano Santarnecchi. 2016. "Non-Invasive Brain Stimulation of the Aging Brain: State of the Art and Future Perspectives." *Ageing Research Reviews* 29: 66–89. https://doi.org/ 10.1016/j.arr.2016.05.006.

"Transcranial Direct Current Stimulation (TDCS) for Depression." 2015, 1–9.

Trebbastoni, Alessandro, Floriana Pichiorri, Fabrizia D Antonio, Alessandra Campanelli, Emanuela Onesti, Marco Ceccanti, Carlo De Lena, and Maurizio Inghilleri. 2016. "Altered Cortical Synaptic Plasticity in Response to 5-Hz Repetitive Transcranial Magnetic Stimulation as a New Electrophysiological Finding in Amnestic Mild Cognitive Impairment Converting to Alzheimer's Disease : Results from a 4-Year Prospective Cohor" 7 (January): 1–10. https://doi.org/10.3389/fnagi.2015.00253.

Vecchio, Fabrizio, Roberta Lizio, Giovanni B. Frisoni, Raffaele Ferri, Guido Rodriguez, and Claudio Babiloni. 2011. "Electroencephalographic Rhythms in Alzheimer's Disease." *International Journal of Alzheimer's Disease* 2011. https://doi.org/10. 4061/2011/927573.

Voss, Patrice, Maryse E. Thomas, J. Miguel Cisneros-Franco, and Étienne de Villers-Sidani. 2017. "Dynamic Brains and the Changing Rules of Neuroplasticity: Implications for Learning and Recovery." *Frontiers in Psychology* 8 (OCT): 1–11. https://doi.org/10.3389/fpsyg.2017.01657.

Zanto, Theodore P, Kelly Hennigan, Mattias Östberg, Wesley C Clapp, and Adam Gazzaley. 2011. "Caregiver Burden in Alzheimer's Disease: Cross Sectional and Longitudinal Patient Correlates" 46 (4): 564–74. https://doi.org/10.1016/j.cortex.2009.08.003.Predictive.

In: Sleep, Cognition and Emotion
Editors: I. Ramírez Salado et al.
ISBN: 978-1-53617-561-5

Chapter 6

EFFECTS OF ENVIRONMENTAL POLLUTION ON SLEEP

Eduardo Sánchez-Islas* and José Samuel Mucio-Ramírez
Departamento de Neuromorfología Funcional, Dirección de Investigaciones en Neurociencias, Instituto Nacional de Psiquiatría Ramón de la Fuente Muñiz, CDMX, México

ABSTRACT

Environmental pollution is a serious global problem for human health; there is an alarming increase in diseases associated with this issue. The Persistent Organic Pollutants (POPs) have similar physical and chemical properties like high chemical and thermal stability, non-flammable, electrically resistant, low volatility at normal temperatures, the varying degree to biological and chemical degradation. Another important aspect of POPs is that have low water solubility and high lipid solubility, so that confers the capacity for bioaccumulation in fatty tissues. The POPs are classified in three categories; pesticides, industrial chemicals, and unintentional production by-products. There are growing studies demonstrating the relationship between sleep disorders and air pollution. It also has been shown that particulate matter and dust particles cause alterations in sleep, as the respiratory diseases associated with it, among the most frequent are snoring and apnea syndrome. In this sense, the present work shows the impact and the effects of POPs, particulate matter and dust onto sleep. These pollutants, directly and indirectly, disrupt rapid eye movements from REM sleep and the different cerebral waves that finally disrupt sleep causing some of the sleep disorders such as apnea syndrome, snoring and respiratory disorders mentioned.

Keywords: PCBs, PBDEs, endocrine disruptors, pesticides

* Corresponding Author's Email: edaxy@imp.edu.mx.

Introduction

Environmental pollution is a serious global problem that currently worries the international community. Not only are the large industrialized cities are the most polluted, also small and far cities, both give us alarming data also, causing major concern about the pollutants and its effects on human health, starting with headache, metabolic diseases and more. The pollution is responsible for an alarming and increasing list of illnesses in humans, animals and plants. This has generated an international interest in this problem. Pollution and health are big topic that we cannot cover in only one chapter; therefore, we will focus on the available and relevant up-to-date knowledge of this issue and the effects of some pollutants on sleep, since there are abundant reports showing the relationship between sleep disorders and human exposition to several pollutants.

Persistent Organic Pollutants, Particulate Material and Dust

There are more than 200 chemical agents considered toxic to living organisms, among them are the Persistent Organic Pollutants (POPs), a variety of organic chemical substances with a base structure made up of carbon and hydrogen and sometimes halogen radical substitutions like chlorine or bromine. The physical and chemical properties of these compounds are in general; they have high chemical and thermal stability, non-flammable, electrically resistant, low volatility at normal temperatures, they have varying degree to biological and chemical degradation, another important aspect is that have low water solubility and high lipid solubility, so that confers the capacity for bioaccumulation in fatty tissues.

As mentioned, POPs are also semi-volatile, so as consequence once released remains almost intact for years and depending on the weather conditions they can move long distances in the atmosphere before deposition occurs. The POPs have been used excessively in several human activities during the past decades and as a result of the releases to the environment, POPs are now widely distributed over large regions (including those where POPs have never been used), that explains the high concentration levels measured in the North Pole or Antarctica. This broad distribution throughout the environment is the result of natural processes involving soil, water and the air.

POPs concentrate in living organisms thanks to its liposolubility leading to a process called bioaccumulation. Even they are not soluble in water, POPs are readily absorbed in fatty tissue, so it is easy to understand that can accumulate in fish tissue, predatory birds, mammals, and humans, when any of them travel, the POPs do same within them. It is known that POPs can be found in people and animals living in regions thousands of kilometers from any major POPs source. Then, the POPs are found at higher concentrations and levels in the food chain; becoming toxic to both humans and wildlife. This extensive contamination of environmental media and living organisms includes many foodstuffs and as consequence of the uninterrupted exposure of many species, including humans, for long periods of time that span generations, resulting in both acute and chronic toxic effects.

POPs Classification

There is a list of proposed and confirmed diseases which have been generated by POPs pollution, specific effects of POPs include diverse skin allergies and hypersensitivity, damage to the central and peripheral nervous systems, reproductive disorders, disruption of the immune system and they can also have developmental and carcinogenic effects. Commercial mixtures of POPs (e.g., Aroclor, PBDEs,) are also considered endocrine disrupters (León-Olea et al. 2014), which by altering the hormonal system, can damage the reproductive and immune systems of exposed individuals as well as their offspring (web site: http://chm.pops.int/default.aspx).

The United Nations Environment Programme (UNEP) being aware that POPs poses major and increasing threats to human health and the environment, created the Stockholm Convention in May 2001 to classify the most toxic pollutant compounds in a list. The first classification included twelve POPs, they have been called "dirty dozen", and recognized to cause severe effects on human's health and the ecosystem (Table 1). There are placed also in three categories; pesticides, industrial chemicals or by-products.

Table 1. The12 Persistent Organic Pollutants. POPs

No.	Pollutant/year	Classification as Stockholm convention
1	Dioxins (1920s)	combustion by-products from plastics or PCBs
2	Furans (1920s)	combustion by-products from plastics or PCBs
3	PCBs (1929)	Several industrial uses, fornitures, plastics, foams etc.
4	DDT (1942)	Pesticides/Insecticide
5	Dieldrin (1948)	Pesticides/Insecticide
6	Heptachlor (1948)	Pesticides/Insecticide
7	Clordane (1945)	Pesticides/Insecticide
8	Hexachlorobenzene (1945)	Fungicide
9	Toxaphen (1948)	Pesticides/Insecticide
10	Aldrin (1949)	Pesticides/Insecticide
11	Endrin (1951)	Pesticides/rodenticide
12	Mirex (1959)	Pesticides/Insecticide

Source: Stockholm Convention 2001, http://chm.pops.int/default.aspx

Pesticides: Aldrin, Chlordane, DDT, Dieldrin, Endrin, Heptachlor, Hexachlorobenzene, Mirex, Toxaphene.

Industrial chemicals: Hexachlorobenzene, Polychlorinated Biphenyls (PCBs).

Unintentional production by-products: Hexachlorobenzene; Polychlorinated Dibenzo-p-dioxins and Polychlorinated Dibenzofurans (PCDD/PCDF), and PCBs.

This list is constantly updated. In fact, those that are classified by their toxicity – sixteen new POPs were listed in Annex A, B, and C to the Stockholm Convention in June 2017 (Table 2).

Table 2. The 16 New Persistent Organic Pollutants. POPs

No.	Pollutant	Classification as Stockholm convention
1	α-hexachlorocyclohexan	Insecticide
2	β-hexachlorocyclohexane	Insecticide
3	Chlordecone	Agricultural pesticides
4	Decabromodiphenyl ether (commercial mixture, c-DecaBDE)	Flame retardant
5	Hexabromobiphenyl	Flame retardant
6	Hexabromocyclododecane	Flame retardant additive
7	Hexabromodiphenyl ether and heptabromodiphenyl ether (commercial octabromodiphenyl ether)	Flame retardant additive
8	Hexachlorobutadiene	Solvent
9	Lindane	Insecticide
10	Pentachlorobenzene	Flame retardant and fungicide
11	Pentachlorophenol and its salts and esters	Herbicide, fungicide and insecticide
12	Perfluorooctane sulfonic acid (PFOS), its salts and perfluorooctane sulfonyl fluoride (PFOSF)	Additive electronic parts
13	Polychlorinated naphthalenes	Plastic additive
14	Short-chain chlorinated paraffins (SCCPs)	Flame retardant
15	Technical endosulfan and its related isomers	Insecticide
16	Tetrabromodiphenyl ether and pentabromodiphenyl ether (commercial pentabromodiphenyl ether)	Flame retardant additive

Source: Stockholm Convention 2017, http://chm.pops.int/default.aspx.

Particulate Matter and Dust

The recent data taken from the World Health Organization's 2018 (WHO) update to its Global Ambient Air Quality Database (https://www.who.int/airpollution/data/cities/en) shows the average annual concentration of particulate matter (PM), that is microscopic solid or liquid matter suspended in the air, with a diameter between 2.5 and 10 micrometers (μm) (PM10), fine particles with a diameter of 2.5 μm or less (PM2.5), ultrafine particles, dust, ashes and soot. Only to have an idea here some numbers from five cities in the world, the values are annual averages in micrograms/m^3 respectively of PM2.5 and PM10:

- Jiaozuo, Central China • PM2.5 = 85.0 and PM10 = 142.0
- Kanpur, North India • PM2.5 = 173.0 and PM10 = 318.5
- Pasakha, Bhutan • PM2.5 = 149.9 and PM10 = 275.0
- Delhi, North India • PM2.5 = 143.1 and PM10 = 292.3
- Greater Cairo, Egypt • PM2.5 = 116.6 and PM10 = 284.0

QUICK OVERVIEW OF SLEEP

When we ask what is sleep lots of definitions are found; "is to rest your mind and body, usually at night when you are lying in bed with your eyes closed" (from Longman Dictionary of Contemporary English; (https://www.ldoceonline.com/dictionary/sleep), another definition points that "sleep is the physiological state where the subject's ability to react to external stimuli is considerably diminished" (Porkka-Heiskanen, Alanko and Stenberg 2007). Since humans spend about one-third of his life sleeping, researches have been studying and demonstrated that sleep is an important part life, is essential to survival similar as food and water taking.

Sleep is a state of mind and body, where consciousness is altered, sensory activity is relatively inhibited, reduced muscle activity and inhibition of nearly all voluntary muscles during rapid eye movement (REM) sleep (Ferri et al. 2008). While we sleep, the brain and body in certain way is active, that includes neuron communication, and is the best way to form or maintain pathways that helps learn and create new memories (Brown et al. 2012). Sleep affects almost all tissue and system in the body, which mean the brain, heart, and lungs, immune function, mood, and disease resistance. Researches in human showed that lack of sleep becomes it harder to concentrate and respond quickly, also poor or bad quality sleep, increases the risk of disorders including high blood pressure, cardiovascular disease, diabetes, depression and alteration in circadian rhythmicity (Balbo, Leproult and Van Cauter 2010; Brown et al. 2012). The internal circadian clock promotes sleep daily at night, so alteration in it will affect general homeostasis, in fact, it has been demonstrated that moderate sleep disturbance leads to stress and homeostasis imbalance (Möller-Levet et al. 2013; Gonnissen et al. 2013). Physiological methods of monitoring and measuring changes while vertebrates sleep include the electroencephalogram (EEG) of brain waves, it is useful to record the brain cortex that had a specific pattern of electric activity. Electrooculography (EOG) of eye movements, and electromyography (EMG) of skeletal muscle activity.

The brain is in high activity during sleep, in this sense; sleep becomes a complex and dynamic physiological process. In the first hour or so of sleep, the brain progresses through a series of stages during which brain waves slow down. Usually sleep is divided into two types: non-rapid eye movement (NREM which has three stages) sleep and REM sleep. The sleep stages appear in a rhythmic, repetitive pattern throughout the night during a basic approximately 90–100-minute rhythmic cycle of NREM and REM sleep. In stage 1 of NREM sleep, a short period of several minutes of relatively light sleep occur, the heartbeat, breathing, and eye movements are slow. In the second stage, there is light sleep of NREM before enter deeper sleep, it is the NREM stage 3. Then occur a transitional period called slow-wave sleep or deep sleep, here the body temperature and heart rate fall in this phase. Over the next half hour or so, brain activity alters drastically, from deep slow wave sleep to REM sleep (usually lasts 10 to 15 minutes), characterized by neocortical EEG waves similar to those observed during waking, also known as paradoxical sleep, represents a smaller portion of total sleep. It is the main occasion for dreams and is associated with desynchronized and fast brain waves, eye movements and loss of muscle tone. In healthy young adults between ages 20 and 40 years, stage 1 NREM occupies approximately 2–5% of total sleep, stage 2 occupies 45–55%, stage 3 occupies 3–8%, and stage REM occupies 10–15% (Pace-Schott 2009).

REM and NREM sleep are linked to specific brain waves and neuronal activity. Several times during a night people cycle through different stages of NREM and REM sleep, with increasingly longer, deeper REM periods occurring toward morning. During NREM 1 sleep is relatively light, breathing, heartbeat and eye movements are slow, alpha waves activity with eyes closed and begin theta waves activity. During NREM 2 sleep of light sleep entering in deeper sleep breathing, heartbeat and eye movements are slower, muscles relaxation is present. Brain waves activity slows. K complex and spindle activity can be observed (12–15 Hz). NREM stage 3 is the period of deep sleep. Heartbeat and breathing slow to their lowest levels during sleep. Brain waves become even slower Presence of delta waves activity. REM sleep occurs about 90 minutes after falling asleep. The eyes move rapidly from side to side behind closed eyelids. Mixed frequency brain waves activity becomes closer to that seen in wakefulness. Breathing becomes faster and irregular, heart rate and blood pressure increases. Here occur the dreams, where muscles from arms and legs become temporarily paralyzed (Porkka-Heiskanen, Alanko and Stenberg 2007).

Circadian rhythm and homeostasis are the two internal biological mechanisms that work together to regulate the wake-sleep cycle. Circadian rhythms regulate changes in the brain and body that occur over the course of a day in wakefulness to body temperature, metabolism, and the release of hormones. Another important regulator is the biological clock (based on a roughly 24-hour day) that controls most circadian rhythms. This latter process is controlled by the suprachiasmatic nuclei (SCN) of the anterior hypothalamus, an anatomical structure considered to be the circadian master clock in mammals. Human research, conducted under a range of experimental conditions (e.g., internal desynchronization of the sleep–wake cycle, forced asynchrony paradigms, fragmented sleep–wake cycles, sleep deprivation, and sleep displacement), has highlighted the existence of a robust drive to maintain wakefulness toward the end of the habitual waking day (Lavie 2001). So, it is clear that affections on circadian rhythm will result in a reduction in sleep quality, decrease in alertness and performance in a normal day and impaired cognitive performance.

Neurotransmitters and Neuropeptides Involved in Sleep

The neurochemistry of Sleep involves a great variety of neurotransmitters and neuropeptides participates in sleep mechanism, the most studied are Adenosine; Cytokines; Nitric Oxide; GABA; Prostaglandins, Acetyl Choline, Melatonin and Vasopressin among others (Porkka-Heiskanen, Alanko and Stenberg 2007; Jouvet 1972). Only for mention some of them we know that Melatonin (N-acetyl-5-methoxytryptamine), plays a major role in regulating circadian rhythms, this a hormone secreted by the pineal gland modulates the sleep-wake cycle, its secretion is enhanced during darkness and suppressed during day light (Claustrat, Brun and Chazot 2005). It has been studied the arginine vasopressin (AVP) in the men and in rhesus monkeys. Born et al. (1992) tested whether AVP affects human sleep, they infused AVP intravenously at a constant rate. Plasma AVP concentrations were determined during sleep to ascertain that stable increases in plasma AVP levels were achieved. Increases in plasma AVP concentrations close to the upper limit of the normal physiological range of variability were associated with a marked reduction in time spent in REM sleep. In addition, AVP was found to increase signs of central nervous arousal (such as time spent awake), with these effects, however, being less consistent than those on REM sleep.

This finding suggests a physiological role for AVP in the regulation of human REM sleep. The group of Chen induced hypoglycemia in ovariectomized rhesus monkeys by intravenous administration of insulin in the presence of continuous intracerebroventricular infusions of the V1 receptor antagonist (deamino-Pen1, O-Me-Tyr2, Arg8)-vasopressin (Chen, Ordög and Knobil 1988). They found that AVP antagonist significantly attenuated the decrease in wakefulness, suggesting that AVP may, directly or indirectly induce somnolence due a moderate degree of hypoglycemia. Nitric oxide (NO) is a gaseous intercellular signaling molecule that regulates physiological and pathophysiological processes in the brain, it is produced by three main isoforms of NO synthase (NOS): neuronal (nNOS, constitutive calcium dependent), endothelial (eNOS, constitutive, calcium dependent) and inducible (iNOS, calcium independent (Dawson and Dawson 1995). Immunopositive neurons nNOS are distributed throughout the brain cortical interneurons, supraoptic and paraventricular nuclei hypothalamic, in the laterodorsal tegmentum (LDT), pedunculopontine tegmentum (PPT) and dorsal raphe. NO in neurons of the pontine tegmentum facilitates sleep. There is evidence that NO has a central role in hypothalamic circuits governing sleep among other functions like cardiovascular, osmoregulation, thyroid status, circadian rhythm, nutrient homeostasis, reproduction and growth (Canabal et al. 2007; Ding et al. 1994; Liu et al. 1998; Rettori et al. 1993). Another interesting point is that NO synthesized within the cholinergic population of neurons of the basal forebrain, while under control of the LDT, might modulate the spectral components of the EEG (Gautier-Sauvigné et al. 2005). In the cortex, NO release, measured with voltammetry, is higher during wakefulness and decreases during NREM and REM sleep (Burlet, Leger and Cespuglio 1999), Burlet recommend the use of NO voltammetric measurements combined with polygraphic recordings and local microinjections of drugs to avoid misleading interpretation, and essentially maintaining an intact NO-synthesizing system is required for normal sleep regulation. There is growing body of evidence about the role of NO in sleep; therefore, a large consensus that the intercellular messenger NO is widely implicated in the modulation of the sleep–wake states (Gautier-Sauvigné et al. 2005).

Pollution and Its Effects on Sleep

There is lot of information that environmental pollution affects breathing and people tend to have more symptoms of dry nose, eye irritation, conjunctivitis, dry throat and, that irritation alters the sleep pattern. The polluted air we breathe daily is composed of very fine and gaseous solid microparticles, as well as combustion products (nitrogen oxides), produced by the burning of fossil fuels, car emissions or the energy industry among others. Although there is a growing interest in the influence of the environment on sleep, there are growing studies that assess the relationship between sleep and air pollution.

It has been shown that the particles cause alterations in sleep, as the respiratory diseases associated with sleep, among the most frequent are snoring and apnea syndrome. For example, there is an association of PM10 with sleep and sleep-disordered breathing in adults from seven U.S. urban areas, they found evidence for the called Sleep Heart Health Study (SHHS). 1) Increases in the respiratory disturbance index are associated with increases in short-term temperature; 2) increases in sleep disorders, as measured by both the respiratory disturbance index and the overnight hypoxemic index, and decreases in the percentage of sleep efficiency; are associated with increasing levels of daily particulate matter in summer months. In this sense, they suggest that elevation in air pollution may increase the risk of poor sleep (Zanobetti et al. 2001).

PESTICITIES AND SLEEP

Pesticides include a wide variety of compounds that serve to control unwanted pests, such as insects, weeds and fungi. They are used in agricultural, residential, commercial, and school settings, both indoors and outdoors. In general, these pesticides are slightly soluble in water and are stable in the presence of sunlight, moisture, air, and heat, which makes them quite persistent in the environment. Human exposure to pesticides can occur through numerous pathways. Dietary ingestion is an important route of exposure to pesticides (Lu et al. 2008; Muñoz-Quezada et al. 2012). In the countries where the use of these compounds has been limited or prohibited, their residues are still frequently found in food, especially of animal origin, precisely because they are very stable in the environment.

The impact of pesticides on human health and sleep has been recognized internationally, which has required immediate action. There is some information of the following: Aldrine, chlordane, 1,1,1-trichloro-2,2-bis(p-chlorophenyl) ethane (DDT), Dieldrin, Endrin, Heptachlor, Mirex, Toxaphene and Hexachlorobenzene. Information on cases and the effects on sleep physiology will be described.

Organophosphorus Compounds

Organophosphorus (OP) compounds are organic substances containing phosphorus (Lewis 1998; Roberts and Reigart 2013). They are mainly used in control of a variety of insects in agriculture as an alternative to chlorinated hydrocarbons that persist in the environment. However, some organophosphorus compounds are highly effective insecticides, although some are extremely toxic to wildlife and man (Roberts and Reigart 2013). OP pesticides are a diverse class of acetylcholinesterase (AChE) inhibitors that are responsible for tremendous morbidity and mortality worldwide, killing approximately 300,000 people annually (Eddleston et al. 2012). OP compounds exert their toxicological clinical effects acting primarily on AChE, enzyme widely located throughout the nervous system, and red blood cells. OPs bind to AChE enzyme and inhibit it, resulting in an increase in acetylcholine levels within the post-synaptic cleft. This cause in overstimulation throughout the central and peripheral nervous system and to cause a variety of neurological conditions such as seizures, long-term cognitive deficits (Almeida et al. 2019; Walker 1998; Mostafalou and Abdollahi 2017; McCarren and McDonough 2016; Bruchajzer, Frydrych, and Szymańska 2015) and sleep disorders as apnea (Gaspari and Paydarfar 2011; Gaspari and Paydarfar 2007; Carey, Dunn and Gaspari 2013; Gaspari and Dunn 2013).

One of these central effects that cause exposure to OP is the apnea, this happens when breathing stops again and again while sleeping. This condition has been documented in several studies in animal models and where it has been suggested that in humans it can be a major cause of mortality after exposure to OP (Gaspari and Paydarfar 2007). There is substantial evidence that points to the importance of specific neuronal groups of the ventral cord. The pre Bötzinger complex (preBötC) contains cholinergic neurons (Shao and Feldman 2000) and is an integral part of normal respiratory function (St-Jacques and St-John 1999). Other sites in the ventral region of the medulla are involved in respiratory control and contain cholinergic synapses, including the ventral surface of the medulla (Nattie et al. 1989), the ambiguous nucleus (Jordan and Spyer 1981) and the retrotrapezoid nucleus (Rosin,Chang and Guyenet 2006). These and other sites of the brainstem may be involved in OP-induced apnea, for example, by affecting respiratory neurons that respond to chemotherapy or that generates rhythm or by inhibiting the inspiratory impulse through the pontin cholinergic circuits (Kok 1993).

Dichlorvos

Dichlorvos (2,2-dichlorovinyl dimethyl phosphate, DDVP) is an OP widely used as a pesticide to control household and agricultural pests (mushroom flies, aphids, spider mites, caterpillars, thrips, and whiteflies), in public health and to protect stored products from insects (IARC 1979; IRIS 1999). It is also used in the milling and grain handling industries and to treat a variety of parasitic worm infections in animals and humans. In some communities, it is fed to livestock to control bot fly larvae in manure (Gillett et al. 1972). It acts against insects as both a contact and a stomach poison. Since its introduction in 1964, Dichlorvos has been used commercially and has become a controversial environmental problem because it has been found in rivers and the fact that its harmful toxicity extends far beyond insects (Das 2013). This pesticide has been banned in the EU since 1998 (ATSDR 1997). However other developing countries continue to use it (Cizmas et al. 2015).

Several studies have documented the clinical effect of OP insecticide, alone or mixed with other insecticides, depending on the type and route administration of the compound to cause a toxic effect; they can induce sleep disorders, such as apnea (Takahashi 1991). Rats treated intravenously and oral with lethal doses of P = O type insecticides (covalent binding category), chlorphenvinphs (68 ng/kg) and Dichlorvos (42 ng/kg), both compound are direct cholinesterase (ChE) inhibitors and orally the P = S type, indirect ChE inhibitors, diazinon and fenthion (both doses 1 mg/kg) (Eto 1979). With all compounds, the animals showed typical signs of anti-ChE intoxication. But the chlorphenvinphos and dichlorvos administration animals exhibited typical anti- ChE poisoning signs in cardio respiration: hypertension and apnea which were antagonized by atropine. This study shows that the mechanism of organophosphorus insecticides producing acute toxicities depends on their chemical form and route of administration. P

= O and P = S insecticides produce their toxicity through AChE inhibition (Takahashi 1991).

Acute exposure to high concentrations of OP may cause rapid and lethal apnea, including other respiratory disorders (Goswamy Chaudhuri and Mahashur 1994; Munidasa et al. 2004). However, because it is an acute exposure, little has been documented in humans. A study using adult rats were administered acute doses of Dichlorvos (100 mg/kg, 3 x LD50) subcutaneously. In this study, a rodent model was used to analyze the dynamics of respiratory and cardiovascular collapse during acute OP poisoning. This study found that poisoning caused a rapidly lethal apnea. The effect consistently occurred soon after the poisoning but pulmonary insufficiency was more variable in both timing and severity. Mortality was reduced in animals that were sustained solely by artificial ventilation. In this animal model, OP poisoning causes a sequential "two hit" insult, with rapid apnea followed by delayed impairment of pulmonary gas exchange with prominent airway secretions (Gaspari and Paydarfar 2007).

With non-human primates, African Green Monkey, were given Dichlovoros at an acute dose of 75 mg/kg orally. OP caused apnea in 10 minutes with a progressive decrease in heart rate. AChE activity in blood decreased to zero in 10 min. The treated monkeys received hydrolase isolated from *Agrobacterium radiobacter* (OpdA, 1.2 mg/kg, i.v.) immediately after Dichlorvos poisoning. The animals treated with OpdA, showed no change in heart and respiratory rates from the beginning during an observation period of 240 minutes. AChE activity decreased slowly, but remained above 25% of the baseline throughout the duration. The toxic effects of Dichlorvos are hydrolysed with OpdA, delays AChE inhibition and prevents lethality (Jackson et al. 2014).

Several toxicological chronobiological studies have shown that OP compounds, Trichlorfon and Dichlorvos, are classified as chronotoxic that can alter circadian rhythms and sleep-related molecules synthesis (Jian and Feng 1990). However, studies of chronotoxic effects in humans are still scarce. Nicolau (1983), reported that trichlorfon chronic exposure to rats (10 ppm/day, for 90 days) produced marked changes in the temporal thyroid and adrenal structure (e.g., the loss of circadian rhythms) as a group phenomenon in the content of thyroid DNA and proteins and adrenal RNA, phase alterations in the rhythms of thyroid RNA and adrenal DNA, marked decrease in the amplitude of adrenal DNA and protein rhythms. On the other hand, exposure to Dichlorvos (5 ppm/day, for 90 days) leads to phase alterations, without a change in the average time of circadian rhythms in the content of DNA, RNA and proteins and a marked decrease in the amplitude of the DNA rhythm, in the thyroid and adrenal. The phase alterations found in this study are different in extent and direction, between and within the organs, which indicates an external and internal desyn-chronization of cellular metabolic and proliferative functions (Nicolau 1983).

Organochlorine Compounds

Organochlorine (OC) pesticides are synthetic compounds widely used throughout the world, principally by developed countries (Jayaraj, Megha, and Sreedev 2016; UNEP 2009; Vijgen et al. 2011). They belong to the group of chlorinated hydrocarbons containing at least one covalently bonded atom of chlorine, which have vast application in the chemical industry and in agriculture. These compounds are known for their high toxicity, slow degradation and bioaccumulation (UNEP 2009; Vijgen et al. 2011). In general, they are considered less acutely toxic than insecticides organophosphorus and n-methyl carbamates, but have a greater potential for chronic toxicity (UNEP 2009; Vijgen et al. 2011). Organochlorines are easily absorbed by inhalation or ingestion and can also be absorbed through the skin. They are usually distributed into fatty tissues, the liver and the nervous system. Most are metabolized by the liver and excreted in the urine. For some pesticides, accumulation in fatty tissue occurs during chronic exposure, so its elimination is slow. From 1940 to the 1970s, a number of other organochlorine compounds, such as Aldrine, Dieldrin, Toxaphene, Mirex, Endrin, Lindane, Heptachlor and Chlordane, were widely used as pesticides. After the recognition of their accumulation in the environment and in living beings, and the observation of some adverse effects on the environment, most have been banned or severely restricted in their use (UNEP 2009; Vijgen et al. 2011; Nayyar et al. 2014; Jayaraj, Megha and Sreedev 2016).

Chlordane

Chlordane (1,2,4,5,6,7,8,8-octachloro-3a; 4,7,7a-tetrahydro-4,7-methanoindane) is a prohibited pesticide OC in all its chemical presentations and uses for being highly toxic to living organisms and the environment as reported by several Food and Agriculture Organization, WHO and National Institute for Occupational Safety and Health (NIOSH). Like any OC, chlordane is persistent and bioaccumulates, with potential adverse effects on man and the environment due to continuous long-term exposure through water, food and other sources (ATSDR 1994). Of particular concern is its proven carcinogenic effect in laboratory rodents and its potential impact on human health due to widespread environmental contamination in the food chain. Chlordane is well absorbed in the lungs and gastrointestinal tract and through the skin. Exposure by any route can cause systemic effects (Lewis-Mikhael et al. 2015; Evangelou et al. 2016). Chlordane is a synthetic compound and was widely used as an insecticide in food crops and as a termiticide in buildings and homes. Since 1988, the use and commercial production of chlordane (except for export) has been banned in the United States and in many other countries. However, chlordane residues are still present due to previous use in many homes and

other places where it was used, as well as in the surrounding soil. There may still be old supplies of Chlordane in locations such as warehouses, garages and landfills. (web site: https://www.atsdr.cdc.gov/MMG/
MMG.asp?id=349&tid=62).

Chlordane pesticide is considered highly toxic to the environment and human health, in the same way as other environmental toxins, little is known about the toxic effect on sleep. The reports that are taken are of cases of personnel exposed in the manufacturing, of industrial accidents and domestic use. In a cross-sectional epidemiological investigation of the health status of 261 people (between 17 and 64 years of age) from 85 private households that were chronically exposed to chlordane for termite control, the possible effects on human health were evaluated. The study reported that the exposed people presented very frequent acute symptoms, such as: headache (22%), sore throat and respiratory infections (16%), fatigue (14%). While the chronic ones were less frequent, such as: sinusitis (21%), bronchitis (1 3%), migraine (8%), among others. It should be noted that in frequent acute symptoms it was also reported that people had difficulty sleeping. Comparing symptoms such as sinusitis, bronchitis and migraine with the measured concentration levels of chlordane in the indoor air of the homes, a dose-response relationship was found after individually and simultaneously controlling age, sex and smoking. In this study, it was evident that the households studied with chronic exposure to chlordane exceeded acceptable concentration limits (Menconi et al. 1988).

Aldrine and Dieldrin

Aldrin and Dieldrin are an OC insecticide that was widely used from 1948 until the 1990s, when it was banned in most countries. However, in 1972, the EPA approved the use of Aldrin and Dieldrin to kill termites. Aldrin is a member of the so-called "classic organochlorines" (COC) pesticides group. Aldrin and Dieldrin are two structurally similar compounds that were once used as insecticides. They are synthetic chemicals and do not occur naturally in the environment (ATSDR 2002).

Aldrin (HHDN): 1,2,3,4,10,10-hexachloro1,4,4α,5,8,8α-hexahydro-1,4-endo,exo-5,8-dimethanonaphthalene and Dieldrin (HEOD): 1,2,3,4,10,10-hexacloro-6,7-epoxi-1,4,4α,5,6,7,8,8α-octahidro-1,4 endo,exo-5,8-dimetanonaftalina (ATSDR 2002). Aldrin decays very slowly once released into the environment. Though it is rapidly converted to Dieldrin by plants and bacteria, Dieldrin maintains the same toxic effects and slow decay of Aldrin (Nakatsugawa, Ishida and Dahm 1965). In this section, the two chemicals are treated together because the Aldrin becomes Dieldrin when it enters the environment or the body. Aldrin is not toxic to insects, until it is oxidized by insects to the Dieldrin form which is the active compound.

Although many toxic effects of Aldrine/Dieldrin have been discovered, the exact mechanisms underlying toxicity have not yet been determined. The only toxic Aldrin/Dieldrin-induced process that has been documented to a large extent is neurotoxicity (Richardson et al. 2006). Aldrin/Dieldrin-induced neurotoxicity has been reported to be due to strong stimulation of the central nervous system, as it can cause hyperexcitation and seizures (Gupta 1975). This effect is exerted through two different mechanisms: One mechanism is by the inhibition of transmembrane proteins, which ATPases move calcium in the brain. Aldrin by inhibiting the release of this ion in the nerve terminal increases its intracellular content. This results in an improved release of neurotransmitters (Mehrotra 1988). The second mechanism, Aldrin blocks the activity of the inhibitory neurotransmitter, gamma-aminobutyric acid (GABA) through the GABAA chloride receptor channel complex, by blocking this receptor, chloride does not move to the synapse, and prevents it from hyperpolarizing (Glotfelty 1978).

The AMA Committee on Toxicology (1960) suggested that the median lethal dose by ingestion in humans is probably about 65 mg/kg and "untoword syptoms" will result from a single dose of 10 mg/kg or more. Evidence from animal studies using high doses of Aldrin and Dieldrin OC (24-167 mg/kg) reported that can cause various neurological, hepatic and carcinogenic effects (AMA 1960). According to Occupational Safety and Health Administration (OSHA), recommends a maximum average of Aldrin/Dieldrin in the work air of 250 micrograms per cubic meter of air ($\mu g/m^3$) during a day of eight hours a day, and forty hours a week. The NIOSH, recommends the same limit during a daily day of ten hours, forty hours per week.

Prior to banning the use of Aldrin/Dieldrin in 1990, clinical studies of accidental and occupational exposure were reported (Ibanez-Peterson and De Franzanetti 1957; Hayes 1957; Kazantzis, Mclaughlin and Prior 1964; Jager 1970). Dieldrin spray operators were exposed to toxic and reported significant neuropsychological alterations such as: a temporary personality change characterized by fear, crying, mania and other alterations behavior. Also, patients reported difficulty sleeping and bad dreams. In addition, the operators had an abnormal performance in several psychological tests applied (Ibanez-Peterson and De Franzanetti 1957; Hayes (1957) reported several cases of occupational spraymen poisoning that were exposed to Dieldrin for prolonged periods (thirty or more weeks) in two countries from South American (Ecuador) and African (Nigeria). The symptoms reported were: blurred vision, diplopia, tinnitis, sweating, nausea, alteration of reflexes, incoordination, nystagmus, and personality changes. Including difficulty to sleep and bad dreams. Bianchini and Blazquez (1956) reported a similar case in another Venezuelan workers group with symptoms of intoxication including sleep disorders and electroencephalog-ram alterations. In 1964, was reported a case of two South African adolescents who accidentally ingested contaminated cocoa with 50% Dieldrin. After 30 minutes, the adolescents showed signs of poisoning such as dizziness and profuse vomiting, quickly going into a coma with continuous seizures. Twenty-four hours later with treatment they returned to normal, however one of them reported sleep rhythm disordered and was manic and irrational for about seven days (Fry 1964). However, Prior and Deacon (1969) reported a double-blind pharmaco-logical study involving healthy voluntary men for three years, one group received a placebo and the other group received a daily low oral dose of Dieldrin for a period of two of the three years (Prior and Deacon 1969). The persons treated did not report any alteration in the sleep habit or any other physiological effect. In another morbidity study of workers who were engaged in the manufacture of Aldrin/Dieldrin between 1979 and 1990, they did not present degenerative disorders of the nervous system (de Jong 1991). However, this study reported that between the ages of <30 years and between 46 and 50 years old, workers had a significant increase in mental disorders. The diseases observed in the workers were classified as short-term stress and depression reactions. Operators exposed to the compounds also reported sleep disorders (de Jong 1991).

Polychlorinated Dibenzo-p-Dioxins (PCDD) and Polychlorinated Dibenzofurans (PCDF)

Polychlorinated dibenzo-p-dioxins (PCDD) and polychlorinated dibenzofurans (PCDF) are environmental contaminants detectable in almost all compartments of the global ecosystem in trace amounts. They constitute a group of relevant persistent organic pollutants with increased chronic toxicity (Shiu et al., 1988). PCDD/F can be emitted by different human activities and industrial processes where they can be present as unwanted by-products. PCDD/F can be emitted from the burning of biomass and fossil fuels and stationary sources such as waste incineration (Fiedler 2003).

PCDD and PCDF can have between one and eight chlorine atoms attached to the dibenzo-p-dioxin or dibenzofuran molecule, respectively (Mackay, Shiu and Ma 1992) and form several congeners. This pattern of substitution of atoms results in eight homologs and 75 congeners for PCDD and 135 for PCDF, a total of 210 compounds. Of which 17 PCDD/PCDF compounds have the chemical composition of 2,3,7,8-chlorosubstituted, are the most bioaccumulative and toxic compounds (Van den Berg et al. 2006). Even within these 17 congeners, bioaccumulation factors, as well as toxic potencies, are not similar. The 2,3,7,8 – TCDD (2,3,7,8-Cl4DD or 2,3,7,8-TCDD) congener is the most toxic and was the main pollutant compound found in 2,4,5-trichloroacetic acid (2,4,5-T), the so-called “Orange Agent”, widely disseminated in the Vietnam War (Schecter, Ryan and Constable 1986) and was also the main polluter in the “Seveso Episode” (Mocarelli 2001). In 1997, the International Agency for Research on Cancer (IARC) listed the PCDD/PCDF as persistent organic pollutants, as a carcinogen for humans (Group 1 carcinogen) (IARC 1997).

It was reported that the biological action mechanism of TCDD/TCDF and dioxin-like phosphochlorides act through a specific cytoplasmic receptor present in all cells: the aryl hydrocarbon receptor (ArR) (Poellinger 2000; Mandal 2005; Lindén 2010). This receptor is a transcription factor that participates in gene expression. In fact, experiments in mice showed that high doses of TCDD/TCDF increase or decrease the expression of several hundred genes related to ArR (Tijet et al. 2006). Examples of these genes are those of enzymes that activate the breakdown of strangers in the body and are often toxic compounds. TCDD increases enzymes that break down, for example, carcinogenic polycyclic hydrocarbons such as benzo(a)pyrene (3,4-bezopyere) (Okey 2007).

The toxic effects of PCDD/F were tested on animals (Nau 2006). However, in human epidemiological studies on the carcinogenic and other (sub) chronic effects of PCDD/F there are few reports, these are from the regions where the toxic compounds were accidentally released, such was the case of Seveso, Italy (Reggiani 1978; Bertazzi et al.1998; Bertazzi et al. 2001; Nau 2006; Zieliński et al. 2014). These studies are mainly based on the effect of 2,3,7,8-TCDD (Nau 2006). The basic determinants of pharmacokinetic behavior are similar in animals and in man. The most toxic effects, at chronic doses of PCDD/F, reported are: tiredness, loss of appetite and libido, infertility, chronic skin damage (chloracne), hepatorenal and endocrine damage, carcinogenic and teratrogenic lesions (Tollefson 1991; Sweeney and Mocarelli 2000; Bertazzi et al. 2001; Nau 2006; Donato and Zani 2010; Apostoli, Bergonzi, and Catalani 2011). The oral bioavailability of PCDD/F depends on the dose, the means and in specific to each congener. PCDD/F can be absorbed from animal dairy products (milk, butter, meat and fat) or reabsorbed from the soil, ashes or smooth particles. Once absorbed, PCDD/F will accumulate mainly in lipoproteins in the blood, liver and fatty tissues (Fiedler 2003; Pereira 2004; Larsen 2006; Bocio and Domingo 2005).

The toxic effects of PCDD/F have been reported in humans, as described with other toxic compounds, all reports are of several dramatic occupational accidents in the industries where these compounds are manufactured or used (Zack and Suskind 1980; Moses et al. 1984). Suskind (1953) reported the effects of a group of workers in an occupational accident at the chemical plant in Nitro, West Virginia, in which they were exposed to organochlorine compounds, 1,2,4,5-tetrachlorobenzene. The workers presented chloracne, the typical sign of toxicity caused by TCDD, in addition to other systemic signs and symptoms. The operators also reported sleep disturbances, insomnia and hypersomnolence.

In another accident in Binghamton, NY, 482 people who were potentially exposed to polychlorinated biphenyls (PCBs), dibenzo-p-dioxins and dibenzofurans from an electric transformer fire. People the most affected by the compounds had a high incidence of chloracne, abortions, congenital malformations, cancer and mortality. While in the less affected they reported milder signs and symptoms (muscle pain, skin color changes). In the same way, as in previous occupational accidents, people reported very nervous and sleep disturbances (Fitzgerald et al.1989).

In Linz, Austria, Neuberger, Kundi and Jäger (1998) conducted a follow-up study of personnel exposed (159) to TCDD/F (mainly 2, 3, 7, 8-TCDD), reported from 1969 to 1975. Some had been exposed during laboratory research for this production of Phenoxy chlorophenols and herbicides in 1969–71, several during cleaning and repair in 1972–73 and some during special operations on waste, contaminated tanks, etc. until 1975 (Kauppinen et al. 1993). The results were compared with people without exposure to the component. The researchers reported that in the exposed workers, the most frequent cause of sick leave was chloracne, which persisted in 32%. Neurological symptoms were reported frequently of 32% headache, 30% neuralgia and a considerable 44% had sleep disorders (like apnea) (Neuberger, Kundi and Jäger 1998).

In 1968, in northern Kyūshū, Japan, people experienced a massive poisoning by polychlorinated biphenyls (PCBs). It was later discovered that 2,3,4,7,8-pentachlorodibenzofuran (PCDF) was the main toxic compound among Yusho patients (Mitoma et al. 2015). Rice bran oil produced by Kanemi in Kyushu was contaminated with PCB (Kanechlor-400) during production. For deodorization, the oil was heated using PCB as a heating medium, circulating through pipes. Due to the holes in the pipes, the PCB leaked into the rice bran oil. Rice bran oil contaminated was sold to poultry farmers for use as a food supplement and to consumers for use in the kitchen (Yoshimura 2003; Suzuki and Kobayashi 1968). Later months, farmers began to report that their birds were dying due to the apparent difficulty in breathing; in total 400,000 birds died. Around 14,000 people who had consumed the contaminated rice oil were affected in Japan (Suzuki and Kobayashi 2007; Tsukimori et al. 2008).

Recently Kondo et al. (2018) conducted a complete study of patients with Yusho (140; 74 men and 66 women, average age: 67 years) who became contaminated with PCBs in 1968. It is the first clinical study that examines in more detail sleep disorders among other symptoms in patients who were exposed to high levels of dioxins (2,3,4,7,8 pentachlorodibenzo-furan; PCDF) and dioxin-like compounds. The researchers reported that 51.8% of the patients showed moderate to severe symptoms of insomnia. Also, the number of patients who had difficulty initiating sleep (DIS), difficulty maintaining sleep (DMS), and waking up too early (WE) was 74 (54.0%), 89 (65.0%), and 70 (51.1%), respectively. The number of patients who had any one of the three symptoms was 110 (80.3%). The number of patients who had moderate to severe DIS, DMS, and WE were 41 (29.9%), 52 (38.0%) and 46 (33.6%), respectively. The number of patients who had any of the three moderates to severe symptoms was 71 (51.8%). Studies in animals exposed to PCDD have reported that it can affect the circadian rhythm and thereby alter genes, proteins and hormone production, such as melatonin, an important molecule in sleep (Pohjanvirta et al. 1989; Garrett 2006; Sato et al. 2008). Pohjanvirta et al. (1989) reported that animals exposed to a TCDD high dose (50 mg/kg, i.p.) for three days and subsequently at different times slaughtered, observed that TCDD significantly depressed serum melatonin levels. This study highlights the effect of this compound on the production of hormones regulated by the circadian rhythm.

α-Hexachlorocyclohexane y β-Hexachlorocyclohexane

Hexachlorocyclohexane (HCH) is formally known, for its synthetic relationship with benzene, as benzene hexachloride (BHC). Like other organochlorine insecticides, it also accumulates in biological tissues. HCH is a polyhalogenated organic compound consisting of a six-carbon ring with a chlorine and hydrogen attached to each carbon. HCH is a mixture of eight or more stereoisomers used in the agricultural area as a pesticide to protect crops from various pests (ATSDR 2005). When HCH is manufactured by benzene photochlorination, a product of variable chemical composition (isomers) is produced, which generally contains 60 to 70% α-HCH, 5 to 12% β-HCH, 10 to 15% γ-HCH, 6 to 10% of δ-HCH, and some other minor components. One of these forms, gamma-HCH (or γ-HCH, commonly called Lindane), is produced and used as an insecticide in fruits, vegetables and forest crops, and in animals and animal premises (Willett, Ulrich, and Hites 1998). This compound is a white solid whose vapor can evaporate in the air. The vapor is colorless and has a slight musty smell when present in 12 or more parts of HCH per million parts of air (ppm). γ-HCH has not been produced in the United States since 1976. However, imported γ-HCH is available in the United States for the use of insecticides such as dust, dust, liquid or concentrate. It is also available as a prescription medication (lotion, cream or shampoo) to treat and/or control scabies (mites) and lice in humans. The production of HCH was banned in 1976, but the compound can still be used in other third world countries. In 2009, γ-, α- and β-HCH were designated as persistent organic pollutants in the Stockholm Convention and, as such, should be phased out in places where it is still used (POPRC 2008). Similar to other organochlorine pesticides, the liver appears to be a primary white organ of HCH; however, a substantial number of toxicological studies have examined the ability of the compound to alter reproductive and developmental characteristics in experimental animals or to induce serious teratogenic effects (ATSDR 2005).

The EPA recommends the restricted use of HCH and has also recommended how much of this compound may be present in drinking water during specific periods without causing health effects. This organization states that children should not have more than 1.2 milligrams of HCH per liter of drinking water (mg/L) for up to 10 days. While in adults, throughout his life, it is recommended that there should be no more than 0.0002 mg/L of HCH.

The Occupational Safety and Health Administration (OSHA) regulates γ-HCH levels in the workplace. The maximum amount allowed in the workplace air during an 8-hour workday in a 40-hour work week is 0.5 mg per cubic meter of air.

The distribution of HCH isomers in humans and animals is primarily into the adipose tissue but also to the brain, kidney, muscle, blood, and other tissues. β-HCH accumulates much greater extent than γ-HCH. The excretion of HCH isomer metabolites is primarily through the urine (Baumann et al. 1980; Siddiqui et al. 1981).

The mode of action of HCH in the nervous system is acting as an antagonist of the chloride channel regulated by GABA. HCH interferes with chloride channels in the synaptic nerve membrane and disrupting the ion transfer and impulse neurotransmission between nerve cells. This antagonistic effect of HCH keeps neurons in a state of continuous activity. Reason why people that are exposed to this component convulse. Therefore, seizures caused by γ-HCH can be antagonized by GABA-A mimetic (ATSDR 2005).

Several studies have reported the carcinogenic, teratological, gastric, neurophysiological and neuropsychological effects caused by the continuous exposure of people in industrial or agriculture to HCH pesticides (Durham 1963; Taddeini and Watson 1968; Allen 1979; Joy 1985; Joy 1982; Dich et al. 1997; Martenies and Perry 2013).

Nigam et al. (1986) and Kashyap (1986) conducted studies in India with industrial personnel where HCH was manufactured, formulated and distributed pesticides. The researchers reported that non-manipulators exposed indirectly and manipulators directly exposed to HCH during the manufacture and formulation of pesticides had mean serum levels of 0.27 ppm (non-manipulators) and 0.6 ppm (manipulators) total HCH. As much as 60-100% of the total serum HCH measured was for the β-HCH isomer. The ranges of serum HCH levels measured in all exposed workers were 0.07-0.72 ppm of β-HCH, 0.004-0.18 ppm of α-HCH, 0-0.17 ppm of γ-HCH and 0-0.16 ppm of δ-HCH. In both studies, both manipulators and non-manipulators presented paresthesias in the face and extremities, headache and vertigo; other symptoms include general malaise, vomiting, tremor, apprehension, confusion, memory problems and loss of libido. In addition, the workers, in both reports, presented sleep loss.

HCH has also been related to causing alterations in circadian rhythms at the level of synthesis of DNA, proteins and hormones in different experimental models (Batth 1972; Schulte-Hermann and Landgraf 1974, 1976; Attia et al. 1990; Attia et al. 1991). Attia et al. (1990) reported the effect of organochlorine insecticide, Lindane (1,2,3,4,5,6-hexachlorocyclo-hexane), in the production of pineal melatonin in male rats exposed to this compound (17,782 mg/kg, gastric route). The researchers report an increase in the activity of pineal N-acetyltransferase, in pineal and serum melatonin levels. They did not observe changes in pineal levels of 5-hydroxytryptophan, serotonin or 5-hydroxyindole acetic acid. The results of the study clearly show that HCH, lindane, modifies the synthesis of pineal melatonin.

Hexabromocyclododecane (HBCD or HBCDD)

Hexabromocyclododecane, which has the molecular formula $C_{12}H_{18}Br_6$ (HBCD or HBCDD). HBCD is an organobromated compound, it is considered a brominated flame retardant due to its physical and chemical characteristics. It consists of twelve carbon atoms, eighteen hydrogens and six bromine atoms attached to the ring. HBCD is mainly used as an additive in extruded polystyrene (XPS) and expanded (EPS) foam that is used as thermal insulation in the construction industry. It is also used in upholstered furniture, automotive interior textiles, car cushions and insulation blocks in trucks, packaging material, cassette recorder casing and electrical and electronic equipment. According to the United Nations Environment Program (UNEP), HBCD is produced in China, Europe, Japan and the United States. The current known annual production is approximately 28,000 tons per year. Most of the market volume is used in Europe and China (UNEP 2009). In 2013, due to its persistence, bioaccumulation, toxicity and ecotoxicity of hexabromocyclododecane, the Stockholm Convention on Persistent Organic Pollutants decided to include it in Annex A of the Convention with specific exemptions for the production and use of expanded polystyrene and extruded polystyrene in buildings (POPRC8.3 2013).

The HBCD, due to its chemical properties and being ubiquitous compounds, has a wide distribution in the environment, animals and humans (Covaci et al. 2006). Animal studies indicate that HBCD is absorbed and excreted rapidly (within 72 hours) after oral exposure. The compound when entering the body is widely distributed in all tissues, the highest concentrations are found in adipose tissue and muscle, followed by the liver, lung, kidney, blood and brain (de Wit 2002). Exposure to HBCD can come from various sources: food, human milk, dust, indoor or outdoor air and soil. Food and human milk remain the major contributors of exposure with an estimated intake of HBCD of 1.12 and 2.1 ng/kg-day respectively (Roosens et al. 2010). Food exposure is significant compared to intake through house dust (0.45 ng/kg-day) or through outdoor and indoor air ($0.25 \times 10{-}3$ to $41 \times 10{-}3$ ng/kg-day) but considering the high concentrations in house dust, dust inhalation is still considered as a major contributor, especially in kids (Driffield et al. 2008).

Animal studies indicate that oral exposure to HBCD induces hepatic cytochrome P450 enzymes in rats (Germer et al. 2006), disrupts the normal absorption of neurotransmitters in the brain of rats (Mariussen and Fonnum 2003) and may produce developmental neurotoxics effects, such as aberrations in spontaneous behavior, learning and memory function after neonatal exposure in rats (Eriksson et al. 2002). There are also indications that it can alter the thyroid hormone system, affect the gene expression mediated by the thyroid hormone receptor (Yamada-Okabe et al. 2005) and induce cancer through a non-mutagenic mechanism (Ronisz et al. 2004).

Very limited information is available regarding the toxicological effects of HBCD in humans and there are no clinical reports concerning sleep. This can be attributed to the lack of sufficient information on human exposure pathways and the bioavailability of the flame retardant through different routes (Law et al. 2008). Recently, Farmahin et al. (2019) employed the sequencing of the complete genome RNA in liver samples of Fischer rats (male and female) exposed for 28 days at different concentrations of technical HBCD (0, 250, 1250 and 5000 mg/kg). The researchers found that exposure to HBCD altered 428 and 250 gene transcripts in males and females, respectively, that was involved in several metabolic and cellular issues among these related to the regulation of circadian rhythm and sleep. These genes are critical for proper glucose regulation and lipid metabolism, sleep and energy balance. The genes related to circadian/sleep regulation that saw significant increases in expression were Per3, Nr1d1 and Nr1d2, and those that decreased significantly in expression were Cry1 and Usp2 in both sexes. Increases in the expression of Per3 and Nr1d1/2 and decreases of Usp2 induced by HBCD can lead to repressed circadian transcriptional activities driven by the CLOCK (Circadian Locomotor Output Cycles Kaput) and B smal1 genes. This study shows that HBDC can be a xenobiotic that can influence circadian rhythms through the regulation of circadian gene expression as suggested with other compounds (Claudel et al. 2007).

The Polychlorinated Biphenyls and Polybrominated Diphenyl Ethers

Persistent organic chemicals used in industrial applications are major contributors to environmental contamination, and are responsible for contributing to a variety of deleterious human health effects the polychlorinated biphenyls (PCBs) are a family of 209 different components (congeners), depending on the number and position of chlorines around the biphenyl ring. The PCBs are fat-soluble substances to which everyone is exposed through ingesting animal fats, inhalation, or dermal contact (Carpenter 2006). PCBs are persistent organohalogen compounds of which chronic exposure can lead to endocrine disruption and neurotoxicity. Experimental studies suggest that Aroclor 1254, a PCB mixture, has the potential to disrupt endocrine functions associated with reproductive and thyroid homeostasis (Coburn, Gillard, and Curras-Collazo 2004; Coburn, Curras-Collazo and Kodavanti 2007; Mucio-Ramirez et al. 2017). Acute effects after high-level exposure to PCBs have been described for some of the organochlorine have caused well documented episodes of mass poisoning called "Yusho" and "Yu Cheng", that occurred in China, Taiwan, and in Japan. (Kundo et al.

2018; Kuratsune et al. 1971). We have been studying the neurotoxicity of Aroclor 1254 and the structurally related brominated flame retardant, DE-71, that belongs to a polybrominated diphenyl ethers (PBDEs). Data from our lab indicate that PCBs and PBDEs significantly reduce physiolo-gically-activated central VP release from rat SON in vitro and that early adult exposure to Aroclor 1254 disrupts osmotically stimulated plasma and central VP (Coburn, Gillard, and Curras-Collazo 2004; Coburn, Curras-Collazo and Kodavanti 2007; Coburn et al. 2015; Mucio-Ramirez et al. 2017). The in utero exposure to Aroclor 1254 produces abnormally reduced nitrergic expression in postnatal rats, showing that NO is affected in early stages of live. We measured the immunereactivity of nNOS in brain tissue, we found that Aroclor 1254- induced disruption of NOS activity and it may represent one mechanism through which PCBs and similarly acting environmental toxicants may affect important regulatory physiological functions governed by the hypothalamus. Given that some PCB congeners have been reported to inhibit hypothalamic NOS in vitro (Sharma and Kodavanti 2002), the experiment we did was designed to determine if PCB exposure in vivo alters NOS activity in SON MNCs during osmotic challenge in order to delineate a potential mechanism of neuroendocrine disruption by PCBs. Upon maturation exposed rats also show reduced stimulated NADPH-d responses, an effect not seen in rats exposed in adulthood. This effect persists into late adulthood, when the reduced NOS activity is coincident with significantly elevated plasma osmolality in response to hyperosmotic stimulation (Coburn et al. 2015). We also examined the differential susceptibility of NOS activity within the SON to developmental and adult exposure to PCBs. As an indicator of NOS activity, we used nicotinamide adenine dinucleotide phosphate diaphorase (NADPH-d) histochemistry, that can be used, in paraformal-dehyde-fixed tissue (Coburn et al. 2015). Our findings indicate that early developmental PCB exposure significantly compromises NOS responses to osmotic activation in the adult supraoptic nucleus, an effect which may lead to PCB-induced disruption of neuroendocrine function and osmoregulation Aroclor 1254 induced disruption of NOS activity that may represent one mechanism through which PCBs may affect important regulatory physiological functions governed by the hypothalamus. Since NO has a central role in hypothalamic circuits governing sleep, its disruption may cause indirectly some sleep diseases.

Based on our previous report showing elevated plasma and reduced somatodendritic vasopressin in response to these pollutants, we hypothesize that these organohalogens may disrupt the AVP transcription and/or protein content in the PVN and SON cells populations in rats with prolonged or acute hyperosmotic challenge. Therefore, the aim of our next research was to elucidate the effects of perinatal exposure to either A1254 or DE-71 (30mg kg/day) on AVP gene and protein responses to osmotic stimulation in the rat PVN and SON hypothalamic nuclei. Perinatal exposure to either A1254 or DE-71 affects the AVP system similarly in males and females; this effect was more evident during the hyperosmotic challenge although significant effects were detected under basal conditions. Exposure dramatically reduces hyperosmotically stimulated responses in AVP content and mRNA expression of MNCs in the PVN and the SON. These PCBs and PBDEs seem to produce similar disruption of neuroendocrine processes. The physiological outcome of the lack of response to increment AVP content in magnocellular cells during hyperosmotic stress in A1254- or DE-71-exposed rats led to a rise in plasma osmolality, an indication of disrupted water and electrolyte balance. Vasopressin, as mention previously not only participates in endocrine regulation of body fluid and cardiovascular function, but also is an important regulator of sleep. Regions such as the hypothalamus, amygdala, and hippocampus are known to contain either AVP neurons or terminals and play a role in cognitive functions and regulating complex social behaviors (Stoop 2012). Our findings of AVP system disruption by Aroclor 1254 and DE-71 highlight the possibility that these organohalogens may affect those other AVP brain functions like sleep.

Conclusion

The information collected to date and the growing appearance of new POPs confirm that these compounds (in particular PCBs and PBDEs; Arochlor or DE-71) as well as PM2.5 particles and dust directly and indirectly disrupt sleep, whether REM NoREM and the different waves recorded in the EEG, that finally disrupt sleep and causes alteration, such as apnea, snoring, and this can cause death. Due to the neurotoxicity of the organohalogens and its constant contact with humans there is a big concern about the lack of adequate legislation in many countries, especially the third world, and permanent monitoring programs to evaluate the degree of contamination in the population, especially in infants. More lines of research and refined studies would be required to know the fine action mechanisms by the pollutants addressed and how they alters cellular signaling, genes involved and brain areas affected involved in sleep. In addition, the high body burdens of PCBs and PBDEs in the general population are of particular concern. Environmental organohalogens, PM and dust should, therefore, be considered a public health threat that must be addressed by the health care institutions of the most polluted countries.

Acknowledgments

This study was supported, in part, by the INPRFM Research Support Fund (NC143290.0), CONACYT founding SEP-CONACYT-MLO-283268.

REFERENCES

Agency for Toxic Substances and Disease Registry (ATSDR). 1994. *"Toxicological Profile for chlordane."* Atlanta, GA: U.S. Department of Health and Human Services, Public Health Service.

Agency for Toxic Substances and Disease Registry (ATSDR). 1997. *"Toxicological Profile for Dichlorvos."* Public Health Service, U.S. Department of Health and Human Services, Atlanta, GA.

Agency for Toxic Substances and Disease Registry (ATSDR). 2002. *"Toxicological Profile for Aldrin/Dieldrin."* Atlanta, GA: U.S. Department of Health and Human Services, Public Health Service.

Agency for Toxic Substances and Disease Registry (ATSDR). 2005. *"Toxicological profile for Hexachlorocyclohexane."* Atlanta, GA: U.S. Department of Health and Human Services, Public Health Service.

Allen, J. R., Hargraves, W. A., Hsia, M. S., Lin, F. S. 1979. "Comparative toxicology of chlorinated compounds on mammalian species." *Pharmacology & Therapeutics* 7(3):513-547. doi: 10.1016/0163-7258(79)90041-x.

Almeida, D. L., Pavanello, A., Saavedra, L. P., Pereira, T. S., de Castro-Prado, M. A. A., de Freitas Mathias, P. C. 2019. "Environmental monitoring and the developmental origins of health and disease." *Journal of Developmental Origins of Health and Disease* 27:1-8. doi:10.1017/S2040174419000151.

AMA Committee on Toxicology. 1960. *"Occupational dieldrin poisoning."* 172:18. P. 2077.

Apostoli, P., Bergonzi, R., and Catalani, S. 2011. "Polychlorinated dibenzofurans and cancer." *Giornale Italiano di Medicina del Lavoro ed Ergonomia* 33(2):185-192.

Attia, A. M., Mostafa, M. H., Soliman, S. A., El-Sebae, A. H., Nonaka, K. O., Withyachumnarnkul, B., Reiter, R. J. 1990. "The organochlorine insecticide 1, 2, 3, 4, 5, 6-hexachlorocyclohexane (lindane) but not 1, 1, 1-trichloro-2, 2-bis (p-chlorophenyl) ethane (DDT) augments the nocturnal increase in pineal N-acetyltransferase activity and pineal and serum melatonin levels." *Neurochemical Research* 15(7): 673-680. doi: 10.1007/bf00973647.

Attia, A. M., Richardson, B. A., Rodriguez, C., Mostafa, M. H., Soliman, S. A., El-Sebae, A. H., Reiter, R. J. 1991. "Lindane may enhance nocturnal pinealN-acetyltransferase activity via β-adrenergic receptors." *Brain Research* 554(1-2):253-256. doi: 10.1016/0006-8993(91)90197-4.

Balbo, M., Leproult, R., and Van Cauter, E. 2010. "Impact of sleep and its disturbances on hypothalamo-pituitary-adrenal axis activity." *International Journal of Endocrinology,* 2010:759234. doi: 10.1155/2010/759234.

Batth, S. S. 1972. "Pesticide effectiveness and biorhythm in the house fly." *Journal of Economic Entomology* 65(4):1191-1193. doi: 10.1093/jee/65.4.1191.

Baumann, K., Angerer, J., Heinrich, R., Lehnert, G. 1980. "Occupational exposure to hexachlorocyclohexane: I. Body burden of HCH-isomers." *International Archives of Occupational and Environmental Health,* 47(2):119-127. doi: 10.1007/bf00716371.

Bertazzi, P. A., Bernucci, I., Brambilla, G., Consonni, D., Pesatori, A. C. 1998. "The Seveso studies on early and long-term effects of dioxin exposure: a review." *Environmental Health Perspectives* 106(suppl 2): 625-633.

Bertazzi, P. A., Consonni, D., Bachetti, S., Rubagotti, M., Baccarelli, A., Zocchetti, C., Pesatori, A. C. 2001. "Health effects of dioxin exposure: a 20-year mortality study." *American Journal of Epidemiology* 153(11):1031-1044. doi: 10.1093/aje/ 153.11.1031.

Bianchini, C., and Blazquez, J. 1956. "Chronic occupational poisoning with dieldrin in humans." *Gaceta Médica de Caracas* 64(1-2): 1-39.

Bocio, A., and Domingo, J. L. 2005. "Daily intake of polychlorinated dibenzo-p-dioxins/polychlorinated dibenzofurans (PCDD/PCDFs) in foodstuffs consumed in Tarragona, Spain: a review of recent studies (2001–2003) on human PCDD/PCDF exposure through the diet." *Environmental Research* 97(1):1-9. doi: 10.1016/j.envres.2004.01.012.

Born, J., Kellner, C., Uthgenannt, D., Kern, W., Fehm, H. L. 1992. "Vasopressin regulates human sleep by reducing rapid-eye-movement sleep." *American Journal of Physiology-Endocrinology and Metabolism* 262(3):E295-E300. doi: 10.1152/ajpendo.1992.262.3.E295.

Brown, R. E., Basheer, R., McKenna, J. T., Strecker, R. E., McCarley, R. W. 2012. "Control of sleep and wakefulness." *Physiological Reviews* 92(3):1087-1187. doi: 0.1152/physrev.00032.2011.

Bruchajzer, E., Frydrych, B., and Szymanska, J. A. 2015. "Organophosphorus Flame Retardants-Toxicity and Influence on Human Health." *Medycyna Pracy* 66(2):235-265. doi: 10.13075/mp.5893.00120.

Burlet, S., Leger, L., and Cespuglio, R. 1999. "Nitric oxide and sleep in the rat: a puzzling relationship." *Neuroscience* 92(2):627-639. doi: 10.1016/S0306-4522(99)00031-7.

Canabal, D. D., Potian, J. G., Duran, R. G., McArdle, J. J., Routh, V. H. 2007. "Hyperglycemia impairs glucose and insulin regulation of nitric oxide production in glucose-inhibited neurons in the ventromedial hypothalamus." *American Journal of Physiology-Regulatory, Integrative and Comparative Physiology* 293(2):R592-R600. doi:10.1152/ajpregu.00207.2007.

Carey, J. L., Dunn, C., and Gaspari, R. J. 2013. "Central respiratory failure during acute organophosphate poisoning." *Respiratory Physiology & Neurobiology* 189(2):403-410. doi: 10.1016/j.resp.2013.07.022.

Carpenter, D. O. 2006. "Polychlorinated biphenyls (PCBs): routes of exposure and effects on human health." *Reviews on Environmental Health* 21(1):1-24. doi:10.1515/reveh.2006.21.1.1.

Chen, M. D., Ördög, T., and Knobil, E. 1998. "Mitigation of the somnolence of insulin-induced hypoglycemia by a vasopressin V1 receptor antagonist in the rhesus monkey." *Psychoneuroendocrinology* 23(5):497-503. doi: 10.1016/S0306-4530(98)00018-3.

Cizmas, L., Ackerman, L., Donnelly, D. A., Donnelly, K. C., and McDonald, T. J. 2015. "Measurement of Organophosphate Pesticides, Organochlorine Pesticides, and Polycyclic Aromatic Hydrocarbons in Household Dust from Two Rural Villages in Nepal." *Journal of Environmental & Analytical Toxicology* 5:261. doi: 10.4172/2161-0525.1000261.

Claudel, T., Cretenet, G., Saumet, A., Gachon, F. 2007. Crosstalk between xenobiotics metabolism and circadian clock. *FEBS Letters* 581(19):3626-3633. doi: 10.1016/j.febslet.2007.04.009.

Claustrat, B., Brun, J., and Chazot, G. 2005. "The basic physiology and pathophysiology of melatonin." *Sleep Medicine Reviews* 9(1):11-24. doi:10.1016/j.smrv.2004.08.001.

Coburn, C. G., Currás-Collazo, M. C., and Kodavanti, P. R. S. 2007. "Polybrominated diphenyl ethers and ortho-substituted polychlorinated biphenyls as neuroendocrine disruptors of vasopressin release: effects during physiological activation in vitro and structure–activity relationships." *Toxicological Sciences* 98(1):178-186. doi: 10.1093/toxsci/kfm086.

Coburn, C. G., Gillard, E. R., and Currás-Collazo, M. C. 2004. "Dietary exposure to aroclor 1254 alters central and peripheral vasopressin release in response to dehydration in the rat." *Toxicological Sciences* 84(1):149-156. doi:10.1093/toxsci/kfi046.

Coburn, C. G., Watson-Siriboe, A., Hou, B., Cheetham, C., Gillard, E. R., Lin, L., Currás-Collazo, M. C. 2015. "Permanently compromised NADPH-diaphorase activity within the osmotically activated supraoptic nucleus after in utero but not adult exposure to Aroclor 1254." *Neurotoxicology* 47:37-46. doi: 10.1016/j.neuro.2014.12.009.

Covaci, A., Gerecke, A. C., Law, R. J., Voorspoels, S., Kohler, M., Heeb, N. V., De Boer, J. 2006. "Hexabromocyclododecanes (HBCDs) in the environment and humans: a review." *Environmental Science & Technology* 40(12):3679-3688. doi: 10.1021/es0602492.

Das, S. 2013. "A review of dichlorvos toxicity in fish." *Current World Environment* 8(1):143. doi:10.12944/CWE.8.1.08.

Dawson, V. L., and T. M. Dawson. 1995. "Physiological and toxicological actions of NO in the central nervous system." In *Nitric oxide, biochemistry, molecular biology and therapeutic implications* edited by L. Ignarro, and F. Murad, 323-42. San Diego: Academic press.

de Jong, G. (1991). Long-term health effects of aldrin and dieldrin. A study of exposure, health effects and mortality of workers engaged in the manufacture and formulation of the insecticides aldrin and dieldrin. *Toxicology Letters* Suppl 1:1-206.

de Wit, C. A. 2002. "An overview of brominated flame retardants in the environment." *Chemosphere* 46(5):583-624. doi: 10.1016/s0045-6535(01)00225-9.

Dich, J., Zahm, S. H., Hanberg, A., and Adami, H. O. 1997. "Pesticides and cancer." *Cancer Causes Control* 8(3):420-43. doi: 10.1023/a:1018413522959.

Ding, J. M., Chen, D., Weber, E. T., Faiman, L. E., Rea, M. A., Gillette, M. U. 1994. "Resetting the biological clock: mediation of nocturnal circadian shifts by glutamate and NO." *Science* 266(5191):1713-1717. doi: 10.1126/science.7527589.

Donato, F., and Zani, C. 2010. "Chronic exposure to organochlorine compounds and health effects in adults: cancer, non-Hodgkin lymphoma. Review of literature." *Annali di igiene: Medicina preventiva e di comunita* 22(4):357-367.

Driffield, M., Harmer, N., Bradley, E., Fernandes, A. R., Rose, M., Mortimer, D., Dicks, P. 2008. "Determination of brominated flame retardants in food by LC–MS/MS: diastereoisomer-specific hexabromocyclododecane and tetrabromobisphenol A." *Food Additives and Contaminants* 25(7):895-903. doi: 10.1080/0265203070 1882999.

Durham W. F. 1963. "Pesticide residues in foods in relation to human health." In *Residue Reviews/Rückstands-Berichte. Residue Reviews/Ruckstands-Berichte* edited by F. A. Gunther, vol. 4, 33-81. New York, NY: Springer.

Eddleston, M., Adhikari, S., Egodage, S., Ranganath, H., Mohamed, F., Manuweera, G., Dawson, A. H. 2012." Effects of a provincial ban of two toxic organophosphorus insecticides on pesticide poisoning hospital admissions." *Clinical Toxicology* 50(3):202-209. doi: 10.3109/15563650.2012.660573.

Eriksson, P., Viberg, H., Fischer, C., Wallin, M., Fredriksson, A. 2002. "A comparison on the developmental neurotoxic effects of hexabromocyclododecane, 2, 2´, 4, 4´, 5, 5´-hexabromodiphenylether (PBDE 153) and 2, 2´, 4, 4´, 5, 5´-hexachlorobiphenylether (PCB 153)." *Organohalogen Compounds,* 57, 389-390.

Eto, M. 1979. *Organophosphorus Pesticides: Organic and Biological Chemistry.* Boca Raton, FL: CRC Press.

Evangelou, E., Ntritsos, G., Chondrogiorgi, M., Kavvoura, F. K., Hernández, A. F., Ntzani, E. E., Tzoulaki, I. 2016. "Exposure to pesticides and diabetes: a systematic review and meta-analysis." *Environment International* 91:60-68. doi:10.1016/j.envint.2016.02.013.

Ferri, R., Manconi, M., Plazzi, G., Bruni, O., Vandi, S., Montagna, P., L. Ferini-Strambi, Zucconi, M. 2008. "A quantitative statistical analysis of the submentalis muscle EMG amplitude during sleep in normal controls and patients with REM sleep behavior disorder." *Journal of Sleep Research* 17(1):89-100. doi: 10.1111/j.1365-2869.2008.00631.x.

Fiedler, H. 2003. "Dioxins and Furans (PCDD/PCDF)." In *Persistent Organic Pollutants. The Handbook of Environmental Chemistry* edited by H. Fiedler. Vol. 30, 47-90, Series: Anthropogenic Compounds. Berlin, Heidelberg: Springer.

Fitzgerald, E. F., Weinstein, A. L., Youngblood, L. G., Standfast, S. J., Melius, J. M. 1989. "Health effects three years after potential exposure to the toxic contaminants of an electrical transformer fire." *Archives of Environmental Health: An International Journal* 44(4):214-221. doi: 10.1080/00039896.1989.9935886.

Fry, D. R. 1964. "Human Dieldrin Poisoning." *Lancet* 1:764.

Garrett, R. W., and Gasiewicz, T. A. (2006). "The aryl hydrocarbon receptor agonist 2, 3, 7, 8-tetrachlorodibenzo-p-dioxin alters the circadian rhythms, quiescence, and expression of clock genes in murine hematopoietic stem and progenitor cells." *Molecular Pharmacology* 69(6):2076-2083. doi: 10.1124/mol.105.021006.

Gaspari, R. J., and Dunn, C. 2013. "Dichlorvos exposure to the Kölliker-Fuse Nuclei is sufficient but not necessary for OP induced apnea." *Neurotoxicology* 39, 132-137. doi: 10.1016/j.neuro.2013.06.009.

Gaspari, R. J., and Paydarfar, D. 2007. "Pathophysiology of respiratory failure following acute dichlorvos poisoning in a rodent model." *Neurotoxicology,* 28(3):664-671. doi:10.1016/j.neuro.2007.02.002.

Gaspari, R. J., and Paydarfar, D. 2011. "Dichlorvos-induced central apnea: effects of selective brainstem exposure in the rat." *Neurotoxicology* 32:206-14. doi: 10.1016/j.neuro.2011.01.005.

Gautier-Sauvigné, S., Colas, D., Parmantier, P., Clement, P., Gharib, A., Sarda, N., Cespuglio, R. 2005. "Nitric oxide and sleep." *Sleep Medicine Reviews* 9(2):101-113. doi: 10.1016/j.smrv.2004.07.004.

Germer, S., Piersma, A. H., Van der Ven, L., Kamyschnikow, A., Fery, Y., Schmitz, H. J., Schrenk, D. 2006. "Subacute effects of the brominated flame retardants hexabromocyclododecane and tetrabromobisphenol A on hepatic cytochrome P450 levels in rats." *Toxicology* 218 (2-3), 229-236. doi: 10.1016/j.tox.2005.10.019.

Gillett, J. W., R. J. Harr, F. T. Lindstrom, D. A. Mount, A. D. Clair, L. J. Weber. 1972. "Evaluation of human health hazards on use of dichlorvos (DDVP), especially in resin strips." In *Residue Reviews. Residue Reviews* edited by F. A. Gunther, and J. D. Gunther. Vol. 44, 115-159. New York: Springer.

Glotfelty, D. E. 1978. "The Atmosphere as a Sink for Applied Pesticides." *Journal of the Air Pollution Control Association* 28(9): 917-921. doi: 10.1080/00022470.1978.11490579.

Gonnissen, H. K., Adam, T. C., Hursel, R., Rutters, F., Verhoef, S. P., Westerterp-Plantenga, M. S. 2013. "Sleep duration, sleep quality and body weight:

Parallel developments." *Physiology & Behavior, 121,* 112-116. doi: 10.1016/j.physbeh.2013.04.007.

Goswamy, R., Chaudhuri, A., and Mahashur, A. A. 1994. Study of respiratory failure in organophosphate and carbamate poisoning." *Heart & Lung: The Journal of Critical Care* 23(6), 466-472.

Gupta, P. C. 1975. "Neurotoxicity of chronic chlorinated hydrocarbon insecticide poisoning: A clinical and electroencephalographic study in man". *The Indian Journal of Medical Research* 63(4): 601–606.

Hayes, W. J. 1957. "Dieldrin poisoning in man." *Public Health Reports* 72(12):1087-1091.

Ibánez, P., and de franzetti, R. P. 1957. "Psychometric Studies and their Clinical Value in respect of Dieldrin Spraymen." *Boletín de la Oficina Sanitaria Panamericana* 43(6): 531-3.

International Agency for Research on Cancer (IARC). 1979. *"IARC Monographs on the Evaluation of the Carcinogenic Risk of Chemicals to Humans: Some Halogenated Hydrocarbons."* Vol 20. World Health Organization, Lyon.

International Agency for Research on Cancer (IARC). 1997. *"Polychlorinated dibenzo-para-dioxins and polychlorinated dibenzofurans. Monographs on the Evaluation of Carcinogenic Risks to Humans."* Vol 69. World Health Organization, Lyon.

Jackson, C. J., Carville, A., Ward, J., Mansfield, K., Ollis, D. L., Khurana, T., Bird, S. B. 2014. "Use of OpdA, an organophosphorus (OP) hydrolase, prevents lethality in an African green monkey model of acute OP poisoning." *Toxicology* 317:1-5. doi: 10.1016/j.tox.2014.01.
003.

Jager, K. W. 1970. *Aldrin, Dieldrin, Endrin and Telodrin: An Epidemiological and Toxicological Study of Long-Term Ocupacional Exposure.* Amsterdam, London, New York: Elsevier Publishing Company.

Jayaraj, R., Megha, P., and Sreedev, P. 2016. "Organochlorine pesticides, their toxic effects on living organisms and their fate in the environment." *Interdisciplinary Toxicology* 2(3-4):90-100. doi: 10.1515/intox-2016-0012.

Jian, T., and Feng, Z. Y. 1990. "Chronotoxicologic studies on dichlorphos in mice and humans." *Progress in Clinical and Biological Research* 341A:503-10.

Jordan, D., and K. M. Spyer. 1981. "Effects of acetylcholine on respiratory neurones in the nucleus ambiguus-retroambigualis complex of the cat." *Journal Physiology* 320:103–11. doi: 10.1113/jphysiol.1981.sp013937.

Jouvet, M. 1972. "The role of monoamines and acetylcholine containing neurons in the regulation of the sleep-waking cycle." In *Neuropysiology and Neurochemistry of Sleep and Wakefulness* edited by Ergebnisse der Physiologie, biologischen Chemie und experiment-tellen Pharmakologie. Vol. 64, 166-307 Berlin, Heidelberg: Springer.

Joy, R. M. 1982. "Mode of action of lindane, dieldrin and related insecticides in the central nervous system." *Neurotoxicology and Teratology* 4(6):813-23.

Joy, R. M. 1985. "The effects of neurotoxicants on kindling and kindled seizures." *Fundamental and Applied Toxicology* 5(1):41-65. doi:10.1016/0272-0590(85)90049-1.

Kashyap, S. K. 1986. "Health surveillance and biological monitoring of pesticide formulators in India." *Toxicology Letters* 33(1-3):107-14. doi: 10.1016/0378-4274(86)90075-5.

Kauppinen, T., Kogevinas, M., Johnson, E., Becher, H., Bertazzi, P-A., Bueno de Mesquita, H.B., Coggon, D., Green, L., Littorin, M., Lynge, E., Mathews, J., Neuberger, M., Osman, J., Pannett, B., Pearce, N., Winkelmann, R., and R.Saracci. 1993. "Chemical exposure in manufacture of phenoxy herbicides and chlorophenols and in spraying of phenoxy herbicides." *American Journal of Industrial Medicine* 23:903-920. doi: 10.1002/ajim.4700230607.

Kauppinen, T., Toikkanen, J., Pedersen, D., Young, R., Kogevinas, M., Ahrens, W., ... Mirabelli, D. 1998. *"Occupational exposure to carcinogens in the European Union in 1990-93."*

Kazantzis, G., McLaughlin, A. I. G., and Prior, P. F. 1964. "Poisoning in industrial workers by the insecticide aldrin." *Occupational and Environmental Medicine* 21(1), 46-51.doi: 10.1136/oem.21.1.46.

Kok, A. 1993. "REM sleep pathways and anticholinesterase intoxication: a mechanism for nerve agentinduced, central respiratory failure." *Medical Hypotheses* 41(2):141–9. doi: 10.1016/0306-9877(93)90061-t.

Kondo, H., Tanio, K., Nagaura, Y., Nagayoshi, M., Mitoma, C., Furue, M., Maeda, T. 2018. "Sleep disorders among Yusho patients highly intoxicated with dioxin-related compounds: A 140-case series." *Environmental Research* 166:261-268. doi: 10.1016/j.envres.2018.
05.033.

Kuratsune, M., Yoshimura, T., Matsuzaka, J., and Yamaguchi, A. 1971. "Yusho, a poisoning caused by rice oil contaminated with polychlorinated biphenyls." *HSMHA Health Reports* 86(12), 1083-91.

Larsen, J. C. 2006. "Risk assessments of polychlorinated dibenzo- p-dioxins, polychlorinated dibenzofurans, and dioxin-like polychlori-nated biphenyls in food." *Molecular Nutrition & Food Research* 50(10):885-96. doi: 10.1002/mnfr.200500247.

Lavie, P. 2001. "Sleep-wake as a biological rhythm." *Annual Review of Psychology* 52(5):277-303. doi:10.1146/annurev.psych.52.1.277.

Law, R. J., Herzke, D., Harrad, S., Morris, S., Bersuder, P., and Allchin, C. R. 2008. "Levels and trends of HBCD and BDEs in the European and Asian environments, with some information for other BFRs." *Chemosphere, 73*(2):223-241. doi: 10.1016/j.chemosphere.2008.02.066.

León-Olea, M., Martyniuk, C. J., Orlando, E. F., Ottinger, M. A., Rosenfeld, C. S., Wolstenholme, J. T., Trudeau, V. L. 2014. "Current concepts in neuroendocrine disruption." *General and Comparative Endocrinology 203:*158-173. doi: 10.1016/j.ygcen.2014.02.005.

Lewis, R. A. 1998. *Lewis' Dictionary of Toxicology.* CRC Lewis. 763. USA: CRC Press LLC.

Lewis-Mikhael, A. M., Olmedo-Requena, R., Martínez-Ruiz, V., Bueno-Cavanillas, A., and Jiménez-Moleón, J. J. 2015. "Organochlorine pesticides and prostate cancer, Is there an association?" A meta-analysis of epidemiological evidence. *Cancer Causes & Control* 26(10):1375-1392. doi: 10.1007/s10552-015-0643-z.

Lindén, J., Lensu, S., Tuomisto, J., Pohjanvirta, R. 2010. "Dioxins, the aryl hydrocarbon receptor and the central regulation of energy balance." *Frontiers in Neuroendocrinology,* 31(4):452-478. doi: 10.1016/j.yfrne.2010.07.002.

Liu, H., Terrell, M. L., Bui, V., Summy-Long, J. Y., and Kadekaro, M. 1998. "Nitric oxide control of drinking, vasopressin and oxytocin release and blood pressure in dehydrated rats." *Physiology & Behavior,* 63(5): 763-769. doi:10.1002/ana.410440510.

Lu, C., Barr, D. B., Pearson, M. A., and Waller, L. A. 2008. Dietary intake and its contribution to longitudinal organophosphorus pesticide exposure in urban/suburban children." *Environmental Health Perspectives* 116(4): 537-542. doi: 10.1289/ehp.10912.

Mackay D., W. Y. Shiu, and K. C. Ma. 1992. *Illustrated Handbook of Physical-chemical Properties and Environmental Fate for Organic Chemicals*. Vol. II. Boca Raton, FL, USA: Lewis Publishers.

Mandal, P. K. 2005. "Dioxin: a review of its environmental effects and its aryl hydrocarbon receptor biology". *Journal Comparative Physiology B* 175(4):221-30. doi: 10.1007/s00360-005-0483-3.

Mariussen, E., and Fonnum, F. 2003. "The effect of brominated flame retardants on neurotransmitter uptake into rat brain synaptosomes and vesicles." *Neurochemistry International* 43(4-5):533–42. doi: 10.1016/s0197-0186(03)00044-5.

Martenies, S. E., and Perry, M. J. 2013. "Environmental and occupational pesticide exposure and human sperm parameters: a systematic review." *Toxicology* 307:66-73. doi: 10.1016/j.tox.2013.02.005.

McCarren, H. S., and McDonough, J. H. 2016. "Anticonvulsant discovery through animal models of status epilepticus induced by organophosphorus nerve agents and pesticides." *Annals of the New York Academy of Sciences* 1374(1):144-50. doi: 10.1111/nyas.13092.

Menconi, S., Clark, J. M., Langenberg, P., Hryhorczuk, D. 1988. "A preliminary study of potential human health effects in private residences following chlordane applications for termite control." *Archives of Environmental Health: An International Journal* 43(5):349-52. doi: 10.1080/00039896.1988.9934947.

Mehrotra, B.D., Ravichandra Reddy S., and D. Desaiah. 1988 "Effect of subchronic dieldrin treatment on calmodulin-regulated Ca2+ pump activity in rat brain." *Journal of Toxicology and Environmental Health* 25(4):461-9. doi: 10.1080/15287398809531224.

Mitoma, C., Uchi, H., Tsukimori, K., Yamada, H., Akahane, M., Imamura, T., Furue, M. 2015. "Yusho and its latest findings-A review in studies conducted by the Yusho Group." *Environment International* 82:41–48. doi: 10.1016/j.envint.2015.05.004.

Mocarelli, P. 2001. "Seveso: a teaching story." *Chemosphere* 43(4-7):391-402. doi: 10.1016/s0045-6535(00)00386-6.

Möller-Levet, C. S., Archer, S. N., Bucca, G., Laing, E. E., Slak, A., Kabiljo, R., Lo J. C., Santhi, N., von Schantz, M., Smith, C. P., Dijk, D. J. 2013. "Effects of insufficient sleep on circadian rhythmicity and expression amplitude of the human blood transcriptome." *Proceedings of the National Academy of Sciences USA* 110(12):E1132-41. doi: 10.1073/pnas.1217154110.

Moses, M., Lilis, R., Crow, K. D., Thornton, J., Fischbein, A., Anderson, H. A., and Selikoff, I. J.1984. "Health status of workers with past exposure to 2,3,7,8-tetrachlorodibenzo-p-dioxin in the manufacture of 2,4,5-trichlorop2enoxyacetic acid: Comparison of findings with and without chloracne." *American Journal of Industrial Medicine* 5(3):161-182. doi: 10.1002/ajim.4700050303.

Mostafalou, S., and Abdollahi, M. 2017. "Pesticides: an update of human exposure and toxicity." *Archives Toxicology* 91(2):549-599. doi: 10.1007/s00204-016-1849-x.

Mucio-Ramírez, S., E. Sánchez-Islas, E. Sánchez-Jaramillo, M. Currás-Collazo, V. R. Juárez- González, M. Y. Álvarez-González, L. E. Orser, B. Hou, F. Pellicer, R. S. Kodavanti, M. León-Olea. 2017. "Perinatal exposure to organohalogen pollutants decreases vasopressin content and its mRNA expression in magnocellular neuroendocrine cells activated by osmotic stress in adult rats." *Toxicology and Applied Pharmacology* 329:173-189. doi: 10.1016/j.taap.2017.05.039.

Munidasa, U. A. D. D., Gawarammana, I. B., Kularatne, S. A. M., Kumarasiri, P. V. R., and Goonasekera, C. D. A. 2004. "Survival Pattern in Patients with Acute Organophosphate Poisoning Receiving Intensive Care. *Journal of Toxicology: Clinical Toxicology* 42(4):343-347. doi:10.1081/CLT-120039539.

Muñoz-Quezada, M. T., Iglesias, V., Lucero, B., Steenland, K., Barr, D. B., Levy, K., Ryan, P. B., Alvarado, S., and Concha, C. 2012. "Predictors of exposure to organophosphate pesticides in schoolchildren in the Province of Talca, Chile." *Environment International* 47:28-36. doi: 10.1016/j.envint.2012.06.002.

Nakatsugawa, T., Ishida, M., and Dahm, P. A. 1965. "Microsomal epoxidation of cyclodiene insecticides." *Biochemical Pharmacology* 14:1853-1865. doi: 10.1016/0006-2952(65)90276-5.

Nattie, E. E., Wood, J., Mega, A., Goritski, W. 1989. "Rostral ventrolateral medulla muscarinic receptor involvement in central ventilatory chemosensitivity." *Journal of Applied Physiology* 66(3):1462–70. doi: 10.1152/jappl.1989.66.3.1462.

Nau, H. 2006. "Impacts and impact mechanisms of "dioxins" in humans and animals." *Deutsche Tierärztliche Wochenschrift* 113(8):292-7.

Nayyar, N., Sangwan, N., Kohli, P., Verma, H., Kumar, R., Negi, V., ... Lal, R. (2014). Hexachlorocyclohexane: persistence, toxicity and decontamination. *Reviews on Environmental Health* 29(1-2): 49-52.

Neuberger, M., Kundi, M., and Jäger, R. 1998. "Chloracne and morbidity after dioxin exposure (preliminary results). *Toxicology Letters* 96-97:347-50. doi: 10.1016/s0378-4274(98)00092-7.

Neuberger, M., Rappe, C., Bergek, S., Cai, H., Hansson. M., Jäger, R., Nicolau, G. Y. 1983. "Circadian rhythms of RNA, DNA and protein in the rat thyroid, adrenal and testis in chronic pesticide exposure. III. Effects of the insecticides (dichlorvos and trichlorphon)." *Physiologie*, 20(2):93-101.

Nigam, S. K., Karnik, A. B., Majumder, S. K., Visweswariah, K., Suryanarayana Raju, G., Muktha Bai, K., Lakkad, B. C., Thakore, K. N., Chatterjee, B. B. 1986. "Serum hexachlorocyclohexane residues in workers engaged at a HCH manufacturing plant." *International Archives of Occupational and Environmental Health* 57(4):315. doi: 10.1007/BF00406186.

Okey, A. B. 2007. "An aryl hydrocarbon receptor odyssey to the shores of toxicology: the Deichmann Lecture, International Congress of Toxicology-XI". *Toxicology Sciences* 98(1):5-38. doi:10.1093/toxsci/kfm096.

Pace-Schott, E. F. 2009. "Sleep Architecture" In *The Neuroscience of Sleep,* edited by R., Stickgold and M. Walker 12. San D*iego:* Academic press.

Pereira, M. 2004. "Polychlorinated dibenzo-p-dioxins (PCDD), dibenzofurans (PCDF) and polychlorinated biphenyls (PCB): Main sources, environmental behaviour and risk to man and biota." *Química Nova* 27(6):934-943. doi:10.1590/S0100-40422004000600018.

Persistent organic pollutants review committee (POPRC). 2008. *"Draft risk management evaluation for alpha hexachlorocyclohexane."*

Poellinger, L. 2000. "Mechanistic aspects-the dioxin (aryl hydrocarbon) receptor". *Food Additives and Contaminants* 17(4):261-6. doi:10.1080/026520300283333.

Pohjanvirta, R., Tuomisto, J., Linden, J., and Laitinen, J. 1989. "TCDD reduces serum melatonin levels in Long-Evans rats." *Pharmacology Toxicology* 65(3):239-40. doi: 10.1111/j.1600-0773.1989.tb01164.x.

POPRC8.3 2013. Available online: http://chm.pops.int/TheConvention/POPsReviewCommittee/ReportsandDecisions/tabid/3309/Default.aspx (accessed 10.02.114).

Porkka-Heiskanen T., Alanko, L., and Stenberg, D. 2007. "Neurochemi-stry of Sleep." In *Handbook of Neurochemistry and Molecular Neurobiology* edited by A. Lajtha and J. D. Blaustein, Boston, MA: Springer.

Prior, P. F., and Deacon, P. A. 1969. "Spontaneous sleep in healthy subjects in long-term serial electroencephalographic recordings." *Electroencephalography and Clinical Neurophysiology* 27(4):422-424.

Reggiani, G. 1978. "Medical problems raised by the TCDD contamination in Seveso, Italy." *Archives of Toxicology* 40(3):161-88.

Rettori V., Belova, N., Dees, W., Nyberg, C., Gimeno, M., and McCann, S. 1993. "Role of nitric oxide in the control of luteinizing hormone-releasing hormone release in vivo and in vitro." *Proceedings of the National Academy of Sciences USA* 90(21):10130-4. doi:10.1073/pnas.90.21.10130.

Richardson, J. R., Caudle, W. M., Wang, M., Dean, E. D., Pennell, K. D., and Miller, G. W. 2006. "Developmental exposure to the pesticide dieldrin alters the dopamine system and increases neurotoxicity in an animal model of Parkinson's disease". *FASEB Journal* 20(10):1695-7. doi: 10.1096/fj.06-5864fje.

Roberts, J. R., and Reigart, J. R. 2013. "Chapter 5. Organophosphate Insecticides. In: *Recognition and Management of Pesticide Poisonings: Overview,* edited by J. R. Roberts, and J. R. Reigart. 43. 6th Edition. Environmental Protection Agency.

Ronisz, D., Finne, E., Karlsson, H., and Förlin, L. 2004. "Effects of the brominated flame retardants hexabromocyclododecane (HBCDD) and tetrabromobisphenol-A (TBBPA) on hepatic enzymes and other biomarkers in juvenile rainbow trout and feral eelpout." *Aquatic Toxicology* 69(3):229-45. doi: 10.1016/j.aquatox.2004.05.007.

Roosens, L., Cornelis, C., D'Hollander, W., Bervoets, L., Reynders, H., Van Campenhout, K., Heuvel, Van Den R., Neels, H., and Covaci, A. 2010. "Exposure of the Flemish population to brominated flame retardants: Model and risk assessment." *Environment International* 36:368–376. doi: 10.1016/j.envint.2010.02.005.

Rosin, D. L., Chang, D. A., and Guyenet, P. G. 2006. "Afferent and efferent connections of the rat retrotrapezoid nucleus." *Journal Comparative of Neurology* 499(4):64–89. doi: 10.1002/cne.21105.

Sato, S., Shirakawa, H., Tomita, S., Ohsaki, Y., Haketa, K., Tooi, O., Santo, N., Tohkin, M., Furukawa, Y., Gonzalez, F. J., Komai. M. 2008. "Low-dose dioxins alter gene expression related to cholesterol biosynthesis, lipogenesis, and glucose metabolism through the aryl hydrocarbon receptor-mediated pathway in mouse liver." *Toxicology and Applied Pharmacology* 229(1):10-9. doi: 10.1016/j.taap.2007. 12.029.

Schecter, A. J., Ryan, J. J., and Constable, J. D. 1986. "Chlorinated dibenzo-p-dioxin and dibenzofuran levels in human adipose tissue and milk samples from the north and south of Vietnam." *Chemosphere*, 15(9-12), 1613-1620.

Schulte-Hermann, R. 1976. "Feeding rhythms and the diurnal rhythm of cell proliferation in pharmacologically induced liver growth." *Archives of Toxicology* 36(3-4):235-45. doi: 10.1007/bf00340531.

Schulte-Hermann, R., and Landgraf, H. 1974. "Circadian rhythm of cell proliferation in rat liver: synchronization by feeding habits." *Zeitschrift fur Naturforschung C* 29(7-8):421-4.

Shao, X. M., and Feldman, J. L. 2000. "Acetylcholine modulates respiratory pattern: effects mediated by M3-like receptors in preBötzinger complex inspiratory neurons." *Journal Neurophysiology* 83(3):1243–52. doi: 10.1152/jn.2000.83.3.1243.

Sharma, R., and Kodavanti. P. 2002. "In vitro effects of polychlorinated biphenyls and hydroxy metabolites on nitric oxide synthases in rat brain." *Toxicology and Applied Pharmacology* 178(3):127-136. doi.org/10.1006/taap.2001.9328.

Shiu, W. Y., Doucette, W., Gobas, F. A. P. C, Andren, A., Mackay, D. 1988. "Physical-chemical properties of chlorinated dibenzo-p-dioxins". *Environmental Science & Technology* 22(6):651-658. doi:10.1021/es00171a006.

Siddiqui, M. K. J., Saxena, M. C., and Krishna Murti, C. R. 1981. "Storage of DDT and BHC in adipose tissue of Indian males." *International Journal of Environmental Analytical Chemistry* 10(3-4):197-204. doi: 10.1080/03067318108071546.

St-Jacques, R., and St-John, W. M. 1999. "Transient, reversible apnea following ablation of the pre-Bötzinger complex in rats." *Journal Physiology* 520 Pt 1:303-14. doi: 10.1111/j.1469-7793.1999.00303.x.

Stoop, R. 2012. "Neuromodulation by oxytocin and vasopressin." *Neuron* 76(1):142-159. doi:10.1016/j.neuron.2012.09.025.

Suskind, R. R. 1953. *"A clinicl and environmental survey, Monsato Chemical Company, Nitro West Virginia."* Report of the Kattering Laboratory.

Suzuki, T., and Kobayashi, H. 1968. *"Contamination of rice bran oil with PCB used as the heating medium by leakage through penetration holes at the heating coil tube in deodorization chamber."* Hatamura Institute for the Advancement of Technology.

Sweeney, M. H., and Mocarelli, P. 2000. "Human health effects after exposure to 2,3,7,8-TCDD." *Food Additives and Contaminants* 17(4):303-16. doi: 10.1080/026520300283379.

Taddeini, L., and Watson, C. J. 1968. "The clinical porphyrias." *Seminars in Hematology* 5:335-69.

Takahashi, H., Kojima, T., Ikeda, T., Tsuda, S., Shirasu, Y. 1991. "Differences in the mode of lethality produced through intravenous and oral administration of organophosphorus insecticides in rats." *Fundamental and Applied Toxicology,* 16(3), 459-468.

Tijet, N., Boutros, P. C., Moffat, I. D., Okey, A. B., Tuomisto, J., and Pohjanvirta, R. 2006. "Aryl hydrocarbon receptor regulates distinct dioxin-dependent and dioxin-independent gene batteries". *Molecular Pharmacology* 69(1):140-153. doi:10.1124/mol.105.018705.

Tollefson, L. 1991. "Use of epidemiology data to assess the cancer risk of 2,3,7,8-tetrachlorodibenzo-p-dioxin. *Regulatory Toxicology and Pharmacology* 13(2):150-69.

Tsukimori, K., Tokunaga, S., Shibata, S., Uchi, H., Nakayama, D., Ishimaru, T., Nakano, H., Wake, N., Yoshimura, T., and Furue, M. 2008. "Long-term effects of polychlorinated biphenyls and dioxins on pregnancy outcomes in women affected by the Yusho incident." *Environmental Health Perspectives* 116(5):626-30. doi: 10.1289/ehp.10686.

U.S. Environmental Protection Agency. 1999. *"Integrated Risk Informa-tion System (IRIS) on Dichlorvos."* National Center for Environmental Assessment, Office of Research and Development, Washington, DC.

UNEP. "Report of the Conference of the Parties of the Stockholm Convention on Persistent Organic Pollutants on the work of its fourth meeting." UNEP/POPS/COP.4/38. 8 May 2009.

Van den Berg, M., Birnbaum, L. S., Denison, M., De Vito, M., Farland, W., Feeley, M., Fiedler. H., Hakansson, H., Hanberg, A., Haws, L., Rose, M., Safe, S., Schrenk, D.,Tohyama, C., Tritscher, A., Tuomisto, J., Tysklind, M., Walker, N., Peterson, R. E., Rose, M. 2006. "The 2005 World Health Organization reevaluation of human and Mammalian toxic equivalency factors for dioxins and dioxin-like compounds". *Toxicological Sciences* 93(2): 223–41. doi:10.1093/toxsci/kfl055.

Vijgen, J., Abhilash, P. C., Li, Y. F., Lal, R., Forter, M., Torres, J., Singh, N., Yunus, M., Tian, C., Schäffer, A., Weber, R. 2011. "Hexachlorocyclohexane (HCH) as new Stockholm Convention POPs-a global perspective on the management of Lindane and its waste isomers." *Environmental Science and Pollution Research* 18(2):152-62. doi:10.1007/s11356-010-0417-9.

Walker, C. H. 1998. "Biochemical biomarkers and potentiation of toxicity." *Biotherapy* 11(2-3):113-7.

Willett, K. L., Ulrich, E. M., and Hites, R. A. 1998. "Differential toxicity and environmental fates of hexachlorocyclohexane isomers." *Environmental Science & Technology* 32(15):2197-2207. doi.org/10.1021/es9708530.

Yamada-Okabe, T., Sakai, H., Kashima, Y., Yamada-Okabe, H. 2005 "Modulation at a cellular level of the thyroid hormone receptor mediated gene expression by 1,2,5,6,9,10-hexabromocyclodo-decane(HBCD), 4,4'-diiodobiphenyl (DIB), and nitrofen (NIP)." *Toxicology Letters* 155(1):127–33. doi: 10.1016/j.toxlet. 2004.09.005.

Yoshimura, T. 2003. "Yusho in Japan." *Industrial Health* 41(3):139-48. doi: 10.2486/indhealth.41.139.

Zack, J. A., and Suskind, R. R. 1980. "The mortality experience of workers exposed to tetrachlorodibenzodioxin in a trichlorophenol process accident." *Journal of Occupational Medicine and Toxicology* 22(1):11-4.

Zanobetti A., Redline, S., Schwartz, J., Rosen, D., Patel, S. O'Connor, G. T., Lebowitz, M., Coull, B. A., Gold, D. R. 2001. "Associations of PM10 with sleep and sleep-disordered breathing in adults from seven U.S. urban areas." *American Journal of Respiratory and Critical Care Medicine* 182(6):819-25. doi: 10.1164/rccm.200912-1797OC.

Zieliński M., Kamińska, J., Czerska, M., Ligocka, D., Urbaniak, M. 2014. "Levels and sources of PCDDs, PCDFs and dl-PCBs in the water ecosystems of central Poland - a mini review." *International Journal of Occupational Medicine and Environmental Health* 27(6):902-18. doi: 10.2478/s13382-014-0336-y.

In: Sleep, Cognition and Emotion
Editors: I. Ramírez Salado et al.
ISBN: 978-1-53617-561-5

Chapter 7

MENTAL AND COGNITIVE PHENOMENA DURING SLEEP AND DREAMS

Ignacio Ramírez-Salado*[*], *PhD
Nadia Alejandra Torres-Marcial
and José Carlos Aguirre-Barrera
Laboratorio de Cronobiología y Sueño,
Dirección de Investigaciones en Neurociencias,
Instituto Nacional de Psiquiatría Ramón de la Fuente Muñiz,
Ciudad de México, México

ABSTRACT

The aim of this chapter is to present current and condensed information regarding the process of dreaming and its relation to Theory of Mind, Social Cognition, Simulation Theory and Default Mode Network. We also provide detailed information about the neuroanatomical and neurobiological substrates that underlie dream activity, such as the prefrontal cortex, precuneus, cingulum, hippocampus, and temporoparietal junction, while trying to show the neurobiological differences of these structures during two behavioral states: conscious wakefulness and REM sleep, considering that what happens during dreams should not be thought of as an "altered state of consciousness" but as a different state of consciousness to the waking state. We will focus on how man experiences dream activity along with different processes that occur during dreaming, such as social interactions, behaviors, actions, character recognition, sensory qualities and processes, story building, and the emotions experienced by the dream-self and those assigned to other dream characters. We will also describe dreams as a possible mechanism that serves as a simulation reality process, and its likely functions as a regulator of emotion and social behavior. A brief review of some associated pathologies,

[*] Corresponding Author's Email: salado@imp.edu.mx.

as well as the relation between sleep, autism, Default Mode Network and theory of mind will also be mentioned.

Finally, we intend to postulate some ideas regarding the identification of emotions, feelings and intentions, in both the dream self and other characters.

Keywords: sleep, dreams, REM sleep, theory of mind, social cognition

INTRODUCTION

The waking state, slow wave sleep (SWS) and rapid eye movement sleep (REM sleep) are distinct physiological states shared by some birds and all mammals, including humans (Kleitman 1963; Hassemberg 1965; Michel Jouvet 1972; Aserinsky and Kleitman 1953). In such species the sleep-wake cycle, which consist on alternating between wakefulness and sleep, depends on several factors such as the light-dark cycle, whether they are altricial or precocial species, prey or predator and, in some cases, on season and geographical location.

The sleep-wake cycle also varies throughout ontogeny. The ratio between the waking state and sleep stages, as well as the temporal organization of such stages, undergo important changes. The ratios and temporal organization are different in newborn humans from those found in the adult and elderly populations. REM sleep percentage is higher during the first few years of life, while waking and SWS increase gradually, reaching their highest ratios during older ages.

Some environmental factors can affect this cycle, particularly the light-dark cycle and temperature. On the other hand, some internal conditions, such as changes in mood or emotional state, affective disorders, anxiety, stress, neurological disorders of central origin, fever, pain and epilepsy can also affect the sleep-wake cycle.

The physiological state of sleep is vital to and necessary for maintaining good mental health. When sleep deprivation lasts more than a few days, irritability (Kleitman 1963; Luby et al. 1960; 1962; Kollar et al. 1966), attention errors (Kleitman 1963), memory failures (Williams, Gieseking, and Lubin 1966) and poor task performance (Kleitman 1963) occur. Hyper-reactive behaviors, illusions (Kollar et al. 1966; 1968; Koella 1967) and hallucinations (Kollar et al. 1969) are also common, and some people can develop psychotic states (Wilkinson 1969; Luby et al. 1962).

It has been proven in animal models that total sleep deprivation leads to general weakness, skin lesions, hyperactivity, aggressive behaviors, weight loss, increased metabolic rate and decreased body temperature, and that these conditions can cause death if deprivation continues (Rechtschaffen et al. 1989). Sleep is a homeostatic (endogenous) process that promotes the release of several hormones, mainly growth hormone; regulates changes in brain temperature (Arnulf et al. 1998; Denoyer et al. 1991) and participates in Central Nervous System (CNS) regeneration and in the consolidation of memories.

The first studies on sleep were based on analysis of behaviors, such as posture changes, in healthy subjects, and on analysis of sleep mentation through introspection. Later studies examined the changes in vegetative phenomena that occurred during sleep, such as blood pressure, body temperature, heart rate and pupil diameter. With the development of the electroencephalogram (EEG), performed by Berger in 1930, subsequent studies on sleep centered on the correlation between sleep behavior and electrical activity of the brain. The possibility of registering the electrical activity of the CNS through the EEG lead to numerous studies that aimed to record and examine the EEG rhythms present during several states of consciousness, including sleep. The first major observation was that sleep is a complex, multi-stage process, each one differentiated by distinct patterns of electrical brain activity with identifiable changes in the voltage and frequency of its oscillations. Sleep was thus divided into different stages according to the amount of slow EEG rhythms that appeared during a given period of time (Loomis, Harvey, and Hobart 1937).

In 1953 Aserinsky and Kleitman observed, in a clinical setting, that children would periodically and quickly move their eyes during sleep, without opening them, and that this movement was accompanied by an EEG activity similar to that seen during the initiation of sleep, or during the waking state (high frequency, low voltage). They named this stage Rapid Eye Movement (REM) sleep.

After corroborating that this phase is also normally found in adult subjects, they were able to prove the cyclical nature of these periods, which usually appear four or five times during each night, alternating with SWS stages. These authors also discovered that, if subjects were awakened during REM sleep, they were able to report what they were dreaming upon awakening. After Aserinsky and Kleitman discovered REM sleep in humans, Dement demonstrated in 1958 that cats have an identical stage of sleep. In 1959 Jouvet and Michel would describe this phase as REM sleep in the cat. Based on these findings, later studies on sleep have focused on defining the different brain structures that participate in the integration and regulation of each of its stages, as well as on which neurotransmitters or humoral factors are able to regulate them.

The discovery of REM sleep allowed for the completion of the classification of the stages of sleep in humans. A common criteria was established in 1968 in an international committee headed by Rechtscaffen and Kales. This classification was based on the consistent temporal relation between changes in EEG, electrooculogram (EOG) and electromyogram (EMG) activity during sleep. Electromyography

DREAMS

From polygraphic studies of sleep in the man, it has been established that from just seconds before the onset of REM sleep (when the subjects are still in SWS or non-REM sleep), dreams can occur, but they represent only 15 percent of total dreams, however its content is consistent and similar to the normal processes of thinking. The other 85 percent of dreams appear related to the rapid eye movements of REM sleep and they were accompanied by perceptual phenomena of hallucinatory type, they are involuntary and with sense of reality, and are often inexplicable, and inconsistent.

Dreams are predominantly perceptual phenomena in which the visual system plays a major role. The visual phenomena are presented in the 100 percent of the dreams, being less frequently the hearing and even more rare tactile and smells phenomena. Dreams in addition are accompanied by emotional changes, reminiscences of personal type and conceptual (anamnesis), as well as vegetative changes.

Sensory	Vegetative	Emotional	Mental
Visual	Bradycardia	Anxiety	Hallucinations*
Auditory	Tachycardia	Distress	Personal reminiscences
Somesthetic	Apnea	Fear	Incongruousness
Tactile	Tachypnea	Pleasure	Incoherence
Olfactory	Sweating	Sadness	Discontinuity
Gustative	Penile erection	Joy	Strangeness
Pain	Ejaculation	Rage	Sensation of reality
Movement	Swelling	Guilt	Lucid dreams
Pleasure		Surprise	

* This term is used due to the similarity with pathological hallucinations (perception of something that does not exist in the surrondings), but they are not considered abnormal phenomena during REM sleep.

Figure 1. Dream components.

The mental activity that happens during dreams can be explained by the desynchronization of the EEG activity of REM sleep, similar to that that occurs during the consciousness waking state. The visual and auditory phenomena, could be partly explained by the spread of the ponto-geniculo-occipital waves (PGO) potentials toward the sensorial systems. Currently it has been proposed that the mental process and its corresponding cortical activation during REM sleep differs from the mental processes and activation of the cortex during wakefulness, possibly because during the waking state the cortex is activated by cholinergic actions, while during REM this activation occurs by means of glutamatergic activation and neurotransmission (Steriade 2000).

For the "hallucinatory", emotional and vegetative components of dreams, and for their memories, the limbic system is involved. It is well known that this system integrates functions related to the regulation of emotional tone and vegetative system, which in turn participates in the process of memory consolidation. It has been demonstrated that some of its structures show changes in its bioelectric activity during REM sleep. The neuronal discharge frequency increases in the amygdala and hippocampus in relation to the frequency during the waking state or SWS. It has been demonstrated that there are monosynaptic connections, in some cases, reciprocal between limbic structures (amygdala, hippocampus and hypothalamus) and the pontine nuclei involved in the generation of the PGO potentials found that PGO potentials spread toward the cingulate circumvolution, the hippocampus and the temporal lobe amygdala, which indicates that these limbic structures are activated phasically during REM sleep.

The amygdala and the hypothalamus play an important role in a wide range of somatic motor responses related to emotion, such as flight, defense, fury and attack, acting directly on the brainstem, through the amygdalofugal ventral bundle. The emergence of the PGO waves during REM sleep is related to temporary changes of the cardiac and respiratory frequencies. The fact that amygdala electrical stimulation increases PGO activity suggests that the amygdala participates in the vegetative emotional components of REM. Moreover, the electrical stimulation of the cingulate gyrus, hippocampus and amygdala in man causes elaborate visual and auditory hallucinations, in addition to the feeling of "being dreaming". The phasic activation PGO of the limbic system, could explain the hallucinatory components of REM sleep.

THEORY OF MIND

The concept of 'theory of mind' (ToM) refers to the social skill of humans [and perhaps other animals] to attribute mental states to others, including intentions, hopes, expectations, beliefs and desires, in order to interpret, explain and predict their behavior (Leslie 2001).

The term was first proposed in 1978 by Psychologist David Permack. During its development several approaches have been proposed. Among these are the module theories, the theory theories, the simulation theories, (Baars and Gage 2010) and the rationality theories (Shanton and Goldman 2010). Like both these publications [and many others] mention, these approaches are not to be considered mutually exclusive, as evidence for several aspects of each has been consistently gathered.

To explore the possibility of a ToM during dreaming we must first define what dreaming is, as well as determine the processes and mechanisms that either support or constitute what is known as a ToM. Dreaming can be defined as any "mental activity occurring in sleep characterized by vivid sensorimotor imagery that is experienced as waking reality despite such distinctive cognitive features as impossibility or improbability of time, place, person and actions [...]" (Hobson, Pace-Schott, and Stickgold 2000).

It is important to note that, despite its specific and measurable physiological and phenomenological characteristics, according to these same authors, dreaming can be thought of as taking place in a continuum consisting of three distinct and measurable dimensions: Activation, Information and Mode. The variability in these dimensions means that dreams can be highly variable, which is why the question of the existence of a theory of mind in dreams must be reformulated as two different questions: is theory of mind capable of appearing in dreams? And, if so, how frequently does it, or its underlying mechanisms and/or processes, appear in dreams?

Mainly because of technical constraints, studies of sleep mentation have historically relied on questionnaires and self-reports, as far as their cognitive characteristics, and sometimes on electroencephalography as a physiological record of brain activity. On the other hand, the studies on ToM have adopted several approaches including research on Developmental Psychology, psychiatric disorders, brain lesions and, during the last two decades, brain imaging.

In addition to the indirect nature of the correlations between sleep mentation and brain activity found using high spatial resolution brain imaging techniques, any correlation between brain activity and dreaming must take into account the fact that neuromodulation is different between the waking state and all stages of sleep (Hobson, Pace-Schott, and Stickgold 2000), which means that brain activity information gathered through imaging during sleep cannot be directly correlated with the same information obtained during the waking state, though it has been proposed that most brain regions, especially those on the cortex, perform similar functions during dreaming to those they perform during the waking state (Hobson, Pace-Schott, and Stickgold 2000).

STUDIES ON THEORY OF MIND AND DREAMING

Based on a simple definition of ToM: "the awareness of what others are thinking and the ability to attribute feelings to others", Kahn and Hobson (2005) found in a study of 35 participants who submitted a total of 320 dream reports, participants thought that at least one dream character was feeling something[1] toward or about him or her in 82.5% of their dreams. Awareness of feelings between dream characters was also reported. On a second study by these authors, published in the same article referenced above, it was found that when the 24 subjects were asked if they noticed whether they were wondering, thinking, or being concerned about a dream character's thoughts or intentions, they answered positively twice as often as they did negatively. Some of them did not recall.

Kahn and Hobson do mention that the medial prefrontal cortex (mPFC), the temporoparietal junction area (TPJ) and the amygdaloid complex, and proper functional connections between the first one and the other two, have been reported as necessary for normal performance on ToM tasks (Frith 2001). Also, that these areas are all active during REM sleep. It is worth noting, however, that subsequent imaging studies have partially redefined the areas thought to be associated with ToM.

---

[1] In this study 'feelings' include intentions, such as "he was trying to lose, so that I can feel better", and "I somehow know that he wants to have sex with me".

More specifically, Mars, et al. (2012) found a significant overlap between the Default Mode Network (DMN) areas and the areas active while performing ToM tasks in healthy subjects in the left angular gyrus and the posterior cingulate cortex (PCC), extending into the precuneus (PCUN).

Also, Scholz, et al. (2009) corroborated, through the use of functional magnetic resonance imaging, that a specific neural region within the right temporoparietal junction (rTPJ) activated distinctly when subjects performed a ToM task, separate from an adjacent area within the rTPJ which activated when participants processed non-social information.

Pace-Schott (2003) proposes that the neurobiology of autism can be seen as a mirror image of the frontal deficits that occur during dreams, in the sense that these patients retain the capacities of the dorsolateral prefrontal cortex (dlPFC) but show deficits in tasks associated with orbitofrontal and ventromedial functions, which is somehow contrary to what happens during REM sleep, in which these regions are at least partially and selectively activated, and thus social reasoning, including a theory of mind, is preserved [in his opinion].

He further argues that the correct functioning of these regions has been shown to be sensitive to sleep deprivation, and that their lesion is associated with either significant augmentation or attenuation of dreaming. Another important point is that people with autism and Asperger's Syndrome (AS) show significantly attenuated dreaming.

While behaviorally this appears to be the case, more recent studies have proposed that the DMN, which includes regions in the amygdala, the orbitofrontal and mPFC, and is related to social cognition and self-referential processing may be "largely intact in Autism Spectrum Disorder (ASD) but underutilized perhaps because of a dysfunction in control systems that regulate its use" (Buckner, Andrews-Hanna, and Schacter 2008), as some regions have been found to be larger in ASD either during childhood, such as the amygdala (Abell et al. 1999; Schumann et al. 2004) or adolescence (Waiter et al. 2004). However, both activity during social cognition tasks and functional connectivity have been found to differ in ASD when compared to controls (Anderson et al. 2011) especially in the ventromedial prefrontal cortex (vmPFC) and PCC (Kennedy, Redcay, and Courchesne 2006).

One Modular Approach to Theory of Mind

Leslie, Friedman and German (2004) propose a modular model of ToM with three principal features, which work through two basic mechanisms. To answer the question of whether ToM is present in dreams, we must look for the existence of these features and mechanisms during the dreaming states.

The three features are: 1) "a representational system powerful enough to represent beliefs and desires as such", also called "metarepresentation"; 2) "successful reasoning with belief and desire metarepresentations recruits an inhibitory selection process (SP)"; and 3) "because people's mundane beliefs are usually true, the best guess about another person's belief is that it is the same as one's own", also known as "true-belief default".

The proposed mechanisms are: 1) the search in memory of either one, or two separate indexes for 'desires' and 'beliefs'; 2) the application of the 'true-belief default', mentioned above; 3) the application of any number of necessary inhibition SPs, applied to the default(s), to arrive at the correct logical conclusion[2].

From the studies of Kahn and Hobson published in 2005 referenced above, we know that people's dreams can, and often do include situations in which a dream character's feelings and intentions seem obvious, or at least likely, to the dreamer, or at least that these perceptions are reported during the waking state while subjects elaborate on the memories of their dreams[3].

This would suggest that the representational system capable of representing beliefs and desires as such, that is, as hypothetical but probable mental states related to, but not causally connected with reality, is active at the time that dreaming occurs, or at least that it is perceived as such during the conscious reconstruction of the dream which takes place during the waking state.

The inhibitory SP also appears to be present, at least with one iteration, in one of the three examples provided by Kahn and Hobson: "but then I felt that he was not playing seriously and he was trying to lose, so that I can feel better", assuming that the default would be to think that the dream character playing, like the dreamer, wanted to win. This is, of course, unless the 'defaults' were changed by the emotionally negatively charged nature of the dream itself, which is often the case (Merritt et al. 1994; Domhoff 1996), or that the specific emotional charge kept toward the dream character, e.g., mistrust, made it so that the 'default' changed to 'he wants to deceive me', though this is only speculation on our part. These interpretations would be in line with activation of the mPFC, TPJ (Eichenlaub et al. 2014) and amygdaloid complex during REM sleep, assuming, of course, that the dreams reported happened during this stage; and with their conclusion that ToM does exist in dreams.

They would also be consistent with the finding that regional cerebral blood flow (rCBF) in the mPFC is the same during REM sleep as it is during the waking state (Maquet, et al. 2005), that an increase in rCBF in the anterior cingulate cortex (ACC) has been consistently reported during REM sleep (Maquet 2000) and that activity in the ACC

---

[2] For example, Sally believes that the marble is in box B, but the marble is actually in box A, so this is a false belief Sally has, whereas the 'default' would be to assume that Sally has the same information as myself. Also, Sally wants to avoid opening the box that contains the marble, so Sally desires to avoid the marble, whereas the default would be to desire to have the marble.

[3] For a succinct review of methodological considerations in the study of dreaming, including the reduction of psychological states to narrative reports, see Hobson, Pace-Schott and Stickgold (2003).

and within the mPFC, including both their dorsal and medial regions, has been associated with performance of ToM tasks (Northoff et al. 2006)

An Organizing Framework for Theory of Mind

According to Simon Baron-Cohen (1995) the development of ToM depends on the successful and independent development of four kinds of skills: detection of intentions of others, detection of eye direction, shared attention and a theory of mind module, which comprises a set of higher level social skills and a repertoire of implicit knowledge about others.

Detection of Intentions of Others

In this context, detection of intentions refers to the ability to "perceive intention or purposeful action in many forms of biological and non-biological movement", or to the tendency to "interpret *action* as intention". This mechanism, at least in its most primitive form, is also present in many mammals (Baars and Gage 2010).

This first skill can be found in at least some dreams, as reported by Kahn and Hobson (2005), and as may seem obvious with simple introspection. Additionally, in support of this observation is the fact that activation of the superior temporal sulcus (STS), an area involved in the perception of biological motion, occurs during dreaming, in both REM and non-REM (NREM) sleep (Siclari et al. 2017). It is important to note, however, that this activation was observed when subjects reported having the sense of moving during their dream, and not in situations when they appeared to be detecting a dream character's intentions, though this was not part of the researchers' objective in this article.

On the other hand, the inferior frontal gyrus (IFG), an area proposed to be crucial for the interpretations of other's intentions, and part of the 'Mirror Neuron System' (Iacoboni et al. 2005), seems to be, bilaterally, the least active area in the prefrontal cortex during REM sleep (Maquet et al. 2005).

However, the intentions mentioned in the Kahn and Hobson study either referred to a dream character's emotions: "I was on a beach with an unidentified companion. It was a romantic setting. The person had feelings for me and I for her", or are guessed, instead of perceived: "but then I felt that he was not playing seriously and he was trying to lose, so that I can feel better"; "there's a man sitting at the first table in the café staring at me. I do not know him, but I somehow know that he wants to have sex with me.", which means that, while more anterior and lateral regions of the PFC, which show hypoactivation during REM sleep, might be needed to accurately perceive and verify the intentions of others, they may not be necessary to create the subjective sensation of representing an intention and having it assigned to some other personage, or their level of activity may need be minimal.

This assertion would also correlate to the fact that the caudal part of the vmPFC, which is active during REM sleep, assists the "feeling of rightness" and the related phenomenon of "feeling of knowing" (Pace-Schott 2010). It could also be that these sleep reports come from NREM sleep mentation, during which an increase in activation of the left IFG has been reported using EEG (Siclari et al. 2017), or that dreamers are able to monitor their own intentions, but attribute them to their dream characters instead.

This last proposition would be consistent with both the notion of the participation of mirror neuron systems during sleep mentation (Nielsen 2007) and with bilateral hypoactivation of the TPJ during REM sleep (Maquet et al. 2005), the right part of which shows activation during ToM tasks which require differentiation of self and others (Mars et al. 2012) [though it also participates in other ToM functions which do not require this differentiation].

On this point, the authors mention that the low activation of the TPJ in the right hemisphere during REM sleep is hypothesized to play a greater role in the differentiation between first and third-person perspectives [as compared to the left TPJ], which is especially seen in dream reports where the dream self can participate in both first and third-person perspectives, sometimes within the same dream. For example, the self can either see and act in the dream narrative, or it can see itself acting.

Detection of Eye Direction

Although primate and human eye detection is involved in several higher order processes [e.g., shared attention], a more primitive version of it can be found in other mammals, such as cats, where direct gaze signifies confrontation, and gaze aversion means a cat has accepted that the other cat is the dominant one (Baars and Gage 2010). Detecting others' eye direction or gaze is thought to be related to activity in the STS and the amygdala (Hooker et al. 2003). In relation to interpreting another person's gaze in social situations, the anterior portion of the STS is thought to code information of the different directions in which another's head moves (Carlin and Calder 2013).

On the other hand, the posterior region of the STS, along with the adjoining TPJ, is thought to have several roles in processing social information related to eye movement and gaze, including expectancy-based coding of other's intentions, gaze direction, joint [or shared] attention, and initial processing of head orientation (Carlin and Calder 2013).

As previously mentioned, Siclari et al. (2017) reported increased high-frequency activity in an area surrounding the right STS during REM sleep. However, it is worth noting that the low spatial resolution of the EEG and the fact that the STS serves several distinct functions for social cognition, make it difficult to assert that detection of gaze, or all of its components, are present during dreaming, at least based on neural activation alone. Additionally, Braun, et al. (1997) found no differences in cerebral blood flow in the STS during REM sleep as compared with SWS, and, as previously mentioned, the temporoparietal region was reported as one of the quiescent areas in the parietal lobe during REM sleep (Maquet et al. 2005).

One possibility is that these brain regions respond selectively to changes in another's eye direction, and that the possibly static nature of images hallucinated during dreaming means that they are not needed to decode eye movement in dreams. However, activation of the STS has also been correlated with the viewing of static direct and averted gaze pictures. This is thought to be because variable facial features, such as the eyes and mouth, can have implied motion even when seen as static pictures (Hooker et al. 2003)

Another possibility is that the profound activation of the limbic core, including the amygdala, during REM sleep (Hobson, Pace-Schott, and Stickgold 2000; Maquet et al. 2005), which has shown activation during "passive viewing of both direct and averted gaze [...] and active detection of eye contact [bilateral] and averted gaze [left]", and the fact that bilateral amygdala damage have difficulty identifying gaze direction (Hooker et al. 2003) creates an activation pattern robust enough to process basic information about facial recognition and variable aspects of the faces of dream characters, since we know that character identification through facial features occurs in dreams (Kahn, Pace-Schott, and Hobson 2002)

In any case, it is clear that more research is needed to establish which aspects of eye direction and gaze detection are present in dreams, although we can speculate that at least the most basic components are, perhaps even including the 'implied' movement of static eyes in faces, or even static faces. Or perhaps, the subjective sensation of others' intentions occurs without the need for these mechanisms.

Shared Attention

It refers to the ability to know that another person's gaze indicates the position, and the object, that they are looking at; that we too can direct our gaze at the same place to look at the same thing, and that the other person will probably know that we are both focusing our attention on the same object. It is not present in most mammals, including most primates, and in humans of less than one year of age (Baars and Gage 2010).

At this point it is unclear whether shared attention is possible during dreams. We can only mention that the mPFC seems to respond preferentially to shared, as compared to non-shared attention (Carlin and Calder 2013) and when responding to joint attention versus initiating it (Redcay, Kleiner, and Saxe 2012), and that, as previously mentioned, its rCBF is the same during REM sleep as it is in wakefulness (Maquet et al. 2005).

Bilateral STS and IFG are also implicated in responding to shared attention but, as previously discussed, they both appear to be hypoactive during REM sleep (Maquet et al. 2005). However, Siclari et al. (2017) reported higher levels of activation in both areas during NREM sleep mentation as compared to when subjects were awakened during NREM but reported no dream experience.

The PCC also participates in shared attention tasks (Redcay, Kleiner, and Saxe 2012) and has been found to be active during NREM sleep as compared to pre-sleep wakefulness (Braun et al. 1997; Siclari et al. 2017) but hypoactive during REM sleep (Maquet and Franck 1997; Pace-Schott 2003).

Theory of Mind Module

This module is, according to Baron-Cohen, the last component of a complete ToM, and refers to "a complex knowledge based containing rules of social cognition that develops by [age] four" (Baars and Gage 2010). This module includes: a) that knowledge that appearance and reality are not necessarily the same; b) the passive behavior of someone does not mean that they aren't 'doing something', for example, thinking, imagining or remembering; c) other people can have mental states; d) other people can have knowledge that I don't have, thus I can be deceived; e) I can have knowledge that other people don't have, thus I can deceive others; f) my mental state fluctuates in time; and g) facial expressions are indicators of mental states (Baars and Gage 2010).

Attribution of Mental States

Regarding the attribution of mental states to others, including emotion and beliefs, it has been consistently reported that the mPFC, the TPJ and the PCUN are all active during tasks relating to these forms of attribution (Green, Horan, and Lee 2015; Völlm et al. 2006; Sebastian et al. 2011). Specifically, the posterior cingulate gyrus and amygdala are related with emotion attribution, sometimes referred to as 'cognitive empathy', while the superior frontal gyrus (SFG) and the middle frontal gyrus (MFG) are more closely associated with belief attribution (Lee et al. 2016).

Limbic activation during REM sleep has been reported extensively (Braun et al. 1997; Hobson, Stickgold, and Pace-Schott 1998; Maquet 2000; Pace-Schott 2003) while both the PCC and the PCUN have been reported as two of the least active regions during REM sleep (Maquet et al. 2005), as previously mentioned. However, Siclari et al. (2017) have reported a decrease in low frequency power in the PCUN and the posterior cingulate gyrus during episodes of both REM and NREM sleep in which subjects reported having a dream experience right after being awakened, as compared to episodes when they did not report such activity, suggesting that these areas, while hypoactive during REM sleep as compared to the waking state, likely participate in sleep mentation. Also, Braun (1997) reported increased activation in the PCC during REM sleep as compared to activation in SWS or NREM sleep.

Activation of the SFG has been reported during REM sleep as part of the midline attentional structures (Pace-Schott 2003; Maquet et al. 2005), while MFG has been reported as hypoactive during REM sleep (Maquet et al. 2005).

Corroborating the evidence of at least partial activation of these structures during REM and NREM sleep mentation is the fact that sleep reports often contain attribution of both emotions and thoughts to dream characters [other than the self] (Kahn and Hobson 2005). Regarding this matter, we can assert that attribution of both feelings and thoughts to dream characters is well documented. However, awareness of feelings in dream characters is reported significantly more often than them having thoughts or intentions [82.5% vs 50.4% of dream reports in two different studies] (Kahn and Hobson 2005).

Perspective-Taking

Another skill closely related to ToM is perspective-taking. This ability "allows us to understand how another person thinks and feels about a painful situation" (Baars and Gage 2010). As its definition implies, it can be characterized along two dimensions: cognitive and affective. Cognitive perspective-taking refers to the ability to infer the thoughts and beliefs of others, while affective perspective-taking can be defined as the ability to infer the emotions or feelings of others (Healey and Grossman 2018). Cognitive perspective-taking engages dmPFC and dlPFC preferentially, while affective perspective-taking shows more activation in the vmPFC. This distinction could lead us to propose that affective perspective-taking, and thus also affective empathy ['I feel what you feel', instead of 'I understand how you feel'] may be more prevalent in dreams. However, these same authors (Healey and Grossman 2018) emphasize the fact that these processes have both shared and non-shared neural correlates, and, as was mentioned before, recalled dreams do contain the subjective experience of inferring, or knowing, other dream character's intentions, which implies that at least some capacity for cognitive perspective-taking and cognitive empathy is preserved in dreams.

In perspective taking tasks, in which subjects are asked to imagine the pain that either they or another person would feel during a physically painful situation, an increase in activation is shown in the mPFC, along with other areas associated with pain perception (Jackson et al. 2006).

More specifically, the paracingulate region (right Brodmann area 32) of the mPFC shows activation during both tasks [imagined self and other pain] (Baars and Gage 2010), and has been reported to increase its activity during REM sleep vs. both waking and NREM states (Hobson, Stickgold, and Pace-Schott 1998; Braun et al. 1997).

The ACC (Brodmann area 24), which registered increased activation when subjects imagined their own pain, was also reported to increase its activity during REM vs. the waking state (Hobson, Stickgold and Pace-Schott 1998), suggesting a possible neural basis for the distinction between self and other-pain during dreams.

Another important aspect of social cognition during dreams is the fact that separate brain regions are associated with judging others either as similar to us vs. distinct from us, according to Buber's scale of I-It (Baars and Gage 2010). One of the most salient implications of this distinction is the fact that we may be able to selectively start [and stop] the attribution to others of, among other things, higher mental states and feelings, when we perceive them as different from us.

Interestingly, regions of the vmPFC were more active during judgements about people which subjects considered similar to themselves, whereas portions of the dmPFC were more active when subjects judged people whom they considered 'dissimilar' (Mitchell, Macrae, and Banaji 2006). Since dorsomedial prefrontal activation has been reported during REM sleep as 'partial' [vs. a 'complete' activation of ventromedial regions] and because the latter have also been strongly suggested to support ToM abilities during dreams (Pace-Schott 2010) we may be inclined to suggest that, during dreams, we: a) tend to dream of more known or familiar characters than unknown ones; b) tend to judge dream characters as more similar to us than we judge other people during waking; and/or c) this pattern of activation may make it so that we identify unknown or poorly defined characters as known or familiar.

The first hypothesis seems to be at least partially corroborated by Kahn et al. (2000) who found that, in a sample of 320 dream reports from 33 adults, 48% of reported dream characters represented a known personage to the dreamer, and 35% were generically identified [e.g., policeman], while only 16% were 'new' or totally unknown.

Regarding the third hypothesis, it is important to note that in this same study, while a high percentage of 'known personages' were identified by their face [45%] or their appearance [32%], a remarkably large proportion [44%] were identified by 'just knowing', which could partly corroborate this hypothesis, especially since caudal regions of the vmPFC is thought to "subserve both 'feeling of rightness' and the related phenomenon of 'feeling of knowing', for which more anterior PFC areas" [which are for the most part inactive during REM sleep] "only later provide cognitive verification" (Pace-Schott 2010).

EMPATHY, MIRROR NEURON SYSTEM AND DREAMS

Adequate social functioning can be described as the ability to detect cognitive and emotional processes in others. The ability to attribute mental states (desires, intentions and beliefs) to other people, or to fictional characters (Mar and Oatley 2008), is explained by the concept of ToM, also called mentalizing.

On the other hand, empathy, a related mechanism also associated with social functioning, is defined as the ability to infer and share another's emotional experiences, and it has also been related to the mirror neuron system. Empathy is essential for the development of an early preconceptual stage of emotion identification, as it can mediate emotional contagion (affective resonance) (Nielsen, Powell, and Kuiken 2013). It has also been related to the processing and execution of motor behaviors such as the execution of a specific action (e.g., taking something), observing an action, imagining an action and imagining that another is performing an action.

ToM and empathy are two processes that underlie social cognition, and although they have some functional equivalence, it has been observed that both activate different neural mechanisms. ToM has been related to the activation of the mPFC, PCUN, PCC and bilateral temporal union in brain imaging studies. The mirror neuron system has been linked to a neural circuit that includes the posterior IFG, premotor cortex and inferior parietal lobe (Molenberghs, Cunnington, and Mattingley 2012).

Dream contents often have an important social component. The expression of these dream experiences can take many forms, from experiences of pure perception to pure thinking, and from simple images to complex narratives which are often experienced as similar to the experiences that characterize the waking state (Siclari et al. 2017). Memory is another key component in the presentation of images and dream situations which provides information about the appearance of characters and scenarios, and the relationships established with those characters (Han et al. 2015)

Qualitative findings on the social content of dream experiences have led to the hypothesis that social interactions that take place during dreams trigger neural mechanisms related to social cognition during wakefulness (Han et al. 2015; Schweickert and Xi 2010). Furthermore, imagined social interactions (such as daydreaming, imagining future scenarios or having wandering and confusing thoughts) have also been linked to dream-like brain activity and with waking social experiences (Han et al. 2015)

Some studies have explored the functioning of the mentalizing network during dreaming, specifically in the mental characteristics attributed to dream characters by dreamers (Kahn and Hobson 2005; Kahn, Pace-Schott, and Hobson 2002; Schweickert and Xi 2010). However, the neurophysiological factors responsible for the emotions and actions of other characters during sleep remain unclear.

Electroencephalographic Findings

Pioneering studies conducted with electroencephalography have allowed us to identify the EEG patterns as well as the graphoelements present during sleep. At this time, these findings allow us to identify each one of the sleep stages and sleep architecture. EEG studies have also allowed us to study qualitative aspects of sleep such as awareness (attention, memory, orientation), emotional states, mood alterations and the presence of sleep mentation, based on the interpretation of waves with characteristic morphology, or through oscillatory patterns.

High frequency neuronal oscillations have been identified mainly during wakefulness and REM sleep. Different authors suggest that the functional equivalence between wakefulness and REM sleep is regulated by thalamocortical activity, to be precise, by means of communication between the thalamus and the cortex. The differences between these two behavioral states have been attributed mainly to neuromodulatory mechanisms (Kahn, Pace-Schott, and Hobson 1997). Some studies have reported that high frequency oscillations ranging from 30 to 80 Hz can be registered during some stages of sleep. These are called gamma frequencies and are considered representative of REM sleep. During REM-sleep dreaming, gamma frequencies have been hypothetically associated with cognitive and perceptual, memory processing, and temporal binding of dream imagery. Desynchronization of gamma oscillations between anterior and posterior brain regions has also been described during this stage (Pace-Schott, 2010; Kahn, Pace-Schott, and Hobson 1997).

On the other hand, a higher proportion of slow frequencies also take place during sleep, this slow oscillations can be observed mainly during the first part of the night, decrease during the second half of the night and are directly related to NREM sleep stages. Particularly, the theta band plays a role in the consolidation of memory, spatial memory, spatial learning and the consolidation of fear memories in the hippocampus, amygdala and prefrontal cortex. This finding has been corroborated by slow wave research in animal models (Cornwell et al. 2008; Burgess, Maguire, and O'Keefe 2002). Several studies using magnetoencephalography and intracranial electroencephalography in humans have indicated that the hippocampus has a basal frequency of theta oscillations, which have been linked to memory and spatial learning (Burgess, Maguire, and O'Keefe 2002). However, theta rhythms also appear when individuals react to emotionally relevant stimuli, suggesting that they could also play a role in social cognition (Tendler and Wagner 2015). Slow oscillations in NREM sleep, such as sleep spindles, delta waves and slow cortical oscillations (<1 Hz) are associated with interactions between the thalamus and the cortex, called "intrinsic thalamocortical rhythms" (Achermann and Borbély 1997; Pace-Schott 2010). In more specific situations, slow theta waves registered on the left temporal region have been linked to visualization of fear, anxiety and threat-provoking stimuli (Tendler and Wagner 2015).

Theta activity has been registered in the hippocampus, amygdala and neocortex, in both wakefulness and REM sleep, in humans and rodents. Theta coherence between the hippocampus and the amygdala is related to the presence of PGO waves and the processing of emotional memory in the hippocampus during REM sleep (Hutchison and Rathore 2015). The increase of 1 to 4 Hz power during REM and NREM sleep stages in bilateral parieto-occipital regions has been described only in sleep reports where participants reported awareness of the experience of dreaming (Siclari et al. 2017). However, a clear connection between the rhythmic theta brain activity present during wakefulness and the processing of social stimuli during sleep has not yet been established.

Other EEG studies have also described some elements of the EEG related to memory encoding and recovery of sleep content in NREM and REM sleep mentation. Many authors have tried to use the memory of dream experiences to understand the content of dreams in general. The ability to remember REM sleep nightmares is related to the spectral power of alpha (10-14.5 Hz) and delta (3-4 Hz) waves in the frontocentral area, these modifications were not observed in polysomnographic recordings when participants did not remember their dream experiences (Marquis et al. 2017). A growing body of research in humans and animals provides behavioral evidence of the role of REM sleep on the modulation of emotional memories, memory consolidation and extinction of fear memory (Marquis et al. 2017; Hutchison and Rathore 2015).

A new theory of the pathogenesis of nightmares stipulates that they originate mainly during an adverse childhood or as a result of traumatic events (Marquis et al. 2017). From a cognitive approach it has been proposed that nightmares as well as dreams facilitate the extinction of fear memory and are related to emotional regulation processes. At the neuronal level it has been described that nightmares are caused by a dysfunction in a hippocampal-amygdala-prefrontal neural circuit (Anderson et al. 2011). Nightmares are considered symptomatic of affective disorders, such as bipolar disorder and alexithymia. These alterations have been explained by abnormalities in the mirror neuron system (Kim et al. 2009; Moriguchi et al. 2009; Beauchemin and Hays 1996; Nielsen, Powell, and Kuiken 2013).

Current models of post-traumatic stress disorder (PTSD) symptoms emphasize the role of REM sleep in the consolidation and extinction of fear memory and in elevated levels of nightmare recall (Pace-Schott 2010; Marquis et al. 2017; Hutchison and Rathore 2015).

Marzano in 2011 supposed that dream recall is a peculiar form of declarative memory and hypothesized that 1) the encoding of dream content during sleep should share some electrophysiological mechanisms with the encoding of episodic memories of the awake brain and 2) recalling of dreams after awakening from NREM sleep and REM sleep should be associated with different brain oscillations (Marzano et al. 2011). Later, DeGenaro proposed that dream recall regards such a peculiar form of episodic information as dream content, which is encoded in declarative memory during sleep, the retrieval of oneiric contents could share some electrophysiological mechanisms with successful episodic memory encoding of the awake brain (De Gennaro et al. 2012)

In the same manner, neurophysiological mechanisms underlying the coding and recovery of episodic memories remain the same between different states of consciousness. The presence and/or absence of dream recall depends on the EEG activity that takes place during sleep and that precedes behavioral awakening. Sleep mentation has been associated with alpha power during sleep, alpha power during a given sleep stage might be a marker of mentation that is recalled following awakening. If a decrease in alpha power were found to be related to recall of mentation from both REM and NREM sleep awakenings, this would be consistent with the interpretation of a similar process associated with dreaming in the two stages (Esposito, Nielsen, and Paquette 2004).

On the other hand, theta activity over the frontal areas seems particularly involved in the retrieval of episodic mnestic traces also in wakefulness (Scarpelli et al. 2015). In particular, it has been found that successful memory retrieval of dreams corresponds to an increase in theta activity before waking (Marzano et al. 2011). One study showed that theta oscillations in the last 5 minutes of REM sleep predicted the retrieval of the memory of subsequent sleep (Scarpelli et al. 2015).

Moreover, some studies also reported a relationship between high frequency range (18–50 Hz, gamma) activity and the reporting of dream memories (Siclari et al. 2017; Voss et al. 2014). Brain activity characterized by increased power in the high frequency range is associated with the presence of memory upon waking during both REM and NREM sleep (Siclari et al. 2017).

Theories about Dream Content

Different descriptive works propose theories that attempt to explain the psychic and social content of dreams as well as their functionality. Such is the case of the Simulation Theory, which states that dreams are a useful way of simulating the world (Perogamvros and Schwartz 2012). It has also been proposed that the reproduction of elements of waking experiences improves memory consolidation during the night given that memory consolidation and reward information are thought to benefit from sleep-related memory consolidation (Perogamvros and Schwartz 2012; Fischer and Born 2009).

There are currently several simulation theories, though they all share the idea that world simulation is the dream's main component. Their differences lie on whether they attribute some evolutionary function(s) to it (Revonsuo 2000; Tart 1987; Foulkes 1985; Hobson, Hong, and Friston 2014; Nielsen 2010; Revonsuo, Tuominen, and Valli 2015; Tuominen et al. 2019).

Social simulation theory proposes that social perception and interaction are more frequent in dreams than in waking, and that most dream characters are people that are close to the dreamer in waking life. It has also been suggested that social interactions are important in dreams because they represent a compensatory mechanism in lonely or isolated people during the day (Tuominen et al. 2019). From the point of view of the simulation theory, Tart proposes that "in both dreaming and waking, an active, complex world simulation process is going on, basically identical in kind….In dreaming….the kind of world, body, and self that be simulated/experienced is vastly riches and more varied than in the waking state" (Tart 1987). On the other hand, Nielsen and Strenstrom (2005) mention that "episodic memories….are altered in such awake that their autobiographical origins are obscured even though the subjective content, within which they appeared is a credible simulation of reality. Dreams seems to take place in real, spatially coherent, environments with which the self-interacts perceptually, for example, by orienting, seeking and assimilating sensory information, much as it does with the real world" (Nielsen and Stenstrom 2005)

On the other hand, Threat simulation theory, indicates that simulation of situations related to dangerous or threatening events help to prepare individuals for future situations presented in waking state (Revonsuo 2000; Valli and Revonsuo 2009). In both theories, it is considered that the main component of these simulations has an adaptive function. The continuity hypothesis postulates that dreaming is a continuation of wakefulness processing without any specific adaptive function (Schredl and Hofmann 2003).

Dreams and Sleep Alterations

These simulations of the waking world sometimes become so intense that dreamers can confuse the dream with the waking state, causing dream enactment behaviors (DEB) during sleep. These behaviors generally represent actions performed by the dream self and/or the dream characters. Likewise, DEBs have also been linked to a system called affective motor resonance which participates in perception and imagination during waking, and it is proposed that the DEBs are regulated by cortical areas similar to those that make up the mirror neuron system (Nielsen and Kuiken 2013).

DEBs appear when dreams are intense enough to overcome the motor inhibition of REM sleep. They are defined as isomorphic behaviors that correspond to the actions of the dreamed self or of other dream characters, and can mainly be observed during the transitions between sleep and wakefulness. Body images that represent us or other characters also tend to appear in dreams (Nielsen and Kuiken 2013). The feeling of presence implies that the experience of identifying the presence of a dream character is distinct from that of just seeing it, and can be defined as a 'body image' of said character. This phenomenon causes the dream to be interpreted more vividly (Nielsen 2007).

Additionally, some authors propose that this mechanism is also orchestrated by the mirror neuron system (Nielsen and Kuiken 2013; Nielsen, Powell, and Kuiken 2013).

The interpretation of external stimuli during sleep is a subject that could also participate in mechanisms underlying ToM (Koulack 1969). Nielsen found that external stimuli can influence the content of dreams through electrical stimulation applied to the wrist or legs during REM sleep (Nielsen 1993). Such stimulation also influenced the actions of the dreamed self and the actions of other dream characters. It was also found that participants are often aware of these stimuli (Kahn and Hobson, 2005).

Some sleep alterations, such as REM behavior disorder, may reflect an abnormal reduction of the right IFG inhibitory activity, leading to both a reduction in muscle atonia and failure to develop separate representations of the dream self and other characters. This area is related to the system of mirror neurons in wakefulness (Nielsen and Kuiken 2013).

The right lower frontal gyrus is a structure linked to the mirror neuron system. During normal REM sleep this area shows similar activity as during waking (Braun et al. 1998) and serves to establish a clear separation between the self and other dream characters (Kahn and Hobson 2005).

Neurobiological Mechanisms of Sleep

Studies on the phenomenology and contents of dreams indicate that dreams reflect interests, desires, likes and personality features of the dreamer. These elements are similar to the experiences and behavior of the dreamer during wakefulness, and are not simple thoughts or abstractions. Dream content has also been studied for the sensory experiences it gives to the dreamer, such as sounds [including speech and conversation], and more rarely tactile, olfactory and gustatory sensations, as well as pleasure and pain (Hobson 2009).

Besides its psychic components, the study of dreaming could also help us infer the organization and function of the brain during sleep. Evidence indicates that the waking brain state (brain organization, brain lesions and neurodevelopment) participates in the content and construction of dreams (Nir and Tononi 2010). Experimental protocols with EEG and neuroimaging have contributed to the study of dreams through the comparison between brain activity during basal wakefulness and REM sleep (Maquet 2000). The main findings indicate a similar activity between active wakefulness and REM sleep with EEG (Hobson, Pace-Schott, and Stickgold 2000; Maquet et al. 2000).

However, not all brain areas are activated in a similar way between wakefulness and REM sleep. Some studies have reported deactivation of the inferior right parietal cortex (Brodmann area 40, related to volition during waking) during REM sleep (Desmurget et al. 2009) and deactivation in brain regions such as the PCC, the inferior parietal cortex,

that it is not impossible that, even when aphasia is not present, the inability to recall dreams may not necessarily imply a cessation of the dreaming phenomenon itself, but an impossibility of the dreamer to bring those dreams to consciousness, either during the waking state and/or during the dream itself.

As mentioned in different sections of this book, dreams seem to indicate brain organization and function, so that the waking brain state after a brain injury has significant repercussions on dreams. Following this neuropsychological approach, some authors propose that cognitive development throughout life also has repercussions on dreams (Nir and Tononi 2010).

Some findings mention that after 7 years of age the reports of dreams become longer and more frequent. The content of dreams after 7 years is given by thoughts and feelings (the child becomes an active participant in the dream), dreams begin to develop a narrative structure and reflect episodic and autobiographical memories (Nir and Tononi 2010; Casey et al. 2005)

Evidence indicates that visuospatial skills depend on the parietal lobes, which are not completely myelinated until the age of 7. Therefore, the study of dreams allows us to make inferences about visuospatial cognitive development and brain maturation in children (Nir and Tononi 2010; Casey et al. 2005). There are no autobiographical episodic memories in dreams before the age of 5, and this phenomenon is thought to be related to problems with conscious episodic memory in waking (Broughton 1968) that is closely related to the development of visual imagination. These results strongly suggest that mental images are the cognitive capacity most related to sleep (Epstein and Simmons 1983). It has also been proposed that the awareness of dreams is largely given by the dreamer's narrative ability (Nir and Tononi 2010).

PGO Waves

The occurrence of REM sleep has been shown to be regularly associated with the deactivation of a neuronal group (REM-off cells) located in the raphe nuclei (serotonergic population) and the noradrenergic group in the locus coeruleus, as well as with the reciprocal activation of one neuronal population (REM-on cells), localized in the Peribrachial area (PBL) (Datta and Hobson 1994; Datta and MacLean 2007). Among the many REM-on cell groups, one selectively active subpopulation is the PGO burst cell complex of the PBL region. Specifically, the pedunculopontine tegmentum is involved in the REM generation through cholinergic cell activation, which in turn is regulated by glutamate (S. Datta 2006). Since REM sleep is physiologically mediated by cholinergic mechanisms (Datta, Calvo, and Hobson 1991), the caudo-lateral part of PBL region is involved in the homeostatic regulation of REM sleep, and is the main structure responsible for PGO generation. A prolonged increase in REM sleep and other PGO related states, following PBL stimulation by carbachol, has been demonstrated (Calvo et al. 1992; Datta et al. 1992). Taken together, these observations indicate a physiological coupling between the PBL cholinergic/PGO trigger zone and the REM generator network that is both strong and enduring.

PGO waves on the cat appear in REM sleep as well as in the immediately preceding SWS episode, the stage known as *"sommeil phasique a ondes lentes"* (SPHOL) (Thomas and Benoit 1967), which occurs just seconds before REM sleep onset. PGO waves of SPHOL occur always in presence of sleep spindles, have a lesser frequency rate, and occur mainly in a simple pattern, unlike those of REM sleep. Whenever the PGO waves increase their incidence, among other signs (muscular atonia, EEG desynchronization and rapid eye movements) a REM sleep episode is initiated.

On the other hand some forebrain areas such as the limbic system, promote the generation of REM sleep, as well as the phasic elements of this phase (Calvo et al. 1987). Anatomical studies have shown reciprocal projections between the PBL and the amygdaloid central nucleus (Nomura et al. 1979; Amaral et al. 1992). Electrophysiological studies have demonstrated that REM PGO waves generated in the pons, but propagated and recorded in the lateral geniculate body, also propagate with a higher latency to the temporal lobe amygdala (40 ms), dorsal hippocampus (25 ms) and to the anterior and posterior regions of the cingulum (30 ms) (Calvo and Fernández-Guardiola 1984).

It is important to mention that the spread of PGO waves toward certain areas, such as the amygdala, can cause intermittent activations in this area. It has been demonstrated that there is activation of the amygdala during REM sleep, which may be the product of a possible endogenous sensorial afferent stimulus, originated in pontine regions. This information could be processed in the amygdala, and this structure may be leading the emotional quality of dreams during REM. It is of interest to note that patients with temporal lobe epilepsy reported more unpleasant and more distressing dreams compared to control subjects. These observations were made by Calvo through experimental awakenings during episodes of REM sleep. This phenomenon may be due to an on-activation of the amygdala during the REM sleep.

It has also been proposed that if the PGO waves spread mainly toward visual structures (lateral geniculate nucleus, occipital cortex) these potentials may be capable of producing hallucinatory visual phenomena. Finally, we know that the rapid eye movements of REM sleep are the final product of PGO waves generation.

Moreover, it has been demonstrated that PGO waves are involved in learning and memory processes. This is known from studies in rats, which after being exposed to learning tasks presented a greater number of P waves. In pontine areas of the rat, the P waves are the equivalent of PGO waves seen in the cat. The increase in this activity does not occur until the animal learns the task. In contrast, animals that are exposed to REM sleep deprivation are unable to learn the same task. It has been proposed that this lack of learning is due to the lack of REM sleep, and mainly to the lack of PGO waves. In the first case, it is possible that the spread of these potentials toward the hippocampus allows greater activation of this structure and, as a product of this spread, the learning process is facilitated.

On the other hand, it has been proposed that the PGO waves are likely to be regulated homeostatically, giving them a character of a biological constant. In cat studies conducted by Jouvet, an average of between 14.500 and 17.000 potentials every 23 hours was obtained. The constant nature of these potentials has been analyzed through studies of sleep deprivation in cats. After total sleep deprivation, PGO waves presents an important rebound of these potentials, as compared to the overall percentage of REM sleep. In humans, these phenomena occur in a similar way, by a rebound of rapid eye movements more than by the total amount of REM sleep.

Dreaming and Autism

Despite the recent advances in technology and methodology, not much is known about the relationship between autism and dreaming. Regarding its relationship to sleep, a case study by (Godbout et al. 1998) on a patient with Asperger's Syndrome (AS) reported no EEG abnormalities and normal sleep latencies, but a significant decrease in SWS (5.1% of total sleep time) and high levels of stage 1 sleep, along with several awakenings. A somewhat more remarkable finding was a very low incidence of sleep spindles which, together with the low percentage of SWS, authors hypothesized could explain the lack of sleep reports produced by the patient after three awakenings following physiological REM sleep. They further elaborate that low spindle density reflects a fundamental malfunction in some selective attention mechanisms in AS and Autism Spectrum Disorder (ASD), which would indicate not necessarily a cessation of dreaming during REM sleep in these groups, but an inability to attend to the dream process and/or to generate a global concept of the experience. It is worth noting, however, that the idea of REM sleep and dreams being two distinct processes initiated/maintained by different brain mechanisms has recently gathered strong evidence (Solms 2000), thus, the assertion by Godbout et al. that normal physiological REM sleep equals sleep mentation in subjects with ASD should be considered with caution, more so because their study included only one patient.

However, a larger study of 28 young adults with ASD, 19 of them with AS diagnosis, found that dream reports by this population had "fewer words[4] and significantly less mention of content elements such as objects, social interactions, characters and activities compared to controls" (Daoust et al. 2008). ASD participants also reported having more

---

[4] 63 *vs.* 168 in controls.

white dreams[5] [6 *vs.* 1 in controls], fewer bad dreams[6], fewer participants with ASD reported experiencing confusion, shyness and sexual arousal, and anger and confusion were less intense when reported. Even when controlling for report length, individuals with ASD reported fewer settings, objects and descriptive elements [e.g., color, size, age, density, intensity, etc.], despite comparable verbal IQ to controls. The authors mention that these findings are in accordance with a series of connectivity and activation peculiarities found in the brains of subjects with ASD, most importantly, with "the 'replacement' of left frontal [verbal and voluntary] activity by activation of visual-associative temporooccipital regions during a working memory task", and with "low activation of the temporooccipital [visual] cortex" during REM sleep in participants with autism (Daoust et al. 2004).

Most of these differences in dream content and reports can also be found in patients with schizophrenia (Poulin et al. 2010). It is possible that some of their shared brain abnormalities, such as reduced activation in the right amygdala, fusiform gyrus and ventrolateral prefrontal cortex during a social cognition task (Pinkham et al. 2008), all of which are active during REM sleep, play a role in these found similarities during the dreaming process, and/or during dream recall. Other similarities at the neurobiological level, such as shared abnormalities in the serotonergic system (Meyer, Feldon, and Dammann 2011) may also be responsible for these particularities during dreaming and dream recall.

Conclusion

Dreams are a complex phenomenon, multiple cognitive factors, emotional states, ontogeny, pathologies, among many factors participate in their construction, specific characteristics and, particularly, in their contents. The aim of the present work was to present relevant data about dream phenomenology and its contents, taking into account proposals and models from cognitive psychology, hypothesis and proposals from renowned sleep specialists, and experimental and basic research information regarding sleep and dream processes.

Mental processes and emotional states during dreams cannot be measured directly using any current technique, however, subjective reports, surveys, EEG studies, imaging studies, among other techniques, have provided information about dream characteristics. Additionally, much is known about brain neurophysiology and its dynamics during REM sleep, as well as its neurotransmission.

---

[5] The subject reports mental activity prior to waking, but cannot recall any of its contents.

[6] Dreams with negative or unpleasant content which do not wake the dreamer up.

In order to summarize, we list the main structures involved in the regulation and generation of dreams and REM sleep: temporal lobe amygdala (emotion, fear); pre-frontal cortex (cognitive processes similar to wakefulness, decision making); temporo-parietal junction (differentiation of the dream self and other characters, self-recognition, dream recall); hippocampus (episodic memory, visual spatial memory); pontine structures (locus coeruleus, attention; peri-locus coeruleus alpha, muscular atonia; nucleus pontis oralis, cortical activation and theta rhythm generation; pedunculopontine tegmentum and laterodorsal tegmentum: REM sleep generation; peribrachial area (PGO waves generation, visual imagery); cingulum, involved in various cognitive functions, including memory, attention, learning, motivation, emotion, and pain perception, and finally, hypothalamus (orexin/hypocretin-producing neurons, sleep regulation, hormones). It's important to note that some of these structures are the neurobiological basis of some processes involved in Theory of Mind, Default Mode Network, Social Cognition and Empathy, including emotional and motor (mirror neuron system). The activation of the aforementioned structures in certain aspects coincides with the processes that present in conscious wakefulness, although not necessarily with the same neuromodulation, neuronal activation or connectivity. This could explain the differences between memories, characters, objects, places, sensations, etc. created during dreams and those present in the waking state, and could be related with the bizarre nature of dreams.

We consider that the study of Theory of Mind in relation to dreams should be carried out using the components (or modules) of this theory separately. Specifically, we refer to detection of intentions, eye movement, shared attention, and the theory of mind module (people can have different information about the same event, knowing that our and other's perceptions do not necessarily reflect reality, capacity for lying, sarcasm, *faux pas*, etc). Personal dreams have a wide range of variability in their emotions, stories, events and contents dependent on several factors such as emotional state, recent events, previous life experiences, drug use, among others aspects. This variability could mean that the presence or absence of specific abilities related to ToM may be dream specific.

The main function of dreams is to prepare us for possible future events. In this sense, dreams are capable of creating virtual scenarios, situations and social interactions, usually related to everyday life and the dreamer's experiences. On the other hand, the hypothesis of continuity tells us that there is a continuity between the experiences that occurred during wakefulness and those we experienced during dreams. In this sense, the content of dreams is decisive for the upcoming experiences during wakefulness and vice versa. From a different perspective, Threat Simulation Theory proposes that we can simulate dangerous situations, which allows us to anticipate or prepare for threatening situations or simulate situations that generate conflict or that represent danger. Nightmares and dreams related to Post-traumatic Stress Disorder allow repetitive exposure to the stimulus and/or aversive event, altering their perception through reworking and facilitating their incorporation into the content of dreams, e.g., sexual assault, war veterans, situations of extreme violence, and natural disasters.

REFERENCES

Abell, Frances., Krams, Michael., John, Ashburner., Richard, Passingham., Karl, Friston., Richard, Frackowiak., Francesca, Happé., Crhis, Frith. & Uta, Frith. (1999). "The Neuroanatomy of Autism: A Voxel-Based Whole Brain Analysis of Structural Scans." *Neuroreport*, *10* (8), 1647–51. https://journals.lww.com/neuroreport/pages/articleviewer.aspx?year=1999&issue=06030&article=00005&type=abstract.

Achermann, P. & Borbély, A. A. (1997). "Low-Frequency (<1 Hz) Oscillations in the Human Sleep Electroencephalogram." *Neuroscience*, *81* (1), 213–22. https://doi.org/10.1016/S0306-4522(97)00186-3.

Amaral, David G., Joseph, L. Price., Asla, Pitkänen. & Thomas Carmichael, S. (1992). "Anatomical Organization of the Primate Amygdaloid Complex." In *The Amygdala: Neurobiological Aspects of Emotion, Memory, and Mental Dysfunction*, 1–66. New York: Wiley-Liss.

Anderson, Jeffrey., Jared, Nielsen., Alyson, Froehlich., Molly, DuBray., Druzgal, T., Annahir, Cariello., Jason, Cooperrider., et al. (2011). "Functional Connectivity Magnetic Resonance Imaging Classification of Autism." *Brain : A Journal of Neurology*, *134*, 3742–54. https:// doi.org/10.1093/brain/awr263.

Arnulf, Isabelle., Jean-Pierre, Sastre., Colette, Buda. & Michel, Jouvet. (1998). "Hyperoxia Increases Paradoxical Sleep Rhythm in the Pontine Cat." *Brain Research*, *807*, 160–66.

Aserinsky, Eugene. & Nathaniel, Kleitman. (1953). "Regularly Occurring Periods of Eye Motility, and Concomitant Phenomena, during Sleep." Science. *US: American Assn for the Advancement of Science*. https://doi.org/10.1126/science.118.3062.273.

Baars, Bernard J. & Nicole, M. Gage. (2010). "Social Cognition: Perceiving the Mental States of Others." In *Cognition, Brain, and Consciousness*, edited by Bernard J. Baars and Nicole M. Gage, 2nd ed., 444–63. Academic Press. https://doi.org/https://doi.org/10.1016/ B978-0-12-375070-9.00014-0.

Baron-Cohen, Simon. (1995). *Mindblindness. An Essay on Autism and Theory of Mind.* 1st ed. Massachusetts: Bradford.

Beauchemin, Kathleen M. & Peter, Hays. (1996). "Dreaming Away Depression: The Role of REM Sleep and Dreaming in Affective Disorders." *Journal of Affective Disorders*, *41* (2), 125–33. https:// doi.org/10.1016/S0165-0327(96)00080-8.

Braun, A. R., Balkin, T. J., Wesensten, N. J., Carson, R. E., Varga, M., Baldwin, P., Selbie, S., Belenky, G. & Herscovitch, P. (1997). "Regional Cerebral Blood Flow throughout the Sleep-Wake Cycle. An H215O PET Study." *Brain*, *120* (7), 1173–97. https://doi.org/10.1093/ brain/120.7.1173.

Braun, A. R., Thomas, Balkin., Nancy, Wesensten., Fuad, Gwadry., Richard, Carson., Mary, Varga., Paul, Baldwin., Gregory, Belenky. & Peter, Herscovitch. (1998). "Dissociated Pattern of Activity in Visual Cortices and Their Projections During Human Rapid Eye Movement Sleep." *Science*, *279*, 91–95. https://doi.org/10.1126/ science.279.5347.91.

Broughton, R. J. (1968). "Sleep Disorders: Disorders of Arousal?: Eneuresis, Somnambulism, and Nightmares Occur in Confusional States of Arousal, Not in 'Dreaming Sleep'." *Science*, *159* (3819), 1070–78. https://doi.org/doi:10.1126/ science.159.3819.1070.

Büchel, C., Price, C., Richard, Frackowiak. & Karl, Friston. (1998). "Different Activation Patterns in the Visual Cortex of Late and Congenitally Blind Subjects." *Brain: A Journal of Neurology*, *121*, (Pt 3, 409–19.

Buckner, Randy L., Jessica, R. Andrews-Hanna. & Daniel, L. Schacter. (2008). "The Brain's Default Network: Anatomy, Function, and Relevance to Disease." *Annals of the New York Academy of Sciences*, *1124*, 1–38. https://doi.org/10.1196/annals.1440.011.

Burgess, Neil., Eleanor, A. Maguire. & John, O'Keefe. (2002). "The Human Hippocampus and Spatial and Episodic Memory." *Neuron*, *35* (4), 625–41. https://doi.org/10.1016/s0896-6273(02)00830-9.

Calvo, J. M., Badillo, S., Morales-Ramirez, M. & Palacios-Salas, P. (1987). "The Role of the Temporal Lobe Amygdala in Ponto-Geniculo-Occipital Activity and Sleep Organization in Cats." *Brain Research*, *403* (1), 22–30. https://doi.org/10.1016/0006-8993(87)90118-1.

Calvo, J. M., Sandeep, R. Datta., Quattrochi, J. & Allan Hobson, J. (1992). "Cholinergic Microstimulation of the Peribrachial Nucleus in the Cat. II. Delayed and Prolonged Increases in REM Sleep." *Archives Italiennes de Biologie*, *130* (4), 285–301.

Calvo, J. M. & Fernández-Guardiola, A. (1984). "Phasic Activity of the Basolateral Amygdala, Cingulate Gyrus, and Hippocampus During REM Sleep in the Cat." *Sleep*, *7* (3), 202–10. https://doi.org/ .1093/sleep/7.3.202.

Carlin, Johan D. & Andrew, J. Calder. (2013). "The Neural Basis of Eye Gaze Processing." *Current Opinion in Neurobiology*, *23* (3), 450–55. https://doi.org/10.1016/j.conb.2012.11.014.

Casey, B. J., Nim, Tottenham., Conor, Liston. & Sarah, Durston. (2005). "Imaging the Developing Brain: What Have We Learned about Cognitive Development?" *Trends in Cognitive Sciences*, *9*, (3 SPEC. ISS.): 104–10. https://doi.org/10.1016/j.tics.2005.01.011.

Cornwell, Brian R., Linda, L Johnson., Tom, Holroyd., Frederick, W Carver. & Christian, Grillon. (2008). "Human Hippocampal and Parahippocampal Theta during Goal-

Directed Spatial Navigation Predicts Performance on a Virtual Morris Water Maze." *The Journal of Neuroscience*, *28* (23), 5983 LP – 5990. https://doi.org/10.1523/JNEUROSCI.5001-07.2008.

Dahan, Lionel., Bernadette, Astier., Nicolas, Vautrelle., Nadia, Urbain., Bernat, Kocsis. & Guy, Chouvet. (2007). "Prominent Burst Firing of Dopaminergic Neurons in the Ventral Tegmental Area during Paradoxical Sleep." *Neuropsychopharmacology: Official Publication of the American College of Neuropsychopharmacology*, *32*, 1232–41. https://doi.org/10.1038/sj.npp.1301251.

Daoust, Anne-Marie., Élyse, Limoges., Christianne, Bolduc., Laurent, Mottron. & Roger, Godbout. (2004). "EEG Spectral Analysis of Wakefulness and REM Sleep in High Functioning Autistic Spectrum Disorders." *Clinical Neurophysiology*, *115* (6), 1368–73. https:// doi.org/10.1016/J.CLINPH.2004.01.011.

Daoust, Anne-Marie., Félix, Antoine Lusignan., Claude, M. J. Braun., Laurent, Mottron. & Roger, Godbout. (2008). "Dream Content Analysis in Persons with an Autism Spectrum Disorder." *Journal of Autism and Developmental Disorders*, *38* (4), 634–43. https://doi.org/ 10.1007/s10803-007-0431-z.

Datta, Sandeep R., Calvo, J. M. & Allan Hobson, J. (1991). "Long-Term Enhancement of REM Sleep Following Cholinergic Stimulation." *Neuroreport*, *2* (10), 619–22. https://doi.org/10.1097/00001756-199110000-00017.

Datta, Sandeep R., Calvo, J. M., Quattrochi, J. & Allan Hobson, J. (1992). "Cholinergic Microstimulation of the Peribrachial Nucleus in the Cat. I. Immediate and Prolonged Increases in Ponto-Geniculo-Occipital Waves." *Archives Italiennes de Biologie*, *130* (4), 263–84.

Datta, Sandeep R. & Allan Hobson, J. (1994). "Neuronal Activity in the Caudolateral Peribrachial Pons - Relationship to Pgo Waves and Rapid Eye Movements." *Journal of Neurophysiology*, *71* (1), 95–109. https://doi.org/10.1152/jn.1994.71.1.95.

Datta, Subimal. (2006). "Activation of Phasic Pontine-Wave Generator: A Mechanism for Sleep-Dependent Memory Processing." *Sleep and Biological Rhythms*, *4* (1), 16–26. https://doi.org/10.1111/j.1479-8425.2006.00202.x.

Datta, Subimal. & Robert, R. MacLean. (2007). "Neurobiological Mechanisms for the Regulation of Mammalian Sleep-Wake Behavior: Reinterpretation of Historical Evidence and Inclusion of Contemporary Cellular and Molecular Evidence." *Neuroscience and Biobehavioral Reviews*, *31* (5), 775–824. https://doi.org/10.1016/j.neubiorev.2007.02.004.

Dement, William. (1958). "The Occurence of Low Voltage, Fast, Electroencephalogram Patterns during Behavioral Sleep in the Cat." *Electroencephalography and Clinical Neurophysiology*, *10* (2), 291–96. https://doi.org/10.1016/0013-4694(58)90037-3.

Denoyer, Michel,, Marcellem Sallanon,, Colettem Buda,, Georgesm Delhomme., Dittmar, A. & Jou. (1991). "The Posterior Hypothalamus Is Responsible for the

Increase of Brain Temperature during Paradoxical Sleep." *Experimental Brain Research*, *84*, 326–34. https://doi.org/ 10.1007/BF00231453.

Desmurget, Michel., Karen, Reilly., Nathalie, Richard., Alexandru, Szathmari., Carmine, Mottolese. & Angela, Sirigu. (2009). "Movement Intention After Parietal Cortex Stimulation in Humans." *Science* (New York, N.Y.), *324*, 811–13. https://doi.org/10.1126/science.1169896.

Domhoff, William G. (1996). Finding Meaning in Dreams. A Quantitative Approach. 1st ed. New York: Springer US. https://doi.org/10.1007/ 978-1-4899-0298-6.

Eichenlaub, Jean Baptiste., Alain, Nicolas., Jérôme, Daltrozzo., Jérôme, Redouté., Nicolas, Costes. & Perrine, Ruby. (2014). "Resting Brain Activity Varies with Dream Recall Frequency between Subjects." *Neuropsychopharmacology*, *39* (7), 1594–1602. https://doi.org/ 10.1038/npp.2014.6.

Epstein, Arthur W. & Simmons, M. S. (1983). "Aphasia with Reported Loss of Dreaming." *American Journal of Psychiatry*, *140* (1), 108–9. https://doi.org/10.1176/ajp.140.1.108.

Esposito, Maria José., Tore, Nielsen. & Tyna, Paquette. (2004). "Reduced Alpha Power Associated with the Recall of Mentation from Stage 2 and Stage REM Sleep." *Psychophysiology*, *41*, 288–97. https://doi.org/ 10.1111/j.1469-8986.00143.x.

Fischer, Stefan. & Jan, Born. (2009). "Anticipated Reward Enhances Offline Learning During Sleep." *Journal of Experimental Psychology. Learning, Memory, and Cognition*, *35*, 1586–93. https://doi.org/ 10.1037/a0017256.

Foulkes, David. (1985). Dreaming: A Cognitive-Psychological Analysis. Michigan: L. Erlbaum Associates. https://books.google.com.mx/ books?id=xQB_AAAAMAAJ.

Frith, Uta. (2001). "Mind Blindness and the Brain in Autism." *Neuron*, *32* (6), 969–79. https://doi.org/10.1016/S0896-6273(01)00552-9.

Gennaro, Luigi De., Cristina, Marzano., Carlo, Cipolli. & Michele, Ferrara. (2012). "How We Remember the Stuff That Dreams Are Made of: Neurobiological Approaches to the Brain Mechanisms of Dream Recall." *Behavioural Brain Research*, *226*, 592–96. https://doi.org/ 10.1016/j.bbr.2011.10.017.

Godbout, Roger., Cybèle, Bergeron., Emmanuel, Stip. & Laurent, Mottron. (1998). "A Laboratory Study of Sleep and Dreaming in a Case of Asperger's Syndrome." *Dreaming*, *8* (2), 75–88. https://doi.org/ 10.1023/B:DREM.0000005898.95212.58.

Gottesmann, Claude. (2011). "The Involvement of Noradrenaline in Rapid Eye Movement Sleep Mentation." *Frontiers in Neurology*, *2*, 81. https://doi.org/10.3389/fneur.2011.00081.

Green, Michael F., William, P. Horan. & Junghee, Lee. (2015). "Social Cognition in Schizophrenia." *Nature Reviews Neuroscience*, *16* (10), 620–31. https://doi.org/10.1038/nrn4005.

Hall, Calvin S. & Robert, L Van De Castle. (1966). The Content Analysis of Dreams. The Content Analysis of Dreams. East Norwalk, CT, US: Appleton-Century-Crofts.

Han, Hye., Richard, Schweickert., Zhuangzhuang, Xi. & Charles, Viau-Quesnel. (2015). "The Cognitive Social Network in Dreams: Transitivity, Assortativity, and Giant Component Proportion Are Monotonic." *Cognitive Science*, *40*. https://doi.org/10.1111/ cogs.12244.

Hassemberg, L. (1965). Ruhe and Schlaf Bei Säugetieren. Wittenberg-Lutherstand: A Ziemsen.

Healey, Meghan L. & Murray, Grossman. (2018). "Cognitive and Affective Perspective-Taking: Evidence for Shared and Dissociable Anatomical Substrates." *Frontiers in Neurology*, *9*, 491. https://doi.org/10.3389/ fneur.2018.00491.

Hobson, J. Allan. (2009). "REM Sleep and Dreaming: Towards a Theory of Protoconsciousness." Nature Reviews. *Neuroscience*, *10*, 803–13. https://doi.org/10.1038/nrn2716.

Hobson, J. Allan., Charles, Chong Hwa Hong. & Karl, J. Friston. (2014). "Virtual Reality and Consciousness Inference in Dreaming." *Frontiers in Psychology*, *5*, (SEP), 1–18. https://doi.org/10.3389/ fpsyg.2014.01133.

Hobson, J. Allan., Pace-Schott, E. F. & Stickgold, R. (2000). "Dreaming and the Brain: Toward a Cognitive Neuroscience of Conscious States." *Behavioral and Brain Sciences*, *23* (6), 793–842. https://doi.org/ 10.1017/S0140525X00003976.

Hobson, J. Allan., Robert, Stickgold. & Edward, F. Pace-Schott. (1998). "The Neuropsychology of REM Sleep Dreaming." *Neuroreport*, *9* (3), R1–14. https://doi.org/10.1097/00001756-199802160-00033.

Hooker, Christine I., Ken, A. Paller., Darren, R. Gitelman., Todd, B. Parrish., Marsel Mesulam, M. & Paul, J. Reber. (2003). "Brain Networks for Analyzing Eye Gaze." *Cognitive Brain Research*, *17* (2), 406–18. https://doi.org/10.1016/S0926-6410(03)00143-5.

Hutchison, Isabel. & Shailendra, Rathore. (2015). "The Role of REM Sleep Theta Activity in Emotional Memory." *Frontiers in Psychology*, *6*. https://doi.org/10.3389/fpsyg.2015.01439.

Iacoboni, Marco., Istvan, Molnar-Szakacs., Vittorio, Gallese., Giovanni, Buccino. & John, C. Mazziotta. (2005). "Grasping the Intentions of Others with One's Own Mirror Neuron System." *PLoS Biology*, *3* (3), 0529–35. https://doi.org/10.1371/journal.pbio.0030079.

Jackson, Philip. Eric, Brunet-Gouet., Andrew, Meltzoff. & Jean, Decety. (2006). "Empathy Examined through the Neural Mechanisms Involved in Imagining How I Feel versus How You Feel Pain." *Neuropsychologia*, *44*, 752–61. https://doi.org/10.1016/ j.neuropsychologia.2005.07.015.

Jouvet, M. & Michel, F. (1959). "Corrélations Electromyographiques Du Sommeil Chez Le Chat Décortiqué et Mésencéfalique Chronique." *Comptes Rendus Des Séances de La Société de Biologie et de Ses Filiales.*, *153* (3), 422–25.

Jouvet, Michel. (1972). "The Role of Monoamines and Acetylcholine-Containing Neurons in the Regulation of the Sleep-Waking Cycle." *Ergebnisse Der Physiologie, Biologischen Chemie Und Experimentellen Pharmakologie*, *64*, 166–307. https://doi.org/10.1007/ 3-540-05462-6_2.

Kahn, David., Stickgold, R., Edward, F. Pace-Schott. & Allan, Hobson J. (2000). "Dreaming and Waking Consciousness: A Character Recognition Study." *Journal of Sleep Research*, *9* (4), 317–25. https://doi.org/10.1046/j.1365-2869.2000.00213.x.

Kahn, David. & Allan Hobson, J. (2005). "Theory of Mind in Dreaming: Awareness of Feelings and Thoughts of Others in Dreams." *Dreaming*, *15* (1), 48–57. https://doi.org/10.1037/1053-0797.15.1.48.

Kahn, David., Edward, F. Pace-Schott. & Allan Hobson, J. (1997). "Consciousness in Waking and Dreaming: The Roles of Neuronal Oscillation and Neuromodulation In Determining Similarities And Difference." *Neuroscience*, *78* (1), 13–38. https://doi.org/10.1016/ S0306-4522(96)00550-7.

Kahn, David., Edward, Pace-Schott. & Allan Hobson, J. (2002). "Emotion and Cognition: Feeling and Character Identification in Dreaming." *Consciousness and Cognition*, *11* (1), 34–50. https://doi.org/10.1006/ ccog.2001.0537.

Kennedy, Daniel P., Elizabeth, Redcay. & Eric, Courchesne. (2006). "Failing to Deactivate: Resting Functional Abnormalities in Autism." *Proceedings of the National Academy of Sciences*, *103* (21), 8275 LP – 8280. https://doi.org/10.1073/pnas.0600674103.

Kim, Eosu., Young-Chul, Jung., Jeonghun, Ku., Jae-Jin, Kim., Hyeongrae, Lee., So, Young Kim., Sun, I. Kim. & Hyun-Sang, Cho. (2009). "Reduced Activation in the Mirror Neuron System during a Virtual Social Cognition Task in Euthymic Bipolar Disorder." *Progress in Neuro-Psychopharmacology and Biological Psychiatry*, *33* (8), 1409–16. https://doi.org/10.1016/J.PNPBP.2009.07.019.

Kleitman, Nathaniel. (1963). Sleep and Wakefulness. Sleep and Wakefulness. Oxford, England: Univ. Chicago Press.

Klinger, Eric. (2009). "Daydreaming and Fantasizing: Thought Flow and Motivation." In Handbook of Imagination and Mental Simulation, edited by K. D. Markman, W. M. P. Klein, and J. A. Suhr, 225–39. Psychology Press. https://doi.org/10.4324/9780203809846.ch15.

Koella, W P. (1967). Sleep: Its Nature and Physiological Organization. Thomas Books. Thomas. https://books.google.com.mx/ books?id=uTdCAAAAIAAJ.

Kollar, Edward J., Norman, Namerow., Robert, O. Pasnau. & Paul, Naitoh. (1968). "Neurological Findings during Prolonged Sleep Deprivation." *Neurology*, *18* (9), 836. https://doi.org/10.1212/WNL.18.9.836.

Kollar, Edward J., Pasnau, R. O., Rubin, R. T., Naitoh, P., Slater, G. G. & Kales, A. (1969). "Psychological, Psychophysiological, and Biochemical Correlates of

Prolonged Sleep Deprivation." *The American Journal of Psychiatry*, *126* (4), 488–97. https://doi.org/ 10.1176/ajp.126.4.488.

Kollar, Edward J., Slater, G. R., Palmer, J. O., Doctor, R. F. & Mandell, A. J. (1966). "Stress in Subjects Undergoing Sleep Deprivation." *Psychosomatic Medicine*, *28* (2), 101–13. https://doi.org/10.1097/ 00006842-196603000-00002.

Koulack, David. (1969). "Effects of Somatosensory Stimulation on Dream Content." *Archives of General Psychiatry*, *20* (6), 718–25. https:// doi.org/10.1001/archpsyc.1969.01740180102010.

Krebs, Ruth., Dorothee, Heipertz., Hartmut, Schütze. & Emrah, Duzel. (2011). "Novelty Increases the Mesolimbic Functional Connectivity of the Substantia Nigra/Ventral Tegmental Area (SN/VTA) during Reward Anticipation: Evidence from High-Resolution FMRI." *NeuroImage*, *58*, 647–55. https://doi.org/10.1016/ j.neuroimage.2011.06.038.

Lee, Junghee., William, P. Horan., Jonathan, K. Wynn. & Michael, F. Green. (2016). "Neural Correlates of Belief and Emotion Attribution in Schizophrenia." *PLoS ONE*, *11* (11), 1–13. https://doi.org/10.1371/ journal.pone.0165546.

Léna, I., Parrot, S., Deschaux, O., Muffat-Joly, S., Sauvinet, V., Renaud, B., Suaud-Chagny, M. F. & Gottesmann, C. (2005). "Variations in Extracellular Levels of Dopamine, Noradrenaline, Glutamate, and Aspartate across the Sleep–Wake Cycle in the Medial Prefrontal Cortex and Nucleus Accumbens of Freely Moving Rats." *Journal of Neuroscience Research*, *81* (6), 891–99. https://doi.org/10.1002/ jnr.20602.

Leslie, A. M. (2001). "Theory of Mind." In *International Encyclopedia of the Social & Behavioral Sciences*, 15652–56. Pergamon. https:// doi.org/10.1016/B0-08-043076-7/01640-5.

Leslie, Alan M., Ori, Friedman. & Tim, P. German. (2004). "Core Mechanisms in 'Theory of Mind.'" *Trends in Cognitive Sciences*, *8* (12), 528–33. https://doi.org/10.1016/j.tics.2004.10.001.

Loomis, Alfred L., Newton Harvey, E. & Hobart, G. (1937). "Cerebral States during Sleep, as Studied by Human Brain Potentials." *Journal of Experimental Psychology*, *21* (2), 127–44. https://doi.org/https:// doi.org/10.1037/h0057431.

Lopes, da Silva. & Fernando, H. (2003). "Visual Dreams in the Congenitally Blind?" *Trends in Cognitive Sciences*, *7* (8), 328–30. https://doi.org/ 10.1016/S1364-6613(03)00155-4.

Luby, Elliot D., Frohman, C. E., Grisell, J. L., Lenzo, J. E. & Gottlieb, J. S. (1960). "Sleep Deprivation: Effects on Behavior, Thinking, Motor Performance, and Biological Energy Transfer Systems." *Psychosomatic Medicine*, *22*, 182–92. https://doi.org/10.1097/ 00006842-196005000-00003.

Luby, Elliot D., James, L Grisell., Charles, E Frohman., Helen, Lees., Bertram, D Cohen. & Jacques, S Gottlieb. (1962). "BIOCHEMICAL, PSYCHOLOGICAL, AND

BEHAVIORAL RESPONSES TO SLEEP DEPRIVATION." *Annals of the New York Academy of Sciences, 96* (1), 71–79. https://doi.org/10.1111/j.1749-6632.1962.tb50102.x.

Maloney, Karen J., Lynda, Mainville. & Barbara, E Jones. (2002). "C-Fos Expression in Dopaminergic and GABAergic Neurons of the Ventral Mesencephalic Tegmentum after Paradoxical Sleep Deprivation and Recovery." *European Journal of Neuroscience, 15* (4), 774–78. https://doi.org/10.1046/j.1460-9568.2002.01907.x.

Maquet, Pierre. (2000). "Functional Neuroimaging of Normal Human Sleep by Positron Emission Tomography." *Journal of Sleep Research, 9* (3), 207–31. https://doi.org/10.1046/j.1365-2869.2000.00214.x.

Maquet, Pierre. & Franck, G. (1997). "REM Sleep and Amygdala." *Molecular Psychiatry, 2* (3), 195–96. https://doi.org/10.1038/ sj.mp.4000239.

Maquet, Pierre., Steven, Laureys., Philippe, Peigneux., Sonia, Fuchs., Christophe, Petiau., Christophe, Phillips., Joel, Aerts., et al. (2000). "Experience-Dependent Changes in Cerebral Activation during Human REM Sleep." *Nature Neuroscience, 3* (8), 831–36. https://doi.org/ 10.1038/77744.

Maquet, Pierre., Jean-Marie, Péters., Joël, Aerts., Guy, Delfiore., Christian, Degueldre., André, Luxen. & Georges, Franck. (1996). "Functional Neuroanatomy of Human Rapid-Eye-Movement Sleep and Dreaming." *Nature, 383* (6596), 163–66. https://doi.org/10.1038/383163a0.

Maquet, Pierre., Perrine, Ruby., Audrey, Maudoux., Geneviève, Albouy., Virginie, Sterpenich., Thanh, Dang-Vu., Martin, Desseilles., et al. (2005). "Human Cognition during REM Sleep and the Activity Profile within Frontal and Parietal Cortices: A Reappraisal of Functional Neuroimaging Data." In *Progress in Brain Research*, edited by Steven Laureys, *150*, 219–27. Amsterdam: Elsevier B.V. https://doi.org/ 10.1016/S0079-6123(05)50016-5.

Mar, Raymond A. & Keith, Oatley. (2008). "The Function of Fiction Is the Abstraction and Simulation of Social Experience." *Perspectives on Psychological Science, 3* (3), 173–92. https://doi.org/10.1111/j.1745-6924.2008.00073.x.

Marquis, Louis-Philippe., Tyna, Paquette., Cloé, Blanchette-Carrière., Gaëlle, Dumel. & Tore, Nielsen. (2017). "REM Sleep Theta Changes in Frequent Nightmare Recallers." *Sleep, 40* (9). https://doi.org/ 10.1093/sleep/zsx110.

Mars, Rogier B., Franz, Xaver Neubert., Mary, Ann P. Noonan., Jerome, Sallet., Ivan, Toni. & Matthew, F. S. Rushworth. (2012). "On the Relationship between the 'Default Mode Network' and the 'Social Brain.'" *Frontiers in Human Neuroscience, 6*, (JUNE 2012), 1–9. https://doi.org/10.3389/fnhum.2012.00189.

Marzano, Cristina., Michele, Ferrara., Federica, Mauro., Fabio, Moroni., Maurizio, Gorgoni., Daniela, Tempesta., Carlo, Cipolli. & Luigi, de Gennaro. (2011). "Recalling and Forgetting Dreams: Theta and Alpha Oscillations During Sleep

Predict Subsequent Dream Recall." *Journal of Neuroscience*, *31* (18), 6674–83. https://doi.org/10.1523/ JNEUROSCI.0412-11.2011.

Merritt, Jane M., Robert, Stickgold., Edward, Pace-Schott., Julie, Williams. & Allan Hobson, J. (1994). "Emotion Profiles in the Dreams of Men and Women." *Consciousness and Cognition*, *3* (1), 46–60. https://doi.org/10.1006/CCOG.1994.1004.

Meyer, Urs., Joram, Feldon. & Olaf, Dammann. (2011). "Schizophrenia and Autism: Both Shared and Disorder-Specific Pathogenesis via Perinatal Inflammation?" *Pediatric Research*, *69*, (5 PART 2), 26–33. https://doi.org/10.1203/PDR.0b013e318212c196.

Miller, Joseph D., Jorge, Farber., Pam, Gatz., Howard, Roffwarg. & Dwight, C. German. (1983). "Activity of Mesencephalic Dopamine and Non-Dopamine Neurons across Stages of Sleep and Waking in the Rat." *Brain Research*, *273* (1), 133–41. https://doi.org/10.1016/0006-8993(83)91101-0.

Mitchell, Jason., Macrae, C. & Mahzarin, Banaji. (2006). "Dissociable Medial Prefrontal Contributions to Judgments of Similar and Dissimilar Others." *Neuron*, *50*, 655–63. https://doi.org/10.1016/ j.neuron.2006.03.040.

Molenberghs, Pascal., Ross, Cunnington. & Jason, B. Mattingley. (2012). "Brain Regions with Mirror Properties: A Meta-Analysis of 125 Human FMRI Studies." *Neuroscience & Biobehavioral Reviews*, *36* (1), 341–49. https://doi.org/10.1016/J.NEUBIOREV.2011.07.004.

Moriguchi, Yoshiya., Takashi, Ohnishi., Jean, Decety., Makiko, Hirakata., Motonari, Maeda., Hiroshi, Matsuda. & Gen, Komaki. (2009). "The Human Mirror Neuron System in a Population With Deficient Self-Awareness: An FMRI Study in Alexithymia." *Human Brain Mapping*, *30*, 2063–76. https://doi.org/10.1002/hbm.20653.

Murri, L., Massetani, R., Siciliano, G., Giovanditti, L. & Arena, R. (1985). "Dream Recall after Sleep Interruption in Brain-Injured Patients." *Sleep*, *8* (4), 356–62. https://doi.org/10.1093/sleep/8.4.356.

Nielsen, Tore. (1993). "Changes in the Kinesthetic Content of Dreams Following Somatosensory Stimulation of Leg Muscles during REM Sleep." *Dreaming*, *3* (2), 99–113. https://doi.org/10.1037/h0094374.

Nielsen, Tore. (2007). "Felt Presence: Paranoid Delusion or Hallucinatory Social Imagery?" *Consciousness and Cognition*, *16* (4), 975–83. https://doi.org/10.1016/ j.concog.2007.02.002.

Nielsen, Tore. (2010). "Dream Analysis and Classification: The Reality Simulation Perspective." Principles and Practice of Sleep Medicine: Fifth Edition, 595–603. https://doi.org/10.1016/B978-1-4160-6645-3.00051-7.

Nielsen, Tore. & Don, Kuiken. (2013). "Relationships between Non-Pathological Dream-Enactment and Mirror Behaviors." *Consciousness and Cognition, 22* (3), 975–86. https://doi.org/10.1016/ J.CONCOG.2013.06.005.

Nielsen, Tore., Russell, A. Powell. & Don, Kuiken. (2013). "Nightmare Frequency Is Related to a Propensity for Mirror Behaviors." *Consciousness and Cognition, 22* (4), 1181–88. https://doi.org/ 10.1016/J.CONCOG.2013.08.012.

Nielsen, Tore. & Philippe, Stenstrom. (2005). "What Are the Memory Sources of Dreaming?" *Nature, 437* (7063), 1286–89. https://doi.org/ 10.1038/nature04288.

Nir, Yuval. & Giulio, Tononi. (2010). "Dreaming and the Brain: From Phenomenology to Neurophysiology." *Trends in Cognitive Sciences, 14* (2), 88–100. https://doi.org/10.1016/j.tics.2009.12.001.

Nomura, Sakashi., Noboru, Mizuno., Kazuo, Itoh., Kojyuro, Matsuda., Tetsuo, Sugimoto. & Yasuhisa, Nakamura. (1979). "Localization of Parabrachial Nucleus Neurons Projecting to the Thalamus or the Amygdala in the Cat Using Horseradish Peroxidase." *Experimental Neurology, 64* (2), 375–85. https://doi.org/10.1016/0014-4886(79)90277-2.

Northoff, Georg., Alexander, Heinzel., Moritz, de Greck., Felix, Bermpohl., Henrik, Dobrowolny. & Jaak, Panksepp. (2006). "Self-Referential Processing in Our Brain-A Meta-Analysis of Imaging Studies on the Self." *NeuroImage, 31* (1), 440–57. https://doi.org/10.1016/ j.neuroimage.2005.12.002.

Pace-Schott, Edward F. (2003). "Postscript: Recent Findings on the Neurobiology of Sleep and Dreaming." In Sleep and Dreaming, edited by Edward F. Pace-Schott, Mark Solms, Mark Blagrove, and Stevan Harnad, 335–50. Cambridge: Cambridge University Press. https:// doi.org/10.1017/cbo9780511615511.012.

Pace-Schott, Edward F. (2010). "The Neurobiology of Dreaming." In Principles and Practice of Sleep Medicine, edited by Meir H. Kryger, Thomas Roth, and William C. Dement, 5th ed., 563–75. Philadelphia: Elsevier Ltd. https://doi.org/doi.org/10.1016/B978-1-4160-6645-3.00048-7.

Perogamvros, Lampros. & Sophie, Schwartz. (2012). "The Roles of the Reward System in Sleep and Dreaming." *Neuroscience & Biobehavioral Reviews, 36* (8), 1934–51. https://doi.org/10.1016/ J.NEUBIOREV.2012.05.010.

Pinkham, Amy E., Joseph, B. Hopfinger., Kevin, A. Pelphrey., Joseph, Piven. & David, L. Penn. (2008). "Neural Bases for Impaired Social Cognition in Schizophrenia and Autism Spectrum Disorders." *Schizophrenia Research, 99* (1–3), 164–75. https://doi.org/10.1016/ j.schres.2007.10.024.

Poulin, Julie., Sylvie, Chouinard., Tania, Pampoulova., Yves, Lecomte., Emmanuel, Stip. & Roger, Godbout. (2010). "Sleep Habits in Middle-Aged, Non-Hospitalized Men and Women with Schizophrenia: A Comparison with Healthy Controls." *Psychiatry Research, 179* (3), 274–78. https://doi.org/10.1016/j.psychres.2009.08.009.

Rechtschaffen, Allan., Bernard, M Bergmann., Carol, A Everson., Clete, A Kushida. & Marcia, A Gilliland. (1989). "Sleep-Deprivation in the Rat .10. Integration and Discussion of the Findings." *Sleep, 12* (1), 68–87. https://doi.org/10.1016/j.yhbeh.2006.06.003.

Redcay, Elizabeth., Mario, Kleiner. & Rebecca, Saxe. (2012). "Look at This: The Neural Correlates of Initiating and Responding to Bids for Joint Attention." *Frontiers in Human Neuroscience, 6,* (JUNE 2012), 1–14. https://doi.org/10.3389/fnhum.2012.00169.

Revonsuo, Antti. (2000). "The Reinterpretation of Dreams: An Evolutionary Hypothesis of the Function of Dreaming." Behavioral and Brain Sciences. United Kingdom: Cambridge University Press. https:// doi.org/10.1017/S0140525X00004015.

Revonsuo, Antti., Jarno, Tuominen. & Katja, Valli. (2015). "The Avatars in the Machine." *Theoretical Philosophy/MIND Group – JGU Mainz, 32*, 1–28. https://doi.org/10.15502/9783958570375.

Scarpelli, Serena., Cristina, Marzano., Aurora, D'Atri., Maurizio, Gorgoni., Michele, Ferrara. & Luigi, De Gennaro. (2015). "State- or Trait-like Individual Differences in Dream Recall: Preliminary Findings from a within-Subjects Study of Multiple Nap REM Sleep Awakenings." *Frontiers in Psychology, 6.* https://doi.org/10.3389/fpsyg.2015.00928.

Scholz, Jonathan., Christina, Triantafyllou., Susan, Whitfield-Gabrieli., Emery, N. Brown. & Rebecca, Saxe. (2009). "Distinct Regions of Right Temporo-Parietal Junction Are Selective for Theory of Mind and Exogenous Attention." *PLoS ONE, 4* (3). https://doi.org/10.1371/ journal.pone.0004869.

Schredl, Michael. & Friedrich, Hofmann. (2003). "Continuity between Waking Activities and Dream Activities." *Consciousness and Cognition, 12* (2), 298–308. https://doi.org/10.1016/S1053-8100(02)00072-7.

Schumann, Cynthia Mills., Julia, Hamstra., Beth, L. Goodlin-Jones., Linda, J. Lotspeich., Hower, Kwon., Michael, H. Buonocore., Cathy, R. Lammers., Allan, L. Reiss. & David, G. Amaral. (2004). "The Amygdala Is Enlarged in Children But Not Adolescents with Autism; the Hippocampus Is Enlarged at All Ages." *The Journal of Neuroscience, 24* (28), 6392 LP – 6401. https://doi.org/10.1523/ JNEUROSCI.1297-04.2004.

Schweickert, Richard. & Zhuangzhuang, Xi. (2010). "Metamorphosed Characters in Dreams: Constraints of Conceptual Structure and Amount of Theory of Mind." *Cognitive Science, 34*, 665–84. https://doi.org/10.1111/j.1551-6709.2009.01082.x.

Sebastian, Catherine., Nathalie, Fontaine., Geoffrey, Bird., Sarah-Jayne, Blakemore., Stephane, De Brito., Eamon, Mccrory. & Essi, Viding. (2011). "Neural Processing Associated with Cognitive and Affective Theory of Mind in Adolescents and Adult." *Social Cognitive and Affective Neuroscience, 7,* 53–63. https://doi.org/10.1093/scan/nsr023.

Shanton, Karen. & Alvin, Goldman. (2010). "Simulation Theory." *Wiley Interdisciplinary Reviews: Cognitive Science*, *1* (4), 527–38. https://doi.org/10.1002/wcs.33.

Siclari, Francesca., Benjamin, Baird., Lampros, Perogamvros., Giulio, Bernardi., Joshua, J. LaRocque., Brady, Riedner., Melanie, Boly., Bradley, R. Postle. & Giulio, Tononi. (2017). "The Neural Correlates of Dreaming." *Nature Neuroscience*, *20* (6), 872–78. https://doi.org/ 10.1038/nn.4545.

Solms, Mark. (1997). The Neuropsychology of Dreams, A Clinico-Anatomical Study. New York: Lawrence Erlbaum Associates, Inc.

Solms, Mark. (2000). "Dreaming and REM Sleep Are Controlled by Different Brain Mechanisms." *Behavioral and Brain Sciences*, *23* (6), 843–50. https://doi.org/10.1017/S0140525X00003988.

Steriade, M. (2000). "Corticothalamic Resonance, States of Vigilance and Mentation." Neuroscience, *101* (2), 243–76. https://doi.org/10.1016/ S0306-4522(00)00353-5.

Tart, Charles T. (1987). "The World Simulation Process in Waking and Dreaming: A Systems Analysis of Structure." *Journal of Mental Imagery*, *11* (2), 145–58.

Tendler, Alex. & Shlomo, Wagner. (2015). "Different Types of Theta Rhythmicity Are Induced by Social and Fearful Stimuli in a Network Associated with Social Memory." *ELife*, *4*, (February). https://doi.org/10.7554/eLife.03614.

Thomas, J. & Benoit, O. (1967). "Individualisation d'un Sommeil à Ondes Lentes et Activité Phasique." *Brain Research*, *5* (2), 221–35. https://doi.org/10.1016/0006-8993(67)90088-1.

Tuominen, Jarno., Tuula, Stenberg., Antti, Revonsuo. & Katja, Valli. (2019). "Social Contents in Dreams: An Empirical Test of the Social Simulation Theory." *Consciousness and Cognition*, *69*, 133–45. https://doi.org/10.1016/ j.concog.2019.01.017.

Valli, Katja. & Antti, Revonsuo. (2009). "The Threat Simulation Theory in Light of Recent Empirical Evidence: A Review." *The American Journal of Psychology*, *122*, 17–38.

Völlm, Birgit., Alexander, Taylor., Paul, Richardson., Rhiannon, Corcoran., John, Stirling., Shane, McKie., John, Deakin. & Rebecca, Elliott. (2006). "Neuronal Correlates of Theory of Mind and Empathy: A Functional Magnetic Resonance Imaging Study in a Nonverbal Task." *NeuroImage*, *29*, 90–98. https://doi.org/10.1016/ j.neuroimage.2005.07.022.

Voss, Ursula., Romain, Holzmann., Allan Hobson, J., Walter, Paulus., Judith, Koppehele-Gossel., Ansgar, Klimke. & Michael, Nitsche. (2014). "Induction of Self Awareness in Dreams through Frontal Low Current Stimulation of Gamma Activity." *Nature Neuroscience*, *17*, https://doi.org/10.1038/nn.3719.

Waiter, Gordon D., Justin, H. G. Williams., Alison, D. Murray., Anne, Gilchrist., David, I. Perrett. & Andrew, Whiten. (2004). "A Voxel-Based Investigation of Brain

Structure in Male Adolescents with Autistic Spectrum Disorder." *NeuroImage*, *22* (2), 619–25. https://doi.org/10.1016/J.NEUROIMAGE.2004.02.029.

Wilkinson, Dr R T. (1969). "Sleep and Sleep Deprivation: Loss of Sleep." *Proceedings of the Royal Society of Medicine*, *62* (9), 903–4. https://doi.org/10.1177/003591576906200917.

Williams, Harold L., Charles, F. Gieseking. & Ardie, Lubin. (1966). "Some Effects of Sleep Loss on Memory." *Perceptual and Motor Skills*, *23*, (3_suppl), 1287–93. https://doi.org/10.2466/ pms.1966.23.3f.1287.

Wittmann, Bianca C., Nico, Bunzeck., Raymond, J. Dolan. & Emrah, Düzel. (2007). "Anticipation of Novelty Recruits Reward System and Hippocampus While Promoting Recollection." *NeuroImage*, *38* (1), 194–202. https://doi.org/10.1016/J.NEUROIMAGE.2007.06.038.

In: Sleep, Cognition and Emotion
Editors: I. Ramírez Salado et al.
ISBN: 978-1-53617-561-5

Chapter 8

AFFECT, REM SLEEP AND DREAMS: TOWARDS EMOTIONAL INSIGHT

Ana Paula Rivera-García*, PhD
Manuel Alejandro Cruz-Aguilar, PhD
and Ignacio Ramírez-Salado, PhD
Laboratorio de Cronobiología y Sueño,
Dirección de Investigaciones en Neurociencias,
Instituto Nacional de Psiquiatría Ramón de la Fuente Muñiz,
Ciudad de México, México

ABSTRACT

Dreams are one of the most enigmatic and fascinating elements of sleep. Rapid Eye Movement (REM) sleep has a close physiological-functional relationship with the networks of emotional generation and regulation. Emotions are complex processes involving different mechanisms, leading to action in response to changes in the environment. Importantly, emotions are not only present during wakefulness but also during sleep, especially during REM sleep, where more intense dreams occur both emotionally and perceptively. The relevance that emotions have in the integrity of the individual and culture (and vice versa) is not new, but today's affective world has gained such relevance that it is now measured, studied and scrutinized from the political, marketing, medical, holistic, pharmacological, educational, biological and psychological fields, among many others. However, we still need to understand much of the basis of emotion: how it is generated, its components, its intensity, its modulation and, above all, the awareness and recognition of our own emotions. Sleep provides a unique scenario each night in which it is possible to reactivate each and every aspect of the affective spectrum, specifically, in REM sleep's dreamed content. Therefore, the oneiric world can

* Corresponding Author's Email: aprivera@imp.edu.mx.

act as a testing ground for our complex affective universe, either by experiencing emotional states or recognizing emotions present during wakefulness. In this chapter we explore the intricate connections between these three phenomena, sleep, emotion and dreams, from the historical point of view to the physiological and cognitive perspective, in order to shed light on the possible function of dreams as harbingers of emotions.

Keywords: emotion, affect, REM sleep, dreams

"Strange as it may seem today to say, the aim of life is to live, and to live means to be aware, joyously, drunkenly, serenely, divinely aware."

Henry Miller, *The Wisdom of the Heart*.

INTRODUCTION

Sleep is associated with the physical, cognitive and emotional integrity of the organism. Dreams, in particular, are mesmerizing states of the mind that are strongly linked to emotion and inner human psyche. REM sleep stage is distinguished from the different stages of NREM sleep, among other elements, by the appearance of "complex" dreams: dreams with greater perceptual, mnemonic richness, content and emotional intensity, which, in turn are associated with an increase in the activity of limbic and para-limbic neural regions (Nielsen 2000). The exploration of dream content provides information on the emotional context in which the person is, revealing elements of his condition that may go unnoticed in wakefulness. Contrary to the popular belief that dreams are hallucinatory experiences confusing and unconnected with the stimuli of the immediate environment, dream world reflects aspects of waking conditions and viceversa. It is considered that 75-80% of the dreams of both healthy and emotionally affected subjects include daily concerns and interests. Thus, the peculiarities of dreams are linked to the characteristics of the dreamer (age, sex, health, habits, personality and interests), as well as their economic, cultural, environmental and health conditions. In this matter, the similarity that the inner world of dreams keeps with waking life is because dreams reflect the organization and functions of a brain impressed by its environment. Likewise, only through the study of sleep reports is it possible to corroborate the correlations that have so far been sought between the neural activity detected in different phases of sleep and the formal characteristics of the dream experience. Thus, studying the symbolic and emotional content of dreams not only allows us to explore how experiences are reflected in brain function and structure, but also to investigate the neural mechanisms that make them possible. The study of the variation of the human experience of sleep and dreams is therefore an essential resource to examine the way in which individuals relate to their context. In this vein, the framework that REM sleep dreams and emotion provide, have a pivotal role in the processing of salient and emotional waking-life experiences, strongly contributing to emotional consolidation and processing, but, furthermore, may be a direct passage with our deeper, and sometimes overlooked, inner world. In this chapter we will take a more in-depth look to the fascinating intersection of two worlds, the worlds of emotion and sleep.

A Glimpse into Emotions

Understanding the Concept of Emotion

Emotion plays an overwhelming and undenyingly central role in our lives. Yet, because of its ethereal nature, emotions are really hard to define. Initially approached from philosophy, the study of the nature of emotions has been tried to be explained from its most basic physiology, up to the reactions that they generate, by the information they provide to others, by the concepts and beliefs influenced by them and finally by the subjective experience they create. Humans are intrigued by this notion, and it is no simple task: the relevance and preeminence of emotions, mainly in mammals, has been demonstrated in several levels (LeDoux 2012). Emotions influence our behavior, our actions and thoughts, are present in most species and, solely through emotions, we can truly understand ourselves and others (Solomon 1993). They are our sempiternal company, both in waking life and during sleep, so the interest of studying them is not just theoretical. Emotion, in a word, gives us meaning. Still, as some of the most quintessential concepts surrounding human nature, our understanding of them were usually subsumed to more attainable ideas. Emotions are within our grasp and at the same time are as elusive as a butterfly, we're almost always subdued by them, but much of their nature is still a mystery, thus the conflict in defining an almost ever-changing phenomenon. Nevertheless, from last century, psychological science has stimulated a new era of emotion theorizing and research, as did debate about the role of cognition in emotion and emotion in cognition (Lazarus 1982; Zajonc 1980; Pollack 2019). To date, some agreement has been achieved in a universal, quintessential, neural, physiological, cognitive, social and cultural phenomenon as it is emotion. So, let us take a trip down memory lane to understand how we humans got to get a grasp on this concept.

The term *emotion* is now of common usage and knowledge, yet, its origin is not as remote as its congener's "passion," "accident of the soul" or "moral sentiment," which were the terms used before 1830 to describe what we now know as emotions (Smith 2016). As it has been extensively described, most of the then-called 'passions' were attributed to either a balance (or lack of) of four elemental substances in the body: blood, yellow vile, black vile and phlegm, or to the influence of demons or supernatural forces over the body. It was until the 17th century that a modern concept of emotion was developed, Thomas Willis described that a sudden surge of joy or, a nervous tremble appeared because of the activity of the brain and not of liquids in the body. Almost two hundred years later, the philosopher Thomas Brown, proposed a brand-new vocabulary for describing the recent findings regarding mechanical processes behind the experience and expression of passions, hence, he suggested the term '*emotion*' to describe and include all of this new sphere of the body (Smith 2016). Nonetheless, emotion was not a new word; it was developed from Old French *emouvoir* (to stir up) and from Latin *emovere* (move out, remove, agitate). Together, *emotion would describe a process, determined in the context of survival, and related to the action or movement of the individual* (Keil and Miskovic 2015).

In its initial usage, the concept of emotion was mainly based in observable phenomena (emotion expressions), leaving the internal affective process aside (Smith 2015; Izard 2007). It was Charles Darwin one of the first naturalists who approached the study of emotion from a scientific point of view. Particularly, he was interested in its universality and evolving nature. As a result of this new paradigm, emotions were (for the first time) considered not as fixed responses, but as *ongoing processes that help us survive in an everchanging environment* (Darwin 1998). The revolutionary premise of Darwin challenged the well-established notion, mainly supported by William James, which stated that emotions were only a by-product of bodily response, thus, were inherited and unmovable. Nevertheless, it was not easy to stop considering emotions as a mere epiphenomenon of a physical response, it took decades to the scientific community to even start scrutinizing the internal variables involved in emotions. It was after Darwin's contribution that Sigmund Freud's work added depth and complexity to the notion of emotions, showing that not all of them are equally conspicuous and, furthermore, that some of them may even go unnoticed by the individual. In this way, Freud's work revealed that emotions are not always palpable as was initially supposed, and that they may have surreptitious ways of manifesting throughout life (Freud 1955; Smith 2015).

In addition to the biological and physiological components, both the environment and culture also had something to add into the construction of emotions (Lim 2016). Recently—within the past few decades—there has emerged a subset of research dedicated to examining how all of the emotional spectrum may be influenced by culture just as much as by biology and psychology (Lim 2016; Uchida 2009; Elfenbein and Ambady 2002; Mesquita and Frijda 1992). It is mainly through language that the biological responses can be influenced by culture and thus, our private experiences can be imprinted as well by the cultural and environmental cues (Shott 1979). Primarily through the naming and conceptualization of determined emotions, historians have uncovered the specificity of more complex affective experiences, showing that cognitive appraisal processes may play a role in the elicitation, experience and expression of emotion (Ellsworth and Scherer 2003; Scherer et al. 2011). This suggests that emotion may not be as universal as initially was suggested, but more specific to species and culture (Scherer and Brosch 2009; Lim 2016; Uchida 2009; Elfenbein and Ambady 2002; Mesquita and Frijda 1992; Scherer et al. 2011).

Due to the anthropological and ethnographic studies of possible cross-cultural differences in thought, language and customs, it was concluded that emotional speech, just like language, should be interpreted as a set of sign structures that are learned and therefore culturally connected (Lutz andWhite 1986; Scherer et al. 2011; Barfield 1997). It was after Mead's studies that the commonality-based approach of biologists and psychologists was questioned and cultural differentiations in emotion conceptualization was developed (Mead 1928). Nevertheless, given the increasingly complex theoretical perspective, empirical work in the universalist tradition continued to be carried out almost exclusively. With this empirical predominance came what seemed to be broad acceptance of a universalist view, dominating the coverage of emotions in the textbook until very recently. Studies on possible cultural influence on transmission of emotions was somewhat skewed in identifying parallels, not differences (Scherer et al. 2011). However, the possibility that emotion may be much more complex than it was generally understood, solely as basic responses, has begun to be explored.

Emotional expression and social studies currently provide what can be seen as support for both sides of the debate (Ekman and Rosenberg 2005; Russell and Fernandez-Dols 1997). However, little effort was made to reconcile these findings— for and against universality— and to clarify the middle ground. Nevertheless, just as the study of emotion (driven mainly by anthropologists and ethnologists) as a culture-specific value system can contribute, the efforts of biologists and psychologists are geared towards identifying parallels in the underlying behavioral mechanisms, so it is the notion of emotional resemblances that can start to shed light on its conceptualization. So, what do emotions have in common?

All emotions involve changes in multiple systems: subjective, physiological, behavioral and relational (Gross and Barret 2011). Emotions are based on neuronal impulses which in turn trigger a series of reactions that incite an organism to get in motion, but despite its complexity, all of them have some basic components: trigger element (s) (stimulus), physiological response (s), subjective experience and, reactive behaviors (expressions) (Mauss and Robinson 2009). It is important to notice that the components of emotions are characterized by having a cause in common, coinciding in time, and being correlated in intensity, to first experience the emotion and then express it (Delgado 1966; Delgado 2002). Thus, emotion can be defined as: *responses to changes that challenge the survival of the individual*, (Babson and Felder 2015); *adaptive action tendencies that occur in response to changes inside or outside the organism, specifically changes that challenge states and systems necessary for survival* (Keil and Miskovic in Babson 2013); or *responses to internal and external stimulation characterized by a valence connotation and a specific intensity or arousal* (Gross and Barret 2011), Yet, a more integral definition of emotion would involve the seriation of actions required to achieve a determined goal, thus, emotions are processes, more specifically: *processes evolved for facilitating appropriate responses to environmental circumstances to reach individual or social goals in quick and effective ways* (Gross and Barret 2011). We therefore adhere more to a complete conceptualization of emotions as the one of Barret and Campos (1987): *Emotion is a process through which individuals attempt to maintain or change their relations with the environment in terms of its significance for wellbeing.*

A Gamut of Emotions

Initially, it was attributed that the function of emotions was to "mediate" social relations, however, now it has been demonstrated that emotions are not reduced to social contact, but can appear without the presence, whether physical or evoked, of other individuals either from the same species or from a different one (Delgado 1966: Delgado 2002). Importantly, since emotions respond to changes that challenge the survival of the individual, the most basic states and systems that emotions are looking after are those underlying sexual reproduction, homeostasis, metabolism, and defense against bodily harm (Babson and Felder 2015). Therefore, the utter most relevant function of emotions would be to regulate, according to patterns, the internal state (memories, threats, needs, conditions, physiological processes, etc.) with the environment (Ekman et al. 1972; Delgado 2002; Delgado 1966; Russel et al. 2003) and besides of its components, emotions are able to (a) set in affective experiences such as feelings, (b) generate cognitive processes such as assessments, labels, evaluations and judgements, (c) activate generalized physiological adjustments to external conditions, and (d) lead to behavior that is often expressive, aimed at an objective and adaptive (Kleinginna and Kleinginna 1981). Hence, inherent to any emotion lies an almost immediate evaluation of the context within the subjectivity of the individual, emotions are consequently in constant change and evolution, according to the framework, background, physical and mental state, so any type of alteration in the processes involved in the regulation of affect are directly associated with difficulties in social relationships, but mostly within the individual (Gaebel and Wolver 1992; Gross 2007). As Damasio (2003) states, one of the most important recent discoveries of brain science was anticipated by the philosopher Spinoza: the critical role of emotions in ensuring our survival and allowing us to think. It turns out that emotion is not the enemy of reason, but an indispensable accomplice, as Spinoza saw it.

How many emotions are and how are they categorized? On its well-known research, Ekman, Friesen and Ellsworth (1972) found evidence of six basic emotions in humans: joy, surprise, fear, sadness, anger and disgust (the latter combined with contempt), and, as the develop, different emotions, or different magnitudes of the same can give rise to a feeling, mood or temperament, and these in turn can be divided into several emotions (Delgado 1966). A fundamental point about the concept of emotion is that it deals with what is positive and negative for the organism, in such a way that it prepares it for action based on a subjective evaluation (Scherer 2004). It is worth noting at this point that most affect theories, delimit emotions to a phasic or brief response, with a well-defined time related to a specific event that raises that response. Concepts such as thoughts, moods and temperaments, on the contrary, can be viewed as affective mechanisms that are distinct from emotions both in the time they evolve (more extensively) and through various events (Delgado, 1966). The experimental paradigm prevailing in the investigation of affective neuroscience postulates that emotions can be divided into discrete and independent categories and that specific neuronal structures and routes serve each of these emotional categories (Ochsner and Gross 2005; Ekman 1992; Panksepp 1998; Tomkins 1963). Thus from the six basic emotions, more complex affective constructs may be developed such as feelings, affects, sentiments, moods and eventually, temperament (Zilmman 2003).

The theory of basic emotions has produced significant advances in the understanding of emotion, yet, in the fields of clinical psychology and psychiatry, it has left many important questions unresolved (Posner et al. 2005). For instance, the basic emotions construct has not explained the almost ubiquitous comorbid diseases among mood disorders, nor has it resolved the confusion about the neurophysiological foundations of affective disorders. In addition, the theory of basic emotion is largely incompatible with recent findings in behavioral genetics and temperament research. Given these empirical and heuristic limitations of the theory of basic emotions, it has been proposed that a change in the conceptual approaches taken for the study of emotion is needed, where it is understood that all affective states arise from the common superposition of neurophysiological systems (Posner et al. 2005).

Although poorly represented in psychiatry, dimensional models have a long history in psychology (Larsen and Diener 1992; Russell 2003; Schlosberg 1952; Watson, Wiese, Vaidya and Tellegen 1999). A particular dimensional approach, called the circular model of emotion, proposes that all affective states arise from two fundamental neurophysiological systems, one related to valence (a continuum of pleasure-displeasure) and the other associated with arousal or alertness (Russell 1980). Each emotion can be understood as a linear combination of these two dimensions (valence and arousal), or as varying degrees of valence and excitement, considering that all emotions are favored by an appetitive motivational system (associated with positive or pleasant emotions), as well as an aversive system (associated with negative or unpleasant emotions). Joy, for example, is conceptualized as an emotional state that is the product of an activation in the neural systems associated with positive valence or pleasure along with a moderate activation in the neural systems associated with arousal (Russell 2003). In this way the affective states arise from the same two neurophysiological systems, differing in the degree or magnitude of the activation. Therefore, specific emotions arise from the activation patterns within these two neurophysiological systems, along with cognitive interpretations and labeling of these central physiological experiences (Posner et al. 2003).

An adjustment to the continuous pleasure model is the Chromatic Model of the Affective System (Díaz and Flores 2001) in which one of the emotional spectra is reflected: the modality (positive or negative), while the arousal model complements when considering the degree of intensity (high or low) of an emotion. In this way the same emotion can be classified according to the degree of appetite or motivation that represents both in the modality and intensity. The Chromatic Model establishes 14 polar axes of human emotion: calm-tension, certainty-doubt, compassion-anger, fun-boredom, happiness-sadness, pleasure-pain, satisfaction-frustration, desire-aversion, love-hate, courage-fear, vigor-exhaustion, enthusiasm-apathy and pride-humiliation, arranged in two axes, a vertical that specifies the pleasant or unpleasant tone and a horizontal associated to the relaxation or excitement level (Díaz and Flores 2001) (Table 1). In this way the Chromatic Model is more compatible with physiological theories, motor expressions and, in general, it is a tool to provide a more systematic vocabulary of emotions and a scheme by which to evaluate and integrate them.

Table 1. Two-dimension model of emotions depicting affective arousal (high/low) and valence (positive/negative)

HIGH NEGATIVE	HIGH POSITIVE
1. unpleasantness	1. pride
2. pain	2. courage
3. aversion	3. happiness
4. hate	4. satisfaction
5. anger	5. enthusiasm
6. doubt	6. fun
7. tension	7. vigor
LOW NEGATIVE	**LOW POSITIVE**
1. humiliation	1. pleasantness
2. fear	2. pleasure
3. sadness	3. desire
4. frustration	4. love
5. apathy	5. compassion
6. boredom	6. certainty
7. exhaustion	7. calm

(Based on Diaz and Flores 2001; Rivera-García et al. 2018).

It has been described that one key element within the affective process is the emotional reactivity: the individual's threshold, peak intensity, rise time and recovery time in response to emotional stimulation (Derryberry and Reed 2003). All these characteristics vary continuously across individuals and are linked to stable personality dimensions, yet, variations in affective reactivity, which results from difficulties in regulating emotions, is a central feature of most psychiatric conditions (Derryberry and Reed 2003). Finally, another key component of the emotional process are the outputs, or manifestations of affect. Studies on the psychophysiological component of emotions have typically focused on autonomic outputs, such as heart rate, blood pressure, electrodermal activity or muscle tension changes. Increasing evidence indicates that some autonomic indices capture predominantly the arousal dimension (e.g., heart rate) yet, others measure the valence dimension (e.g., facial EMG).

Emotion's Machinery

Emotions are mediated at the neuronal level both by emotional modality-specific structures (either positive or negative) and by regions common to both (Damasio 1994; Cacioppo et al. 2000). The thalamus, hypothalamus, the middle brain and the medial prefrontal cortex stand out among the latter (Newman 1995; Ochsner and Gross 2005). In particular, the medial prefrontal cortex region has been identified as having a predominant role in the regulation of both modalities of emotion as well as in the conscious experience of emotion (Ochsner and Gross 2005), inhibition of emotional expression (Scherer et al. 2001) and in decision-making associated with emotions (Ochsner and Gross 2005). Similarly, the anterior thalamus is involved in behavioral integration and autonomic emotional expression, with the visual relief areas of the lateral geniculate nucleus being a priority for the emotional response to visual stimuli (Newman 1995).

Unlike the regions involved in emotional multimodal activation, the temporal lobe amygdala has been more closely associated with negative emotional generation and experience (Cardinal et al. 2002). Specifically, negative emotions such as sadness and fear have been associated with a particular activation in the left amygdala, either when the person experiences or perceives these emotions. While the activation of the amygdala in negative emotions is significantly greater, an association between amygdala facilitation was also described before positive stimuli compared to neutral stimuli (Ketter et al. 1996; Morris et al. 1996; Wik et al. 1993). On the other hand, substantial activation has been shown in the bilateral occipito-temporal cortex in negative emotions of high anxiety such as phobias and fear (Wik et al., 1993; Fredrikson et al. 1995). Notably, these changes may be reflecting the projections that the amygdala holds with posterior regions involved in visual regulation, calibrating the attention of the visual systems towards threatening stimuli, reflecting intensity (arousal) rather than emotional valence. In positive emotions, an activation of the prefrontal dorsomedial cortex, posterior cingulum, precuneus, bilateral temporal lobes, bilateral temporoparietal junction and right amygdala has been described (Britton et al. 2006; Frewen et al. 2011). This, in turn, is linked to the facilitation of the neighboring ventral striatum, involving regions associated with the reward and pleasure circuit (Cardinal et al. 2002). On the contrary, it has been identified that the activity in these regions is significantly reduced in patients with affective alterations such as major depressive disorder (MDD) (Burton 1989). Likewise, the insula and the anterior orbitofrontal and cingulate cortices are associated with anticipation of reward, decision-making, empathy, and recognition of one's emotions, as opposed to emotions perceived in others, or only evoked (Breiter et al. 1996). It is thus intended to briefly expose part of the complex system that in wakefulness gives rise to generation, experience, and emotional expression, however, as can be seen below, it is during REM sleep stage, that much of the neuronal activation associated with emotion is specifically facilitated, so that emotionality is not limited to wakefulness, but appears and, more importantly, is regulated during sleep (Zohar et al. 2005; van Der Helm et al. 2010; van Der Helm et al. 2011).

What's Sleep Got to Do with It?

Sleep, like emotion, is characterized by the occurrence of electrophysiological, vegetative, cognitive and motor changes. According to the pattern of these components, sleep has been divided into NREM (N1, N2, N3) and REM sleep. REM sleep is characterized by: a) fast and low voltage EEG activity with alternating alpha, beta and theta rhythms (Silber et al. 2007), b) muscular atonia of the anti-gravity muscles, c) rapid eye movements and, at a cognitive level, by d) dreams with higher emotional load, cognitive content and better memory at awakening (complex dreams) (Corsi-Cabrera 1994; Hobson 2011: Hobson 1988). If dreams are defined simply as "any mental activity reported upon awakening from sleep" (Walker 2017), then dreams are as common during REM sleep as during NREM sleep. Yet, if we add to that definition that dreams are hallucinogenic, motoric, emotional and bizarre experiences that occur during sleep (Walker 2017), then we should be narrowing them much more likely to REM sleep dreams. Dreams evoke personal experiences with (or without) emotional tone, and still we may be completely oblivious of them. Dreams can directly impact in wakefulness and are not only affected by our waking life. We now know many of the aspects that determine, generate, modulate and are influenced by dreams, yet, in order to achieve that amount of information, the history behind the study of dreams goes a long way back.

Historical Aspects of the Study of Dreams

Nowadays dreams have a central role among the interests surrounding sleep, however, the curiosity in knowing and understanding dream is far from new. Although there are no indications about the interest of primitive man in dreams, it is assumed that due to its enigmatic nature, since the dawn of humanity, dreams have puzzled dreamers. It is with reflexive writing that the first recorded manifestations of interest in dream activity began. In this way, dreams were not only considered as a means of interpretation of the individual future, but that of entire societies. Examples of the above are the Egyptian Joseph's interpretations of the Pharaoh's dreams of seven fat and seven gaunt cows from which he predicted and avoided major famine (Calvo 1993; Schredl 2018), Artemidorus Daldis ' *Oneirocritica* (the predecessor of dream symbol books) or the Tibetian Dream Yoga (one of the first lucid dream interpretation techniques). The importance of humanity's dreams is such, that even religions were kindled because of them: Muhammad got the revelation of his divine mission in a dream, and Buddha's mother predicted the fate of her son because of a dream. Dreams have also been used for healing purposes in several indigenous cultures (Schredl 2018). Furthermore, since early times, sorcerers have used dream content as means to communicate with gods or spirits. In the late eighteenth and early nineteenth centuries, dreams were used by fortune tellers and charlatans to predict the success or misfortune of people. This practice would lead to the publication of several books where certain meaning was assigned to certain symbols present in dreams, it is worth mentioning that these types of interpretations are still used in the present day. Thus, either in art, mysticism, medicine or science, dreams have occupied an important part of humanity's concerns and relieves.

Medical science began exploring the quantification of vegetative variables during sleep in the 18th and 19th centuries (Schredl 2018). The steady decrease in heart and respiratory rates during sleep was accurately described, although doctors paid little attention to dreams, finding them to be unusual symptoms due to poorly oxygenated brain. It was the French doctor Alfred Maury who made the first attempts in 1862 to explain dream's mental content. To him is due the beginning of dream's systematic and intentional experimentation. Maury was especially interested in visual images and dream-related events that occur at the beginning of sleep, the so-called hypnagogic hallucinations, a term he coined. In an attempt to explore the possible generation of key elements of dreams and its transformation into oneiric visual images, Maury also introduced specific stimuli prior to sleep, and concluded that dreams arose from external stimuli, instantaneously accompanying such impressions as they acted upon the sleeping person (Hobson 1988).

Together with Maury, his contemporary, the Marquis Marie Jean Léon Hervey of Saint-Denis, practiced systematic self-observation, achieving one of the largest dream diaries in history, compiling the dream activity of five years in his work *Les Rêves et les moyens de les diriger (Hobson 1988)*. In his book, Hervey concludes, among other observations, that during the dream will is not interrupted, contradicting what Maury stated. Hervey's merit lies in the important contributions he made to the study of dream phenomenon, such as the one he raises about the imaginative abundance during the dream process, or the importance he granted to the hyper visual quality of dreams (Hobson 1988). Together, Maury and Saint-Denys, were two of the first *oneirologists* in the world.

Freud (1955), on the other hand, would establish in a primordial position the role of dreams in the unconscious life of the human, based on the content of dreams to elaborate his theory about thought processes. Freud distinguished two main components in dreams, the manifest content (the one expressed by the patient) and the latent or subconscious content (Freud 1955). Freud's contribuition is also relevant for being one of the pioneers in the search for the origin of the mental phenomena of dreams, and above all, for being the first to be interested in the moment in which they appear. He proposed that dreams occur continuously during sleep or during the transition from sleep to awakening (Freud 1955). Although it is currently known that the dream activity is not continuous, the second premise is justified, given that dreams we usually remember are those that occur just before waking up (Dement and Wolpert 1958).

Despite the methodological criticisms of Freud's studies on the states of sleep and wakefulness, his work allowed progress on dream's references from wakefulness. That is, the influence of emotionally significant experiences lived during wakefulness and its relation to the content of the dream. Also, Freud gives memory a priority role in dreams, leading him to affirm that during these, memory portraits everything retained, and repressed during wakefulness (Schredl 2018). At the time of Freudian theory, in which the brain was barely listed as the generating organ of mental activity, somatic stimuli were considered as one of the sources of dream activity in sick people, therefore, it was assumed that patients suffering from heart ailments suffered from nightmares because they incorporated into their dream scenes the pain of *angina pectoris* (Schredl 2018). Freud would rightfully reject the theory of the origin of dreams that regarded the vegetative sensation as the forger of dreams, giving the primary role to memory (Freud 1955).

In addition to Freud's contributions, the French doctor Nicholas Vaschide published his extensive observations on sleep and dreams in the book: *Le sommeil et les rêves (1914)*. Vaschide in turn had been inspired by the works of the French psychiatrist Jean-Étienne-Dominique Esquirol, who through the observation of his patients while they slept, described facial expressions and movements in an attempt to predict the content of their dreams. From pulse and breathing patterns, movements of the body and pronounced words occurring during sleep, he concluded that it was almost always possible to determine whether an individual was dreaming or not (Hobson 1988). Esquirol also argued that the main themes of the content of a dream could be predicted most of the times (Hobson 1988). With this background, Vaschide conducted the same study in healthy subjects and described alternate periods of vasoconstriction and vasodilation. Although Vaschide does not determine the exact duration of this vascular cycle, it was probably the first observation of the ninety-minute NREM-REM cycle. Vaschide also woke up the subjects periodically to collect sleep information. Importantly, he observed that when subjects began to fall asleep, there were lapses of attention associated to a loss of synergy of eye movements, then vivid "hallucinations" occurred. (Vaschide 1914; Hobson 1988). Vaschide believed that dreams had a positive function, with its own mechanisms and that they were much more than a mere interruption of brain activity, as many of his contemporaries believed. Vaschide also considered that sleep was an instinctive function of the brain, therefore, a preprogrammed process and not a response to fatigue or the result of brain depression caused by any substance formed during wakefulness. He, therefore, considered sleeping as an active protection of evolutionary origin against threats of wakefulness (as toxic substances) (Vaschide 1914; Hobson 1988). Thus, Vaschide's theory is one of the first to integrate the contributions of Charles Darwin to the study of sleep (Darwin 1998), who argues that evolution is the product of a change in the genetic profile of a population of individuals according to the adaptation to the surrounding environment.

Up to this point, the study of dreams was approached from its mental aspect, but its connection to some functional cerebral state had not yet been taken into account, primarily because of the Cartesian dualism in force at the time, which considered the mind to be separated from the brain. It was until 1925 when Hans Berger developed the first human EEG, that it was possible to link mental states to concurrent brain activity (Dement and Kleitman 1957; Berger and Oswald 1962).

The scientific analysis of dreams began with Aserinsky and Kleitman's discovery of REM sleep in 1953, followed by Dement and Kleitman's own finding in 1957 that rapid eye movements (REMs) corresponded with dream occurrence. Therefore, it was understood for the first time exactly when the dreams took place and that they were linked to a particular physiological cerebral condition. After these pioneer findings, a new wave of scientific quests looking for physiological variables in function of dreams emerged. Specifically, with the collection of verbal reports from several dreams for one night, its components related to various biological variables could be characterized. Initially, a correlation between the number of REMs and dream's emotional intensity was found (Dement and Kleitman 1957). Furthermore, in 1962, Berger and Oswald found a relationship between the intensity of the dream's emotional content and the amplitude of REMs. Likewise, the emergence of more complex dreams was linked to changes in the skin's electrical resistance and abrupt heart and respiratory rate changes (Orem and Barnes 1980). Therefore, dreams were no longer dissociated from the rest of the body, but they had unique visual, vegetative, emotional and mental elements which distinguished them from the rest of sleep (Gruen et al. 1997).

REM Sleep, Dreams and Emotion

The identification of dream mentation in both REM and NREM sleep has been confirmed by many researchers despite the fact that in the beginning, it was considered that dreams were almost exclusive of REM sleep stage. Nevertheless, REM sleep dreams are quite unique: they are more intricate than NREM sleep dreams, they are better remembered, and their emotional and cognitive content is higher than NREM sleep dreams (Nielsen 2000). In the pioneer polysomnographic studies in humans conducted by Foulkes (1962), subjects were asked to classify their experience according to a series of variables that allowed to differentiate the stories of REM from NREM dreams. He found that dreams of REM sleep were characterized by greater perceptual clarity, greater distortion of reality, greater number of visual images, scenes, characters and movement, greater affective participation and were generally estimated as longer. On the other hand, NREM sleep dreams were more conceptual than perceptual, with less drama and physical and emotional participation. NREM sleep dreams were also less complex, and the relationship with recent events in the subject's real life and the continuation of the same theme in different awakenings was more frequent (Foulkes 1962; Nielsen 2000). Thus, a much richer psychophysiological world lies beneath REM sleep dreaming as opposed to NREM sleep dreams.

A key component of REM sleep that has been associated with dream content is motor activity. Despite the well-described muscular atonia characteristic of REM sleep, different patterns of muscle activity have been associated to emotional content of dreams (Gassel et al. 1964; Bliwise et al. 1974; Datta and MacLean 2008; Baldridge et al.1965; Gardner et al. 1983; Nielsen et al. 2009; Nielsen et al. 1989; Gerne and Strauch 1985; McGuigan and Tanner 1971; Shimizu and Inoue 1986; Wolpert 1960; Rivera et al. 2011; Rivera et al. 2018). At this point it should be noted that the excitability to muscle stimulation, whether tonic (sustained) or phasic (brief), is significantly diminished during REM sleep, unlike during NREM, where the response to stimulation remains intact (Hobson 1988). Yet, a paradoxical motor activation (in the midst of a selective inhibition) appear in different muscle groups, which it has been described that may be associated with dream content, serving as a motor execution of dreams (Gassel et al. 1964; Bliwise et al. 1974; Datta and MacLean 2008; Baldridge et al.1965; Gardner et al. 1983; Nielsen et al. 2009; Nielsen et al. 1989; Gerne and Strauch 1985; McGuigan and Tanner 1971; Shimizu and Inoue 1986; Wolpert 1960). Beyond this very interesting activation, it is important to point out the fact that muscle activation during REM sleep is much more significant in the motor groups specifically associated to emotional expression, either in vocalization (Lavigne et al. 2001), movement of the mouth and lips (Shimizu and Inoue 1986), and facial expression (Rivera et al. 2011; Rivera et al. 2018; Clé et al. 2019). Moreover, it has been described that this specificity in muscular activation is associated to emotional dreamed content, functioning as emotional expressions during dreams

(Rivera et al. 2018; Clé et al. 2019). Taking up the Facial Feedback Hypothesis, that posits that affective expressions influence emotional experience (Davis et al. 2009), we propose that motor activity associated to emotional expression during REM sleep might be one of the elements that facilitates emotional feedback.

Several other elements of this sleep stage have shown to play a pivotal role in the processing of salient and emotional waking-life experiences (Scarpelli et al. 2019) and it is through the study of affective disorders that many of this intrincate relations have been portrayed. Since the middle of the 20th century the association between emotional disorders and different components of sleep has been explored (Cartwright and Wood 1994). A pattern of overt early indicators during REM has been described in MDD, such as reduction in latency to this sleep stage, increase in the duration of sleep as well as in the number of REMs. Dream reports in MDD are also afflicted with greater anxiety, anguish and sadness (Cartwright and Wood 1994; Wichniack 2013; Benca et al. 1997). It is important to mention that these changes are reversed as the severity of the MDD decreases (Benca et al. 1997) which indicates a functional association between several REM sleep components and affective aberrant states.

The decrease in emotional reactivity associated with sleep is crucial for daily functioning (Repetti et al. 2002; Sil et al. 2003) and dysfunctional emotional processing associated with sleep after an aversive event is a high risk factor for mental disorders such as post-traumatic stress disorder (PTSD) (Edge 2010; Germain 2013). Under these parameters arose the *Sleep to forget, sleep to remember* (SFSR) theory (Walker and van Der Helm 2009) which associates REM sleep with emotional processing in such a way that we sleep to *forget* the emotional tone, yet sleep to *remember* the tagged memory of that episode, reinforcing of the declarative memory of the emotion while diminishing the degree of affective reactivity (Walker and van Der Helm 2009). The tentative explanation for this process is based on the specific functional activation during REM sleep of the same neuronal regions activated by emotional stimuli during wakefulness (as has been excellently described in Chapter 4), which would facilitate the reactivation and neuronal integration of emotional events (Cahill 2000; McGaugh 2004). Specifically, the increase in activity of the hippocampus and amygdala together with the predominance of Theta rhythm (4-7 Hz) and the decrease of aminergic neurotransmission are suggested as part of the mechanism underlying this association (Maquet et al. 1999; Marquis et al. 2017). Altogether, these features in common indicate a specific role between REM sleep mechanisms and those of emotions, and furthermore, suggest that both this stage (REM) and this process (emotion) may have another common component: complex dreams. According to Revonsuo (2009), the biological function of dreams is to simulate threatening events and rehearse the perception of the threat and avoid it. By reproducing these situations, the theory states that dreams help us process emotions and adapt to the threat (Valli and Revonsuo 2009). Thus, dreaming during REM sleep could play a key role in emotional regulation, particularly by experiencing different ranges of the emotional spectrum but, first of all, by evidencing the affective state of the dreamer.

DREAMS AS HERALDS OF EMOTION

Dreaming is a fascinating experience linked to psychological processes that can be viewed as a gateway to "entry" in the dreamer's inner world (Freud 1955 in Scarpelli et al. 2019). As described earlier, REM sleep plays a major role in promoting various types of memory, namely procedural, episodic and particularly emotional. Yet dreams can play a specific facilitating role in adapting quickly and effectively to changes in the dreamer's context, particularly those involved in affect (Blagrove et al. 2011). Several studies have suggested a cognitive processing sequence ranging from NREM-2 to REM sleep, where the integration of sensorimotor information occurring during NREM stages (especially in NREM-2) is required for subsequent processing of affective character information during REM sleep (Roussy et al. 1996; Gais and Born 2004; Giuditta et al. 1995). Therefore, without the proper distribution of NREM sleep episodes in the first stage of the night, emotional regulation, present in greater quantity during the second part of the night, would be altered.

But it is a central component of REM sleep that often goes unnoticed that can be essential in the specificity between sleeping and affect: its dreams. REM sleep dreams, as opposed to NREM dreams, are particularly flexible in terms of content, as it has been described that are greatly influenced by waking events. Unlike less complex NREM sleep dreams, the stimulation of the limbic, paralimbic and subsequent cortical areas during REM sleep could be associated to the assignment of desires, thoughts and particularly emotions to oneself and to the characters and situations in the dream (Maquet et al. 2005). It is the emotional insight during dreams one of the key components in affective processing: the perception of emotions during dreams can promote internal conflict resolution, which in turn can lead to an emotional catharsis (Deseilles and Duclos 2013; Deseilles et al. 2011) as suggested by Nielsen et al. (1989).

As immediate as emotional processing may be in wakefulness, dreams are directly shaped by context, creating an alternate reality in which the stimulus, the physiological response, the expression (Rivera et al. 2018), and more importantly, the subjective experience of emotions is present, only without voluntary control over the subjective experience. Consequently, one important role of dreams may be its ability to bring to light the dreamer's own affective state (regardless of its suitability), either revealing its current affective status or a much deeper concern. Dreamed emotions can work as proxies, yes by processing, rearranging if necessary, understanding, modulating and regulating emotions, but, first of all by recognizing our own affective status. It is only by paying attention to our feelings that we can truly understand each other, and dreams are the most precise, proximal, inevitable herald of our affective world.

Conclusion

In the present chapter we outlined that dreaming during REM sleep could play a key role in emotional regulation, particularly by experiencing different ranges of the emotional spectrum but, first of all, by evidencing the affective state of the dreamer. The fact that most regions involved in emotional encoding are highly activated during REM sleep can explain the emotional intensity of dreams reported during this sleep stage, but furthermore, may be associated to one of the most important elements of the affective process: the subjective experience of emotion itself. The fact that every night we may have an endogenous source of affective information could be a specific facilitator for experiencing and essaying alternate realities, but first of all, for putting us in contact with our own affective status. In this way, and even though the relation between dreaming, emotional processes and neurobiological correlates is still a work in progress, the exploration of dream content can provide invaluable information on the affective context in which the dreamer is, showing elements of his context that may go unnoticed in wakefulness. Contrary to the popular belief that dreams are hallucinatory experiences confusing and unconnected with the stimuli of the immediate environment, the dream world reflects aspects of the waking conditions. In this way, the peculiarities of dreams are linked to characteristics of the subject such as age, sex, health, habits, personality and interests, as well as their economic, cultural, environmental and health conditions, but, most of all, dreams reflect its affective concerns. Only through the study of sleep reports it is possible to corroborate the correlations that until now have been established between the neural activity detected in different sleep stages and the formal characteristics of the dream experience. Thus, studying the symbolic and emotional content of dreams not only allows us to explore how experiences are reflected in brain function and structure, but also to investigate the neural mechanisms that make them possible. The conscientious study of dreams is therefore an essential resource to examine the way in which individuals relate to their context, but most of all, to their most inner affective state.

Acknowledgments

This work was supported by project NC16020.0 from the Instituto Nacional de Psiquiatría Ramón de la Fuente Muñiz. We thank Maury Martínez and Dulce Islas for their help in this manuscript.

References

Aserinsky, Eugene, Nathaniel Kleitman. 1953. "Regularly occurring periods of eye motility and concomitant phenomena during sleep." *Science* 118:273-274. doi:10.1176/jnp.15.4.454.

Babson, Kimberly A. and Matthew T. Feldner. 2015. *Sleep and Affect: Assessment, Theory, and Clinical Implications*. Elsevier Academic Press.

Baldridge, B., R. Whitman, and M. Kramer. 1965. "The Concurrence of Fine Muscle Activity and Rapid Eye Movements During Sleep." *Psychosomatic Medicine* 27:19-26. doi:10.1097/00006842-196501000-00003.

Barfield, T. 1997. *The dictionary of anthropology.* Oxford, UK: Blackwell.

Benca, R. M., M. Okawa, M. Uchiyama, S. Ozaki, T. Nakajima, K. Shibui, and W. H. Obermeyer. 1997. "Sleep and mood disorders." *Sleep Medicine Reviews* 1:45-56. doi:10.1016/s1087-0792(97)90005-8.

Berger, R. J., I. Oswald. 1962. "Eye movements during active and passive dreams." *Science* 137:601. doi:10.1126/science.137.3530.601.

Blagrove, Mark, Nathalie C. Fouquet, Josephine A. Henley-Einion, Edward F. Pace-Schott, Anna C. Davies, Jennifer L. Neuschaffer, and Oliver H. Turnbull. 2011. "Assessing the dream-lag effect for REM and NREM stage 2 dreams." *PLoS ONE* 6:e26708. doi:10.1371/ journal.pone.0026708.

Bliwise, D., R. Coleman, B. Bergmann, M. Z. Wincor, R. T. Pivik, and A. Rechtschaffen. 1974. "Facial muscle tonus during REM and NREM sleep." *Psychophysiology* 11:497-508. doi:10.1111/j.1469-8986.1974. tb00578.x.

Breiter, Hans C., Nancy L. Etcoff, Paul J. Whalen, William A. Kennedy, Scott L. Rauch, Randy Buckner, Monica M. Strauss, Steven E. Hyman and Bruce R. Rosen. 1996. "Response and habituation of the human amygdala during visual processing of facial expression." *Neuron* 17:875-887. doi:10.1016/S0896-6273(00)80219-6.

Britton, Jennifer C., K. Luan Phan, Stephan F. Taylor, Robert C. Welsh, Kent C. Berridge, and I. Liberzon. 2006. "Neural correlates of social and nonsocial emotions: An fMRI study." *Neuroimage* 31:397-409. doi:10.1016/j.neuroimage.2005.11.027.

Burton, R. 1989. *The anatomy of melancholy*. Oxford, UK: H. Cripps.

Cacioppo, J. T., G. G. Berntson, J. T. Larsen, K. M. Poehlmann, and T. A. Ito. 2000. "The psychophysiology of emotion." In *The Handbook of Emotion,* edited by R. Lewis and J. M. Haviland-Jones, 173-191. New York, US: Guilford Press.

Cahill, L. 2000. "Neurobiological mechanisms of emotionally influenced, long-term memory." *Progress in Brain Research* 126:29–37. doi:10. 1016/S0079-6123(00)26004-4.

Cahill, L. 2000. "Neurobiological mechanisms of emotionally influenced, long-term memory." *Progress in Brain Research* 126:29–37. doi:10. 1016/S0079-6123(00)26004-4.

Calvo, J. M. (1993). "The limbic system and sleep." *Annals of the Mexican Institute of Psychiatry* 4:47-55. http://repositorio.inprf.gob.mx/handle/ 123456789/6514.

Cardinal, Rudolf N., John A. Parkinson, Jeremy Hall, and Berry J. Everitt. 2002. "Emotion and motivation: the role of the amygdala, ventral striatum, and prefrontal cortex." *Neuroscience & Biobehavioral Reviews* 26:321-352. doi:10.1016/S0149-7634(02)00007-6.

Cartwright, R. D., and E. Wood. 1993. "The contribution of dream masochism to the sex ratio difference in major depression." *Psychiatry Research* 46:165e73. doi:10.1016/0165-1781(93)90018-c.

Clé, Marion, Jean-Baptiste Maranci, Sebastian Weyn Banningh, Jade Lanfranchi, Marie Vidailhet, and Isabelle Arnulf. 2019. "Smiling asleep: A study of happy emotional expressions during adult sleep." *Journal of Sleep Research* 28:e12814. doi:10.1111/jsr.12814.

Corsi, C. M. 1994. "The brain, the organ of dreams." *Latin Journal of Thought and Language* 3: 197-225.

Damasio, A. 2003. "Feelings of Emotion and the Self." *Annals of the New York Academy of Sciences* 1001:253-61. doi:10.1196/annals.1279.014.

Damasio, A. R. 1994. *Descartes' Error.* New York: G. P. Putnam.

Darwin, C. 1998. *The expression of emotion in man and animals.* Oxford, UK: Oxford University Press.

Datta, S., and R. R. MacLean. 2008. "Neurobiological Mechanisms for the Regulation of Mammalian Sleep-Wake Behavior: Reinterpretation of Historical Evidence and Inclusion of Contemporary Cellular and Molecular Evidence." *Neuroscience and Biobehavioral Reviews* 31:775–824. doi:10.1016/j.neubiorev.2007.02.004.

Davis, Joshua Ian, Ann Senghas, and Kevin N. Ochsner. 2009. "How Does Facial Feedback Modulate Emotional Experience?" *Journal of Research in Personality* 43:822–829. doi:10.1016/j.jrp.2009.06.005.

Delgado, José María 1966. Emotions. Self-Selection Psychology Textbook. Dubuque Iowa: William C. Brown Company Publishers.

Delgado, José María 2002. *Is the face the mirror of the soul? Physiology of facial expression.* Mexico: Elements: Science and culture.

Dement, W. C., and E. A. Wolpert. 1958. "The relation of eye movements, body motility and external stimuli to dream content." *Journal of Experimental Psychology* 55:543-553. doi:10.1037/h0040031.

Dement, W. C., E. A. Wolpert. 1958. "The relation of eye movements, body motility and external stimuli to dream content." *Journal of Experimental Psychology* 55:543-553. doi:10.1037/h0040031.

Dement, William C., Nathaniel Kleitman. 1957. "The relation of eye movement during sleep to dream activity." *Journal of Experimental Psychology* 53:339-346. doi:10.1037/h0048189.

Derryberry, Douglas, Marjorie A. Reed, and Carolyn Pilkenton-Taylor. 2003. "Temperament and Coping: Advantages of an Individual Differences Perspective." *Development and Psychopathology* 15:1049-66. doi:10.1017/S0954579403000439

Desseilles, M., and C. Duclos. 2013. "Dream and emotion regulation: insight from the ancient art of memory." *Behavior and Brain Sciences* 36:634-659. doi:10.1017/S0140525X13001271.

Desseilles, M., V. Sterpenich, T. T. Dang-Vu, and S. Schwartz. 2011. "REM sleep and emotion regulation." In *Rapid eye movement sleep: Regulation and function, edited by* B. N. Mallick, S. R. Pandi-Perumal, R. W. McCarley, and A. R. Morrison, 427–436. Cambridge, UK: Cambridge University Press.

Díaz, José Luis, and Enrique Flores. 2001. "The structure of human emotion: A chromatic model of the affective system." *Mental Health* 24: 20-35.

Edge, L. C. 2010. "The role of emotional brain processing during sleep in depression." *Journal of Psychiatric and Mental Health Nursing* 17:857–861. doi:10.1111/j.1365-2850.2010.01598.x.

Ekman, Paul, and Erika L. Rosenberg. 2005. *What the face reveals: Basic and applied studies of spontaneous expression using the Facial Action Coding System (FACS)*. New York, NY: Oxford University Press.

Ekman, Paul, Wallace V. Friesen, and Phoebe Ellsworth. 1972. *Emotion in the human face*. Elmsf Pergamon.

Ekman, Paul. 1992. "An argument for basic emotions." *Cognition and Emotion* 6:169–200. doi:10.1080/02699939208411068.

Elfenbein, Hillary Anger, and Nalini Ambady. 2002. "On the universality and cultural specificity of emotion recognition: a meta-analysis." *Psychological Bulletin* 128:203–235. doi:10.1037/0033-2909.128.2. 203.

Ellsworth, Phoebe C., and Klaus R. Scherer. 2003. "Appraisal processes in emotion." In *Series in affective science. Handbook of affective sciences,* edited by Richard J. Davidson, Klaus R. Scherer, and H. Hill Goldsmith, 572–595. New York: Oxford University Press.

Foulkes, W. D. 1962. "Dream reports from different stages of sleep." *The Journal of Abnormal and Social Psychology* 65:14-25. doi:10.1037/ h0040431.

Fredrikson, Mais, Gustav Wik, Peter Annas, Kaj Ericson, and Sharon Stone-Elander. 1995. "Functional neuroanatomy of visually elicited simple phobic fear: additional data and theoretical analysis." *Psychophysiology* 32:43-48. doi:10.1111/j.1469-8986.1995.tb03404.x.

Freud, S. 1955. "*The Interpretation of Dreams*." New York, NY: Basic Books.

Frewen, Paul A., David J. Dozois, Richard W. Neufeld, Maria Densmore, Todd K. Stevens, and Ruth Lanius. 2011. "Neuroimaging social emotional processing in women: fMRI study of script-driven imagery." *Social Cognitive and Affective Neuroscience* 6:375-92. doi:10.1093/ scan/nsq047.

Gaebel, Wolfgang, Wolfgang Wölver. 1992. "Facial expression and emotional face recognition in schizophrenia and depression." *European Archives of Psychiatry and Clinical Neuroscience* 242:46-52. doi:10.1007/BF02190342.

Gais, S., and J. Born. 2004. "Multiple processes strengthen memory during sleep." *Psychologica Belgica* 44:105–120.

Gardner, R., W. I. Grossman, H. P. Roffwarg, and H. Weiner. 1983. "The relationship of small limb movements during REM sleep to dreamed limb action." *Psychosomatic Medicine* 37:147-159. doi:10.1097/ 00006842-197503000-00005.

Gassel, M. Michael, Pier Lorenzo Marchiafava, and Ottavio Pompeiano. 1964. "Phasic changes in muscular activity during desynchronized sleep in unrestrained cats. An analysis of the pattern and organization of myoclonic twitches." *Archives Italiennes de Biologie* 102:449-470. doi:10.4449/aib.v102i3.1837.

Germain, A. 2013. "Sleep disturbances as the hallmark of PTSD: Where are we now?" *American Journal of Psychiatry* 170:372–382. doi:10.1176/appi.ajp.2012.12040432.

Gerne, M., and I. Strauch. 1985. "Psychophysiological indicators of affect pattern and conversational signals during sleep." In *Sleep,* edited by W. P. Koella, E. Rüther, and H. Schulz, 367-369. Stuttgart: Gustav Fischer Verlag.

Giuditta, A., M. V. Ambrosini, P. Montagnese, P. Mandile, M. Cotugno, G. Grassi Zucconi, and S. Vescia. 1995. "The sequential hypothesis of the function of sleep." *Behavioral Brain Research* 69:157–166. doi: 10.1016/0166-4328(95)00012-i.

Gross, James J. 2007. *Handbook of Emotion Regulation.* New York, NY: Guilford Press.

Gross, James J., and Lisa Feldman Barrett. 2011. "Emotion generation and emotion regulation: One or two depends on your point of view." *Emotion Review: journal of the International Society for Research on Emotions* 3:8–16. doi:10.1177/1754073910380974.

Gruen, Isabel, Angélica Martínez, Carlos Cruz-Olloa, Félix Aranday, José María Calvo. 1997. "Characteristics of emotional phenomena in dreams of patients with temporal lobe epilepsy." *Mental Health* 20: 8-15.

Hobson, J. A. 1988. *The Dreaming Brain.* New York: Basic Books.

Hobson, J. A. 2011. *Dream Life: An Experimental Memoir.* Cambridge, MA: MIT Press.

Izard, Carroll E. 2007. "Basic Emotions, Natural Kinds, Emotion Schemas, and a New Paradigm." *Perspectives in psychological science: a journal of the Association for Psychological Science* 2:260-280. doi:10.1111/j.1745-6916.2007.00044.x.

Keil, Andreas, and Vladimir Miskovic. 2015. "Human emotions: A conceptual overview." In *Sleep and affect: Assessment, theory, and clinical implications,* edited by Kimberly A. Babson and Matthew T. Feldner, 23–44. Elsevier Academic Press.

Ketter, Terence A., Paul J. Andreason, Mark S. George, Chulhee Lee, Dbra S. Gill, Priti I. Parekh, Mark W. Willis, Peter Herscovitch, and Robert M. Post. 1996. "Anterior paralimbic mediation of procaine-induced emotional and psychosensory experiences." *Archives of General Psychiatry* 53:59-69. doi:10.1001/archpsyc.1996.01830010061009.

Kleinginna, Paul, and Anne M. Kleinginna. 1981. "A categorized List of Emotion Definitions, with Suggestions for a Consensual Definition." *Motivation and Emotion* 5:263. doi:10.1007/BF00992553.

Larsen, Randy J., and Edward Diener. 1992. "Promises and problems with the circumplex model of emotion." In *Review of personality and social psychology,* edited by M. S. Clark, 25–59. Newbury Park, CA: Sage Publications.

Lavigne, Gilles J., Pierre H. Rompre, Gaétan Poirier, H. Huard, Takafumi Kato, and Jacques Y. Montplaisir. 2001. "Rhythmic masticatory muscle activity during sleep in humans." *Journal of Dental Research* 80:443-448. doi: 10.1177/00220345010800020801.

Lazarus, R. S. 1982. "Thoughts on the relations between emotion and cognition." *American Psychologist* 37:1019–1024. doi:10.1037/0003-066X.37.9.1019.

LeDoux, Joseph. 2012. "Evolution of human emotion: a view through fear." *Progress in Brain Research* 195:431–442. doi:10.1016/B978-0-444-53860-4.00021-0.

Lim, N. 2016. "Cultural differences in emotion: differences in emotional arousal level between the East and the West." *Integrative Medicine Res*earch 5:105–109. doi:10.1016/j.imr.2016.03.004.

Lutz, Catherine, and Geoffrey M. White. 1986. "The anthropology of emotions." *Annual Review of Anthropology* 15:405–436. doi: 10.1146/annurev.an.15100186.002201.

Maquet, Pierre, Jean-Marie Péters, Joël Aerts, Guy Delfiore, Christian Degueldre, André Luxen, and Georges Franck. 1996. "Functional neuroanatomy of human rapid-eye-movement sleep and dreaming." *Nature* 383:163–166. doi:10.1038/383163a0.

Marquis, L. P., T. Paquette, C. Blanchette-Carrière, G. Dumel, and T. Nielsen. 2017. "REM Sleep Theta Changes in Frequent Nightmare Recallers." *Sleep* 40:zsx110. doi:10.1093/sleep/zsx110.

Maury, A. 1862. *Le sommeil et le 224evés* [*Sleep and sleep*]. Paris: Didier et Cie.

Mauss, I. B., and M. D. Robinson. 2009. "Measures of emotion: A review." *Cognition and Emotion 23:*209–237. doi:10.1080/026999308 02204677.

McGaugh, J. L. 2004. "The amygdala modulates the consolidation of memories of emotionally arousing experiences." *Annual Review of Neuroscience* 27:1–28. doi:10.1146/annurev.neuro.27.070203.144157.

McGuigan, F. J., and Robert G. Tanner. 1986. "Covert oral behavior during conversational and visual dreams." *Psychonomic Science* 23:263-264. doi: 10.3758/BF03336104.

Mead, M. 1928. *Coming of age in Samoa.* New York, NY: William Morrow.

Mesquita, Batja, and Nico H. Frijda. 1992. "Cultural variation in emotions: a review." *Psychological Bulletin* 112:179–204. doi:10.1037//0033-2909.112.2.179.

Morris, J., C. D. Frith, D. I. Perrett, D. Rowland, A. W. Young, A. Calder, and R. Dolan. 1996. "A differential neural response in the human amygdala to fearful and happy facial expressions." *Nature* 383:812-815. doi:10.1038/383812a0.

Newman, J. 1995. "Thalamic contributions to attention and consciousness." *Conscious Cognition* 4:137. doi:10.1006/ccog.1995. 1024.

Nielsen, T. A. 2000. "A review of mentation in REM and NREM sleep: "covert" REM sleep as a possible reconciliation of two opposing models." *The Behavioral and Brain Science* 23:851–866. doi:10.1017/s0140525x0000399x.

Nielsen, T. A., D. Kuiken, and D. McGregor. 1989. "Effects of Dream Reflection on Waking Affect: Awareness of Feelings, Rorschach Movement, and Facial EMG." *Sleep* 12:277-286. doi:10.1093/sleep/12 .3.277.

Nielsen, Tore A. 2000. "A Review of Mentation in REM and NREM Sleep: 'Covert' REM Sleep as a Possible Reconciliation of Two Opposing Models." *Behavioral and Brain Sciences* 23:851–866. doi:10.1017/S0140525X0000399X.

Nielsen, Tore, Connie Svob, and Don Kuiken. 2009. "Dream-Enacting Behaviors in a Normal Population." *Sleep* 32:1629-1636. doi:10.1093/sleep/32.12.1629.

Ochsner, Kevin N., and James J. Gross. 2005. "The cognitive control of emotion." *Trends in Cognitive Sciences*, 9:242-249. doi:10.1016/j.tics. 2005.03.010.

Orem, J., C. D. Barnes. 1980. *Physiology in sleep*. New York: Academic Press.

Panksepp, J. 1998. *Affective neuroscience: The foundations of human and animal emotions*. New York: Oxford University Press.

Pollak, Seth D., Linda A. Camras, and Pamela M. Cole. 2019. "Progress in Understanding the Emergence of Human Emotion." *Developmental Psychology* 55:1801-1811. doi:10.1037/dev0000789.

Posner, Jonathan, James A. Russell, and Bradley S. Peterson. 2005. "The circumplex model of affect: An integrative approach to affective neuroscience, cognitive development, and psychopathology." *Development and Psychopathology* 17:715–734. doi:10.1017/S095 4579405050340.

Repetti, R. L., S. E. Taylor, and T. E. Seeman. 2002. "Risky families: Family social environments and the mental and physical health of offspring." *Psychological Bulletin* 128:330–366. doi:10.1037/0033- 2909.128.2.330.

Rivera-García Ana, Ignacio Ramírez-Salado, María Corsi-Cabrera, and José María Calvo. 2011. "Facial muscle activation during sleep and its relation to the rapid eye movements of REM sleep." *Journal of Sleep Research* 20:82–91. doi: 10.1111/j.1365-2869.2010.00853.x.

Rivera-García, Ana Paula, I. E. López Ruiz, Ignacio Ramírez-Salado, J. González-Olvera, F. Ayala-Guerrero, and A. Jiménez-Anguiano. 2018. "Emotional facial expressions during REM sleep dreams." *Journal of Sleep Research* 28:e12716. doi: 10.1111/jsr.12716.

Roussy, Francine, Claude Camirand, David Foulkes, Joseph De Koninck, Malea Loftis, and Nancy H. Kerr. 1996. "Does early-night REM dream content reliably reflect presleep state of mind?" *1113 Dreaming* 6:21–130. doi:10.1037/h0094450.

Russell, J, J. A. Bacharowski, and J. M. Fernández-Dols. 2003. "Facial and Vocal expressions of emotion." *Annual Reviews Psychology* 54:329-349. doi:10.1146/annurev.psych.54.101601.145102.

Russell, James A. 1980. "A circumplex model of affect." *Journal of Personality and Social Psychology* 39:1161–1178. doi:10.1037/ h0077714.

Russell, James A. 2003. "Core affect and the psychological construction of emotion." *Psychological Review* 110:145–172. doi:10.1037/0033-295X.110.1.145.

Russell, James, and José Miguel Fernandez-Dols. 1997. *The psychology of facial expression.* New York, NY: Cambridge University Press.

Scarpelli, Serena, Chiara Bartolacci, Aurora D'Atri, Maurizio Gorgoni, and Luigi De Gennaro. 2019. "The Functional Role of Dreaming in Emotional Processes." *Frontiers in Psychology* 10:459. doi:10.3389/ fpsyg.2019.00459.

Scherer, K. R., A. Schorr, and T. Johnstone. 2001. *Appraisal Processes in Emotion: Theory, Methods, Research.* New York: Oxford University Press.

Scherer, Klaus R. 2004. "Feelings Integrate the Central Representation of Appraisal-Driven Response Organization in Emotion." In *Feelings and emotions The Amsterdam Symposium,* edited by Antony S. R. Manstead, Nico Frijda, Agneta Fischer, 136-157. New York, NY: Cambridge University press.

Scherer, Klaus R., and Tobias Brosch. 2009. "Culture-specific appraisal biases contribute to emotion dispositions." *European Journal of Personality* 23:265–288. doi:10.1002/per.714.

Scherer, Klaus R., E. Clark-Polner, and M. Mortillaro. 2011. "In the eye of the beholder? Universality and cultural specificity in the expression and perception of emotion." *International Journal of Psychology 46*:401–435. doi:10.1080/ 00207594.2011.626049.

Schlosberg, Harold 1952. "The description of facial expressions in terms of two dimensions." *Journal of Experimental Psychology* 44:229–237. doi:10.1037/ h0055778.

Schredl, Michael 2018. *Researching dreams: the fundamentals.* Switzerland, AG: Palgrave Macmillan.

Shimizu, Akira, and Tsuyoshi Inoue. 1986. "Dreamed speech and speech muscle activity." *Psychophysiology* 23:210-214. doi:10.1111/j.1469-8986.1986.tb00620.x.

Shott Susan. 1979. "Emotion and social life: a symbolic interactionist analysis." *American Journal of Sociology* 84:1317–1334.

Silber, Michael H., Sonia Ancoli-Israel, Michael H. Bonnet, Sudhansu Chokroverty, Madeleine M. Grigg-Damberger, Max Hirshkowitz, Sheldon Kapen, Sharon A. Keenan, Meir H. Kryger, Thomas Penzel, Mark R. Pressman, and Conrad Iber. 2007.

"The visual scoring of sleep in adults." *Journal of Clinical Sleep Medicine* 14:121-131.

Silk, J. S., L. Steinberg, and A. S. Morris. 2003. "Adolescents' emotion regulation in daily life: Links to depressive symptoms and problem behavior." *Child Development* 74:1869–1880. doi:10.1046/j.1467- 8624.2003.00643.x.

Smith, Tiffany Watt. 2016. *The Book of Human Emotions: An Encyclopaedia of Feeling from Anger to Wanderlust*. Profile Books Limited.

Solomon, Robert C. 1993. *The Passions: Emotions and the Meaning of Life*. Indianapolis: Hackett Publishing Company.

Tomkins, S. S. 1963. *Affect, imagery, consciousness. I & II.* New York, NY: Springer.

Uchida, Y., S. Kitayama. 2009. "Happiness and unhappiness in East and West: themes and variations." *Emotion* 9:441–456. doi:10.1037/ a0015634.

van der Helm, Els, Ninad Gujar, and Matthew P. Walker. 2010. "Sleep deprivation impairs the accurate recognition of human emotions." *Sleep* 33:335–342. doi:10.1093/sleep/33.3.335.

Vaschide, Nicolas. 1914. *Le sommeil et les revés* [*Sleep and dreams*]. Paris: Flammarion.

Walker, M. 2009. "The role of sleep in cognition and emotion." *Annals of the New York Academy of Science* 1156:168–197. doi:10.1111/j.1749-6632.2009.04416.x.

Walker, Matthew 2017. *Why We Sleep: Unlocking the Power of Sleep and Dreams*. New York, NY: Simon & Schuster Audio.

Walker, Matthew P., and Els van der Helm. 2009. "Overnight therapy? The role of sleep in emotional brain processing." *Psychological Bulletin* 135:731–748. doi:10.1037/a0016570.

Watson, David, David Wiese, Jatin Vaidya, and Auke Tellegen. 1999. "The two general activation systems of affect: Structural findings, evolutionary considerations, and psychobiological evidence." *Journal of Personality and Social Psychology* 76:820–838. doi:10.1037/0022-3514.76.5.820.

Wichniak, A., A. Wierzbicka, and W. Jernajczyk. 2013. "Sleep as a biomarker for depression." *International Reviews in Psychiatry* 25:632-645. doi:10.3109/09540261.2013.812067.

Wik, G., M. Fredrikson, K. Ericson, L. Eriksson, E. S. Stone, and T. Greitz. 1993. "A functional cerebral response to frightening visual stimulation." *Psychiatry Research: Neuroimaging* 50:5-24. doi:10.1016/0925-4927(93)90020-I.

Wolpert, E. A. 1960. "Studies on psychophysiology of dreams: II Electromyographic study of dreaming." *Archives of general Psychiatry* 2:231-241. doi:10.1001/archpsyc.1960.03590080107014.

Zajonc, R. B. 1980. "Feeling and thinking: Preferences need no inferences." *American Psychologist* 35:151–175. doi:10.1037/0003-066X.35.2.151.

Zohar, D., O. Tzischinsky, R. Epstein, and P. Lavie. 2005. "The effects of sleep loss on medical residents' emotional reactions to work events: a cognitive-energy model." *Sleep* 28:47–54. doi:10.1093/sleep/28.1.47.

In: Sleep, Cognition and Emotion
Editors: I. Ramírez Salado et al.

ISBN: 978-1-53617-561-5

Chapter 9

ANALYSIS OF SEXUAL ONEIRIC EXPERIENCES BY GENDER AND SEXUAL ORIENTATION

Ignacio Ramírez-Salado*[*]*, PhD
Manuel Alejandro Cruz-Aguilar, PhD
and Ana Paula Rivera-García, PhD
Laboratorio de Cronobiología y Sueño,
Dirección de Investigaciones en Neurociencias,
Instituto Nacional de Psiquiatría Ramón de la Fuente Muñiz,
Ciudad de México, México

ABSTRACT

Every night during rapid eye movement sleep we replace our waking reality for a vastly more fantastic and strange reality. In each oneiric episode, we experience hallucinatory perceptions with extensive and unlimited components, multiple situations and actions, thus the expression of eroticism and sexual activities can often be present in our dreams. The main objective of the present study is the analysis of erotic and sexual oneiric experiences by gender and sexual orientation. Using an instrument designed in our laboratory, it was possible to obtain a total of 182 women and 120 men. 94 heterosexual men and 26 homosexual men participated in the study. Likewise, 164 heterosexual women and 18 homosexual women participated. The questionnaire quantified different items and categories related to various elements present in every dream episode. The data obtained was compared and differences between men and women for the number of partners and about the kinds of characters that appeared were observed. Differences between different sexual orientations were observed as well. Results indicate that, in general, these dreams are remembered and experienced as pleasant events, full of positive emotional components and highly intense sensory

[*] Corresponding Author's Email: salado@imp.edu.mx.

components. The range of emotions that can be experienced proved to be extensive and diverse, positive emotions were reported in greater numbers. However, when these events were rated as negative emotions such as disgust, fear, anger and anguish, among others, were reported. The type of characters in these negative experiences included strangers, well-known people with whom the dreamer had a close relationship, monsters and animals.

Keywords: dreams, rapid eye movement sleep, erotic dreams, oneiric sexual experiences, sexual orientation, gender

INTRODUCTION

Dreaming is a state of consciousness in which internally generated sensory, motor, verbal, emotional and cognitive experiences can develop into actions and become events that give rise to imaginary plots. In a previous study, where more than 3,500 reports of home-made sleep were evaluated by men and women it was found that, in both genders, approximately 8% of these daily dreams contained some type of sex-related activity. The dream contents referred to situations that included sexual encounters, interactions and behaviors. Sexual intercourse was the most reported type of behavior within the content of sexual dreams, followed by sexual propositions, kisses, fantasies and masturbation (Zadra 2007). Although it was reported that both genders dream equally about sex, their usual contents are significantly different. On the one hand, women tend to dream about their current partner or previous sexual partners, while men are more likely to dream about imaginary sexual characters, or known people with whom they have not had any kind of emotional bond, even though women have as many sexual dreams as men. In this same study, the surveys of 109 women and 64 men, aged 20 to 89 years, who had kept dream diaries for a month, were analyzed. In women, up to 18% of sexual dreams were related to unwanted sexual experiences and 9% to celebrities, showing percentages much higher than those for men in these categories (Zadra 2007).

In this sense, the main dream theories propose that dreams may represent the obtaining of unconscious desires, or that they may be the cognitive derivative of a specific cerebral activation. The psychodynamic theory proposes that desires are frequently associated with impulses of a sexual nature which can manifest in latent content, which is composed of the desires and experiences that motivate the dream, though this content would be hidden from rational understanding through censorship, oblivion and, in some cases, misinterpretation. The manifest content of dreams is the narrative material, the story and events that occur in it. On the other hand, the message (latent content) is expressed through a language of symbols (manifest content), or in a latent message, as would frequently be the case with dreams of a sexual nature (Maggiolini and Codecà 2016).

From a neurobiological point of view, the phenomenon of dreams is explained by the neuromodulation of the structures and cognitive components active during the generation of dreams. A neurocognitive model proposed by Domhoff and Fox (2015) considers dreams as simulations or schemes of the real world, and are hence more faithful to reality than what is considered under other models. This model proposes continuity with wakefulness and, notably, it establishes that dream activity is nothing more than a consequence of neuroanatomic evolution and consciousness, and reflects only the ways in which the individual thinks and processes abstract knowledge. Therefore, dreams of a sexual nature could be explained from a multidisciplinary point of view which takes into account the neuroanatomic processes that generate them, as well as the context under which they emerge, during sleep with rapid eye movements (REM) (Aserinsky and Kleitman 1953).

During REM sleep, which occurs approximately every 90 minutes, the eyes move back and forth with the eyelids closed, the muscles become paralyzed and the body temperature decreases as the dream develops. The Freudian notion of active censorship that can transform the original or "latent" content of a dream into an "apparent" content lacks neurobiological support. On the other hand, unlike the Freudian model of dreams, it has been proposed that dreams may be potentially related to novel phenomena or some probable future, and not just the past. There is evidence of the presence of past and current concerns, or concerns that arise in wakefulness, in the content of dreams (Cartwright et al. 2006; Schredl and Hofmann 2003). However, dreams are generally novel constructions, and when they happen to be reproductions of past events, such reproductions are not literal (Meier 1993). It is important to mention that, during dreams, the brain shows significant levels of activation, and we know that although we cannot speak of a conscious state during dreams, it is also true that we retain the ability to perform different brain processes associated with memory and with the perception or evocation of emotions (Datta and Maclean 2007; Nir and Tononi 2010). Recent studies with functional magnetic resonance imaging (fMRI) reveal that the reward centers and pleasure circuits in the brain are active during REM sleep (Dahan et al. 2007; Lena et al. 2005; Perogamvros and Schwartz 2012).

Men commonly dream about taking part in fights with other men, while women often dream that they are talking animatedly to friends or people they know (these dreams refer to social interactions). The dreams of both genders can include romantic goals; however, men often describe attacks on potential competitors, while women compete with others socially (Zadra 2007). Compared to women, men dream more about other men, while women dream about men and women in equal proportions. Men participate more frequently in physical aggression against other men, while women engage in non-physical forms of aggression, which may consist of verbal rejections and exclusion of others (Zadra 2007).

There is evidence that sex hormones increase during REM sleep. Prolactin, which allows mothers to produce milk and stimulates the testicles in men, raises quickly when sleeping, reaching maximum peaks between 3 a.m. and 5 a.m., when REM sleep predominates. It has been shown that a prolactin release accompanies ejaculation in men, and it has been proposed that prolactin release is associated with the feeling of sexual satisfaction. Oxytocin and vasopressin are two other chemicals released during orgasm, both of which are also associated with sleep (their release often accompanies melatonin). Oxytocin, which is linked to attachment during sexual intercourse, and testosterone, linked to sexual desire, show maximum release peaks at approximately 4 a.m., when REM sleep is most frequent (Abbott 2015). There is also evidence that men experience prominent erections as soon as a period of REM sleep begins (Karacan 1966) and that during this same phase of sleep, women experience congestion or "swelling" of the clitoris (clitoral tumescence) and in some cases vaginal lubrication (Abel et al. 1979).

Regarding emotional ties and their relationship to dreams, the theory of attachment has suggested some categories in which people can be classified: a) those who are happy in their relationship, which is why they feel firmly attached to their partner b) those who are not in a relationship, but who desperately want to be in one, but unfortunately, that is what keeps them worried and anxious, and c) those who are indifferent about the importance of relationships and, therefore, avoid them (Zborowski and McNamara 1998; McNamara et al. 2001). Dylan Selterman, social psychologist at the University of Maryland, has found that dreaming of social interactions with romantic goals significantly predicts romantic interactions with others the next day. For example, if a subject dreamed of having sex with their partner, they were more affectionate, loving and intimate the next day. On the contrary, if that same subject dreamed of having conflicts with their partner, more conflicts were experienced in their relationship the next day (Selterman, Apetroaia and Waters 2012).

These results suggest that dreams can influence our behaviors during wakefulness. Therefore, dreams with sexual content can be associated with attempts to materialize our conscious desires; they could contribute by revealing latent conflicts, or with the resolution of problems associated with our sexuality. In addition, if it is possible to simulate novel sexual events during dreams, this simulation might provide the dreamer with new ideas and behaviors that could strengthen their sexual strategies. We assume the structure of our dreams as something real during their course; however, within dreams we can behave differently than we do in sexual practice during wakefulness. Therefore, it is common to dream of illogical, absurd, violent or aggressive and even funny and fantastic erotic practices. These sexual contents can confuse and worry those who experience them; however, it is important to consider that they are not necessarily the reflection of confused emotional or behavioral states associated with conscious waking life, but perhaps the product of a limbic system and very active pontine areas in combination with a frontal and orbitofrontal cortex with decreased activity.

However, knowledge about the cognitive characteristics of dreams in different gender identities is scarce. The main objective of this study was to describe and quantify emotions, number of sexual partners, type of characters, actions, locations and intensities perceived in dreams with sexual situations in people with different genders and sexual orientations.

METHODS

Participants

We used a sample of 182 women and 120 men. The ages for both genders were between 19 and 62 years (Mean = 34.0, SD = 10.95 years). 94 heterosexual men and 26 homosexual men participated in the study. Likewise, 164 heterosexual women and 18 homosexual women participated. All participants were Mexican.

Materials

An electronic survey was designed in our laboratory to quantify the frequency of emotions, number of sexual partners, type of characters, actions, locations and sexual intensities perceived in dreams by people with different sexual orientations. The survey was made on the *google forms* plataform. The URL to access the survey is displayed:

https://docs.google.com/forms/d/e/1FAIpQLSeoo9Fe7b5KOx4RqMa0mFXvjr_oWB j0eRQxuT-ZJF3kjtyTtw/viewform

This instrument allowed us to analyze and describe the content of dreams with sexual themes and to quantify the characteristics of these sexual dream experiences. Figures 1a and 1b show the survey used in the present study.

Procedure

To obtain the sample, using informative flyers, all members of the College community from different institutions in Mexico were invited to participate. The flyers specified the objectives of the survey and the internet address to access and participate in the study. People did not need to reveal their identity or interact with researchers, they

Erotic Dreams

This survey attempts to gather information about and describe the contents of erotic dreams. It has been reported that only 8% of our dreams have any sexual content. We'd like you to answer the survey as many times as you want, but we ask you to focus on only one dream experience each time. The link is permanent, so you can come back anytime. All gathered data is completely anonymous, and is collected only to better understand the contents of our dreams.

1. Sex
2. Age
3. Sexual preference
   a) Heterosexual
   b) Lesbian
   c) Gay
   d) Bisexual
   e) Transsexual

4. Which of the following emotions describe your dream?
   Remember: answer each time referring to one dream only.
   a) Pain
   b) Passion
   c) Pleasure
   d) Boredom
   e) Apathy
   f) Sexual arousal
   g) Fear
   h) Hatred
   i) Tenderness
   j) Falling or being in love
   k) Rejection
   l) Guilt
   m) Rage
   n) Affection
   o) Fun
   p) Desire
   q) Satisfaction
   r) Mistrust
   s) Joy
   t) Tiredness
   u) Compassion
   v) Sadness
   w) Frustration
   x) Love
   y) Like
   z) Dislike
   aa) Humiliation
   ab) Orgasm
   ac) Aggression
   ad) Empathy

5. How many sexual partners appeared in your dream?
   a) 2
   b) I don't remember
   c) 3 or more
   d) 1

6. The partner(s) who appear in your dream are:
   a) Former sexual partner
   b) Fantastical beings (i.e. fairies, goblins, deities, etc.)
   c) Celebrity (actors, athletes, politicians, artists, etc.)
   d) Stranger
   e) Known person
   f) Current sexual partner
   g) Unpleasant beings (monsters, entities from horror films, etc.)

7. If celebrities appeared, who are they?

8. If unpleasant beings appeared, what were those?

Figure 1a. Survey using during the study.

9. Which of these acts took place during the dream?
   a) Masturbation
   b) Kissing
   c) Fantasies
   d) Sexual propositions
   e) Sex
   f) Flirting

10. Where did the sexual/erotic encounter take place?
   a) Unknown place
   b) Known place
   c) Public place

11. In a scale of 1 to 5, how erotic was your dream? a) Not erotic at all → b) Very erotic

12. Which of the following emotions did you feel DURING the dream?

   a) Joy, excitement, pleasure, happiness
   b) Confusion, surprise, estrangement, indecision
   c) Anguish, worrying, anxiety, fear
   d) Sadness, loneliness, hopelessness, depression
   e) Rage, anger, annoyance, irritation

Thank you! Remember that by participating you are giving very important information that will help us understand not only yours, but everyone's dreams.

Figure 1b. Survey using during the study.

simply had to go to the website and fill out the form. The work was carried out in strict accordance with the Code of Ethics of the World Medical Association for studies involving humans (Declaration of Helsinki, World Medical Association 1964). Once someone decided to participate and access the survey, the instructions asked each subject to evoke or remember a single dream which contained sexual content. With that subjective memory in mind, subjects had to assess what was perceived or experienced during that dream. To know the emotions present in this type of dreams, the subjects had to choose from a list of words the emotions that best described their dream episode. The intention was that subjects did not value these experiences in an overview fashion. The instrument does not contain any mandatory question; however, the following answers were necessary to perform the analysis: gender, age, sexual orientation. This data was necessary for the subject to be included in the study. If at least one of these itmes was not reported, the subject was not included in the results. Due to this condition 19 surveys were eliminated. The data analyzed in this study was gathered between January and June 2016.

Data Analyses

The data obtained was compared, looking for differences between men and women for number of partners and about the kinds of characters that appeared in their dreams, as well as for differences between sexual orientations. For the data analysis, results were transformed to absolute and relative frequencies (mean expressed in percentage).

RESULTS

The most frequently reported emotions in all subjects, regardless of their gender or sexual orientation were: pleasure (mean = 21.5%), desire (mean = 18.25%), passion (mean = 12.75%), orgasm (mean = 7.75%), satisfaction (mean = 10.0%) and love (mean = 5.5%). After the general analysis, the frequencies of these emotions were compared between men and women, and between their sexual orientations. Men and women, regardless of their sexual orientation, showed similar emotions percentages (Figure 2).

Regarding the number of sexual partners present during the dream, subjects were asked to select one of the following options: “one partner,” “two partners,” “three or more partners” or “I do not remember.” 79.9% (mean) of all subjects reported “one sexual partner”; “two partners” was reported in 11.7% (mean); “3 or more partners” was reported in only 4.7% (mean), and the option of “I do not remember” appeard in only 3.7% (mean) of reports. We found no significant differences when comparing men and women. Heterosexual men reported “one partner” in 76.0% (mean) of reports;

heterosexual women did so in 83.0% (mean); homosexual men, 60.0% (mean), and homosexual women reported them in 100% of cases. Homosexual men reported "more than one sexual partner" more frequently than any other group, reporting "two partners" in 20.0% (mean), while heterosexual men did so in 15.0% of cases, and heterosexual women in only 12.0%(mean) (Figure 3).

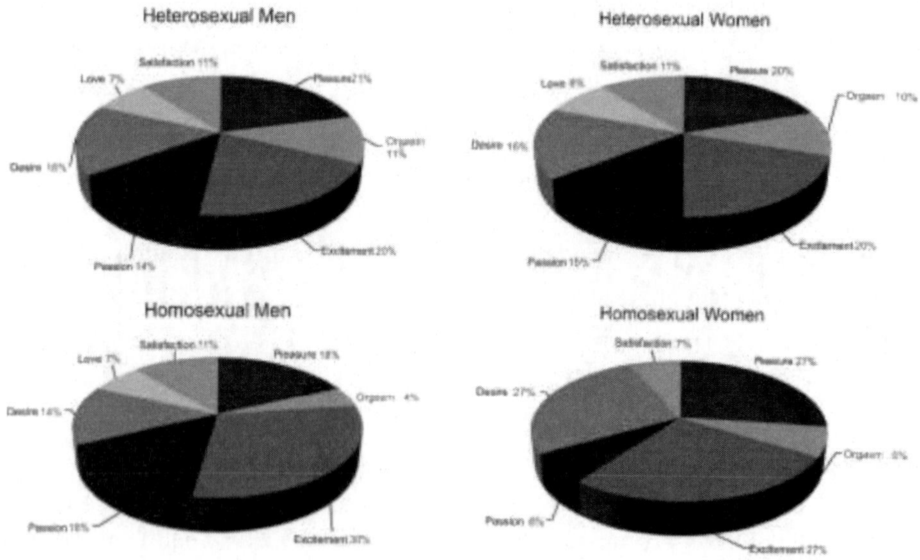

Figure 2. Emotions that best described erotic dreams through the reports obtained.

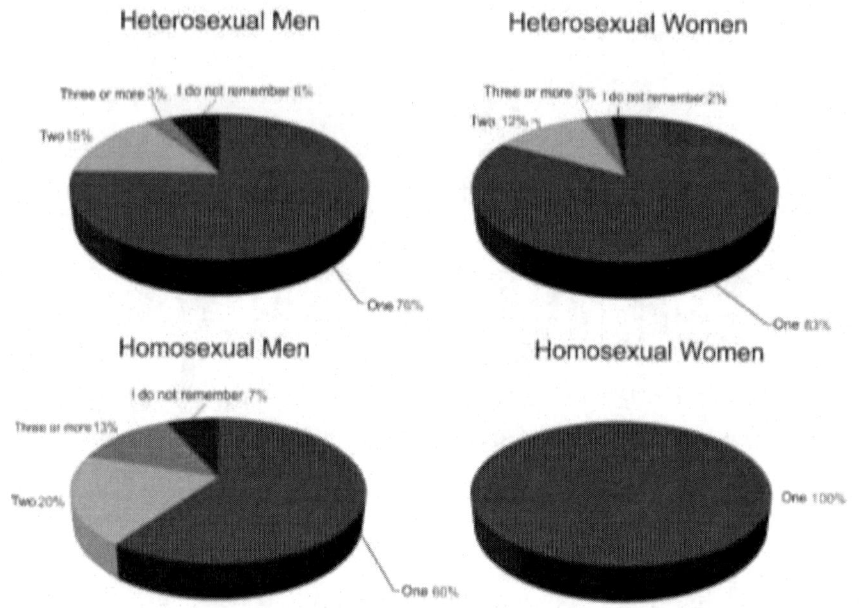

Figure 3. Number of partners evoked during erotic dreams.

To explore the characters present in their dreams, subjects had to answer the following question: “Who was the sexual partner that appeared in your dream?” The following options were available: current sexual partner, past sexual partner, known person, stranger, celebrity, fantastic being and unpleasant being. Significant differences were found when compering men to women. Heterosexual women reported that their sexual partner (in the dream) was their current (real life) partner 30.0% of the time, while heterosexual men did so in only 17.0% of cases. The results indicate that heterosexual men incorporate known people more frequently, especially those with whom they have had no formal or emotional ties. Hetereoseuxal men also reported the partner as “known” 53% of times, while for heterosexual women this percentage was only 38.0%. As for how often they dreamed of public figures, heterosexual men and women reported this in 2.0% of cases, *vs.* 6.0% for homosexual males (Figure 4).

Regarding the type of “present behaviors” in the total of participants, we found that the most reported behavior was the sexual act (mean= 38%). This means that most participants evoke or remember having experienced the sexual act during dreamed sexual behavior. Kisses were reported in 23% (mean), flirting (mean = 13%), sexual propositions (mean = 8%) fantasies (mean = 9%) and masturbation (mean = 9%) were all reported in smaller amounts. When comparing the data according to sexual orientation, homosexual men reported the sexual act more frequently (50.0%) when compared with heterosexual men (35.0%) and heterosexual women (29.0%). In this same group, the frequency of kisses (18.0%) decreased when compared with men (24.0%) and heterosexual women (25.0%) (Figure 5).

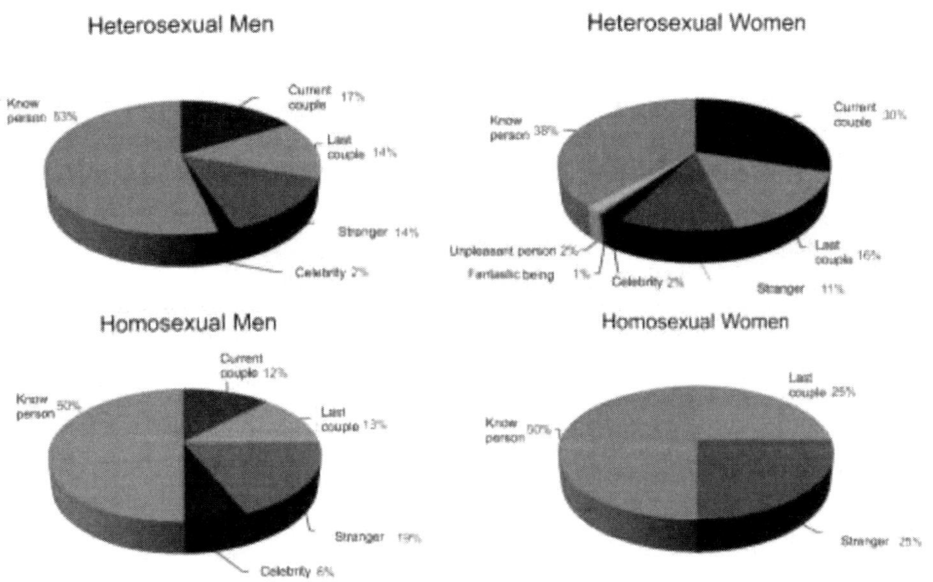

Figure 4. Sexual partners present during erotic dreams.

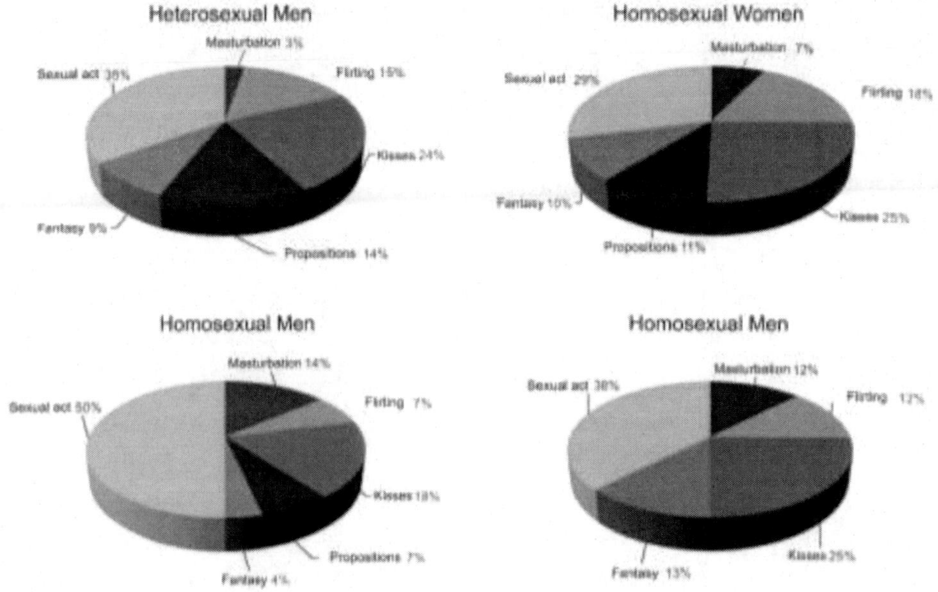

Figure 5. Behaviors and/or actions.

With respect to the "location where the dream sexual encounters took place" the *unknown location* was observed most frequently (mean = 65.7%), followed by *known* (mean = 22.2%) and finally, *a public place* (12.2%). We did not find differences between men and women; however, homosexual women reported 100% of place as unknown (Figure 6).

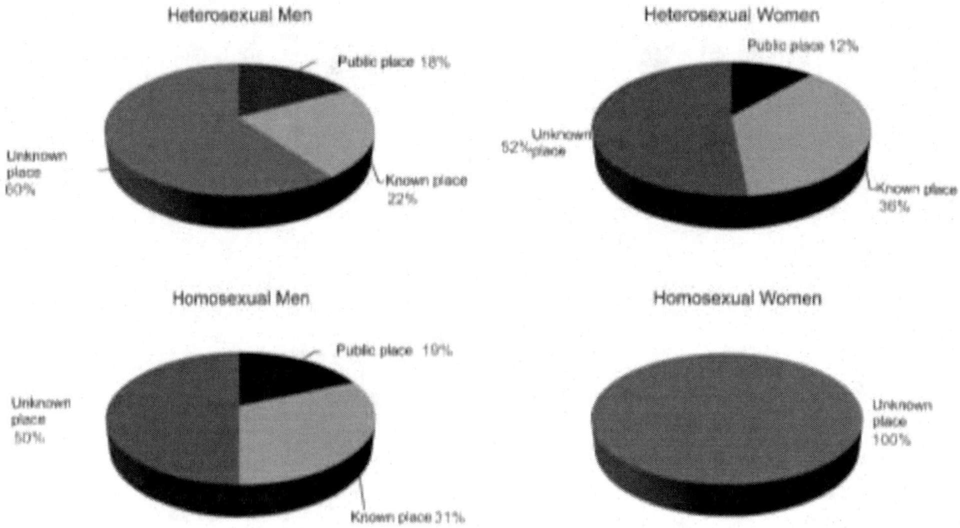

Figure 6. Places where dream sexual encounters occur.

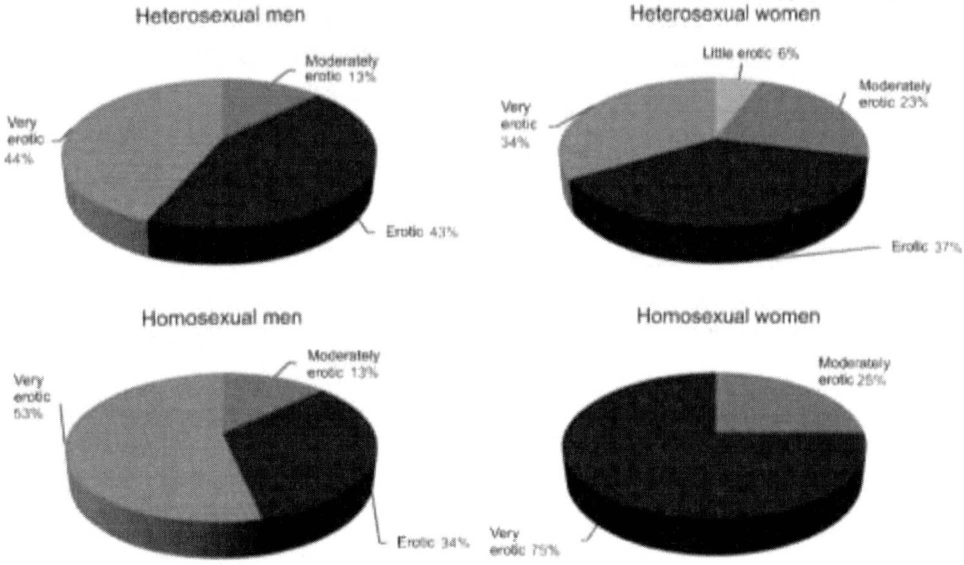

Figure 7. Perceived intensity.

To assess the perceived intensity of the erotic dreams, subjects were asked to answer this question on a scale of one to five: "How erotic was your dream?" where 1 was not erotic and 5 was very erotic. Very erotic was reported in 32.7% (mean), not erotic was not reported and, a little erotic was only reported in heterosexual women. When compareing heterosexual men with heterosexual women we found that men perceive these dreams with greater intensity, reporting very erotic in 44.0% *vs.* 34.0% for women. In the homosexual population, men chose "very erotic" 53.0% of the time, while for homosexual women this proportion rose to 75.0% (Figure 7).

DISCUSSION

The results of our survey indicate differences between the contents of erotic dreams in men and women. The main differences were found in the characters, as women reported dreaming about their current partner more frequently. Results also indicate that, in general, these types of dreams are remembered as pleasant events, full of positive emotional components and with high sensory intensities. These all seem to be normal characteristics of dreams with sexual and/or erotic content. However, they can change dramatically when there is a traumatic event related to sexuality during life, such as rape, negative and/or unpleasant sexual experiences (Duval and Zadra 2010). In these cases, the content of the dream will surely acquire totally negative characteristics, and dreaming about these events could contribute to perpetuating the memory of said event, and could also reinforce it. In addition, the characters will adopt totally negative characteristics

(Duval and Zadra 2010). We make this assertion because, even though only a small number of dreams reported were perceived as unpleasant or as having negative content, some of them included emotions or sensations such as disgust, rejection, guilt, insecurity, anger, annoyance, anguish, fear, sadness, distrust and pain, among others. Some of the characters were "my stepfather's brother," monsters (e.g., alien), "some entity" (e.g., spirits) and animals.

Regarding the perception of love and affection, there's currently not enough evidence to say whether or not they can be experienced during dreams. Our survey showed that the perception of "love" (thought this definition was dependant on the consideration of each subject) was present in some subjects surveyed. Some authors have proposed that some of the characteristics of "theory of mind" may be present during dreams, which would allow the dreamer to recognize emotional and cognitive states, actions and facial expressions in dream characters and in the dream-self during the dream (Kahn and Hobson 2005). However, the identification and description of subtle emotions in dreams is a complex subject.

We can experience different degrees of sensory and emotional excitement during REM sleep, and these are determined by the content of the dream episode. Some vegetative and electrophysiological components registered during this stage are clear manifestations of this excitation. One such example could be nocturnal sexual emissions, which are usually accompanied by erotic dreams. According to some theories, these emissions are the result of the stimulation caused by erotic dreams, though some men report no sexual content during the corresponding dream. In contrast, some women tend to wake up suddenly because of the experience of an orgasm, and more commonly report their corresponding dream experience to have included sexual situations. During sleep, we can experience mental states and situations very similar to those we experience when we are awake (Kahn and Hobson 2005). However, it is important to mention that all dream components, characters, and actions are generated endogenously, without any external stimulus, even though everything we see and everything that appears during our dream episodes is assumed to be real, at least during dream behavior.

From a neurobiological perspective, functional magnetic resonance imaging (fMRI) studies have shown significant activation of the prefrontal cortex (PFC) during dreaming. This region fulfills higher order functions such as planning, decision making, regulation of emotions, and attention, and participates in the elaboration of social judgments that allow us to intuitively differentiate what is right from wrong (Damasio 1994; Ostrosky and Vélez 2008). Such activation, however, is significantly lower during dreaming when compared to the wakeful state. This lower activation could explain our diminished capacity to make social judgements during dreams, which could be reflected in a lower capacity for discrimination between facts and actions that are morally or socially correct from those that are not.

During our conscious states, moral behavior is linked to the interests or welfare of societiy in general, and of people in particular. The word moral refers to the consensus of acceptable customs and behaviors within a social group, or to the inclination of the people in that group to behave in a certain way. It seems possible that the perception and discrimination of morality is diminished during erotic dreams. Thus, any event that takes place during an erotic dream and is perceived as both pleasant and socially unacceptable should not be of any concern, and should not be judged with the same parameters that we would use for an event that took place in waking life, as we are in no way responsible for what happens in our dreams, or even for the 'decisions' that we perceive to have made during them.

The amygdala of the temporal lobe is another structure that has been consistently reported as active during dreams (Corsi-Cabrera et al. 2016), as well as during sexual behaviors. This structure is responsible for the regulation of emotions and instinctive behaviors, and participates in the recognition of facial expressions; controls 'fight or flight' behaviors, discriminates dangerous stimuli and participates in the regulation of sexual behaviors (Armony 2013; Phelps and LeDoux 2005). Its activity is greater if the emotional components of a dream are of high intensity. An intense emotional content, accompanied by a lower activation of anterior brain areas, can give the sensation or perception of events as illogical, bizarre and strange. These strange dreams are the result of a different state of consciousness, reflecting distinct brain activation patterns between the waking state and the REM sleep state.

As mentioned earlier, what happens during dreams is usually perceived as a 'real' experience, even though physiologically, the dream state is linked to hallucinatory processes. Although our erotic dreams may be far from becoming reality, the above depends on the degree of similarity between the dreamed facts (imagined situations) and the facts or events in our daily lives (real situations and events). However, radically different situations may arise from those in everyday life experiences; in this sense, erotic dreams could provide certain behavioral strategies that might favor creativity and freedom in the dreamer. New behaviors and experiences could be facilitated and some psychological aspects associated with insecurity, shyness or fear could be reduced. Results indicate that, in general, these dreams are remembered and experienced as pleasant events, full of positive emotional components and high sensory intensities. The range of emotions that can be experienced proved to be extensive and diverse, and positive emotions were reported in greater numbers. However, most events reported as negative were associated to negative emotions, such as disgust, fear, anger and anguish, among others. The most common type of character in these negative situations turned out to be a stranger, well-known people with whom dreamers had a close link, monsters and animals.

It is possible that men dream more about aggressive interactions with other men because, in our evolutionary past, this was the preferred way to gain access to fertile females. Similarly, women may dream of participating in verbal aggressions more often than men because, in the past, gossip and slander against competitors turned out to be a successful reproduction strategy for them (Zadra 2007; McNamara 2014). On the other hand, dreaming about the current sexual partner can strengthen the emotional bond the next day and could also lead to greater attachment. In contrast, dreaming of someone other than the current partner could be associated with adaptive and evolutionary processes involved in partner selection. Empirically, we postulate that the people present in men's dreams are probably identifiable or recognizable people, and with whom they may have been in direct contact, but with whom there is not a strong affective link. On the contrary, the survey indicates that women dream more often of people with whom they have solid or at least real emotional ties.

In conclusion, erotic dreams can and should be a totally liberating experience. In general, these episodes are lived as pleasant phenomena, full of gratifying sensory experiences. Contrary to what is reported in the Psychoanalytic literature, rather than a repressive experience, these dreams, which occur involuntarily, can foster and stimulate sexual desire; offer opportunities to experience new sexual possibilities, and give the dreamer complete freedom to fulfill wishes that they possibly can't satisfy while awake. In general, we propose that the contents of this type of dream are formed and can come from all of our previous experiences, both distant and recently lived.

References

Abbott, J. Chemical messengers: how hormones help us sleep. The Conversation 2015; September https://theconversation.com/chemical-messengers-how-hormones-help-us-sleep-44983.

Abel, G. G., Murphy, W. D., Becker, J. V., and Bitar, A. 1979. "Women's vaginal responses during REM Sleep." *Journal of Sex & Marital Therapy* 5(1):5-14.doi:10.1080/00926237908403713.

Armony, J. L. 2013. "Current emotion research in behavioral neuroscience: the role(s) of the amygdala." *Emotion Review* 5: 104–115.doi: 10.1177/1754073912457208.

Aserinsky, E., and Kleitman, N. 1953. "Regularly occurring periods of eye motility, and concomitant phenomena, during sleep." *Science* 118: 273-274.doi: 10.1126/science.118.3062.273.

Cartwright, R., Agargun, M, Y., Kirkby, J., and Friedman, J, K. 2006. "Relation of dreams to waking concerns." *Psychiatry Research* 141:261-270. doi:10.1016/j.psychres.2005.05.013.

Corsi-Cabrera, M., Velasco, F., Del Rio-Portilla Y, Armony J, L., Trejo-Martinez, D., Guevara M, A., and Velasco L, A. 2016. "Human amygdala activation during rapid eye movements of rapid eye movement sleep: an intracranial study." *Journal of Sleep Research* 25, 576–582.doi: 10.1111/jsr.12415.

Dahan, L., Astier, B., Vautrelle, N., Urbain, N., et al. 2007. "Prominent burst firing of dopaminergic neurons in the ventral tegmental area during paradoxical sleep." *Neuropsychopharmacology* 32:1232-1241. doi:10.1038/sj.npp.1301251.

Damasio, A, R. 1994 *El error de Descartes* [*Descartes's mistake*]. Chile: Andres Bello.

Datta, S., and Maclean, R, R. 2007. "Neurobiological mechanisms for the regulation of mammalian sleep-wake behavior: reinterpretation of historical evidence and inclusion of contemporary cellular and molecular evidence." *Neuroscience & Biobehavioral Reviews* 31:775-824. doi:10.1016/j.neubiorev.2007.02.004.

Domhoff, G, W., and Fox, K, C.2015. "Dreaming and the default network: A review, synthesis, and counter intuitive research proposal." *Consciousness and Cognition* 33:342-53. doi: 10.1016/j.concog.2015.01.019.

Duval, M., and Zadra, A. 2010. "Frequency and Content of Dreams Associated with Trauma." *Sleep Medicine Clinics* 5 249–260. doi:10.1016/j.jsmc.2010.01.003.

World Medical Association. 1964. Declaration of Helsinki.

Kahn, D., and Hobson, A. 2005. "Theory of Mind in Dreaming: Awareness of Feelings and Thoughts of Others in Dreams." *Dreaming* 15(1):48-57.doi: 10.1037/1053-0797.15.1.48.

Karacan, I. 1966. "Erection cycle during sleep in relation to dream anxiety." *Archives of General Psychaitry* 15, 183-189. doi: 10.1001/archpsyc.1966.01730140071012.

Lena, I., Parrot, S., Deschaux, O., Muffat-Joly, S. et al. 2005. "Variations in extracellular levels of dopamine, noradrenaline, glutamate, and aspartate across the sleep-wake cycle in the medial prefrontal cortex and nucleus accumbens of freely moving rats." *Journal of Neuroscience Research* 81:891-899. doi:10.1002/jnr.20602.

Maggiolini, A., and Codecà, L. 2016. "The typical contents of Freud's and Jung's dreams." *International Journal of Dream Research* 9(1): 1-6.doi:10.11588/ijodr.2016.1.19865.

McNamara, P., Andresen J., Clark, J., Zborowski, M. et al. 2001. "Impact of attachment styles on dream recall and dream content: a test of the attachment hypothesis of REM sleep." *Journal of Sleep Research* 10:117-127.doi:10.1046/j.1365-2869.2001.00244.x.

McNamara, P. *How sex rules our dreams*. Aeon 2014; April 25 http://aeon.co/magazine/altered-states/was-freud-right-about-dreams-all-along/.

Meier B. 1993. *Speech and thinking in dreams*. New York: Harvester, Wheatsheaf.

Nir, Y., and Tononi, G. 2010. "Dreaming and the brain: from phenomenology to neurophysiology." *Trends in Cognitive Sciences* 14:88-100.doi: 10.1016/j.tics.2009.12.001.

Ostrosky, F., and Vélez A, E. 2008. "Neurobiología de la Sensibilidad Moral." ["Neurobiology of Moral Sensitivity"] *Neuropsicología Neuropsiquiatría y Neurociencias* 8:115-126.

Perogamvros, L., and Schwartz, S. 2012. "The roles of the reward system in sleep and dreaming." *Neuroscience & Biobehavioral Reviews* 36:1934-1951.doi: 10.1016/j.neubiorev.2012.05.010.

Phelps, E. A., and LeDoux, J. E. 2005. "Contributions of the amygdala to emotion processing: from animal models to human behavior. *Neuron* 48: 175–187.doi: 10.1016/j.neuron.2005.09.025.

Schredl, M., and Hofmann, F. 2003. "Continuity between waking activities and dream activities." *Consciousness and Cognition* 12(2):298-308. doi:10.1016/s1053-8100(02)00072-7.

Selterman, D., Apetroaia, A., and Waters, E. 2012 "Script-like attachment representations in dreams containing current romantic partners." *Attachment & Human Development* 14:501-515. doi: 10.1080/14616734.2012.706395.

Zadra A. "1093: Sex dreams: what do men and women dream about?" *SLEEP* 2007 30, Abstract Supplement, A376.

Zborowski, M., and McNamara, P. 1998. "Attachment hypothesis of REM sleep: Toward an integration of psychoanalysis, neuroscience, and evolutionary psychology and the implications for psychopathology research." *Psychoanalytic Psychology* 15: 115-3.

In: Sleep, Cognition and Emotion
Editors: I. Ramírez Salado et al.

ISBN: 978-1-53617-561-5

Chapter 10

SLEEP DISTURBANCES IN AUTISTIC SPECTRUM DISORDER (ASD) PATIENTS

***Fructuoso Ayala-Guerrero**, PhD,**
Graciela Mexicano, PhD
and Erik Leonardo Mateos, PhD
Laboratorio de Neurociencias, Facultad de Psicología,
Universidad Nacional Autónoma de México, Ciudad de México, México

ABSTRACT

Sleep is a fundamental process for the whole of living beings, therefore the chronic lack of this process, or its poor quality, increases the risk of physical and mental health disorders. Sleep is not a unitary process, since it is integrated by two different stages: Rapid eye movement (REM) sleep and NREM sleep that are characterized by presenting different behavioral and electrophysiological characteristics. Circadian and homeostatic mechanisms act together to regulate wake-sleep oscillations throughout the 24-hr period.

Both quantity and quality of sleep are critical for brain maturation and synaptic connectivity, therefore sleep disorders may originate detrimental effects on mental abilities Autism spectrum disorder (ASD) is a heterogeneous, behaviorally defined neurodevelopmental syndrome. The neurological and social deficits observed in autistic patients contribute to varying levels of impairment. ASD patients display a high prevalence of sleep problems, such as difficulty falling asleep, fragmented sleep and reduction of nocturnal sleep time. These sleep disturbances contribute to the presence of autistic behavior during the day. Recent advancements in autism research have improved the understanding of this disorder; however, numerous questions about the pathophysiology of ASD remain without precise answers. On the other hand, about a third of people with autism also suffer from epilepsy. The relationship between epilepsy

*Corresponding Author's Email: fayala@unam.mx.

and ASD is poorly understood. It is not known whether epileptic activity in these patients is a secondary phenomenon or whether, in some situations, it is responsible for some of the symptoms. Various educational and behavioral treatments have been implemented to the management of ASD patients arriving to a general agreement that their treatment should be individualized. Pharmacological treatment has been significantly increasing from the past decades. Different types of medical management include numerous categories of drugs.

Society is waiting for new treatments that allow to ASD patients being auto-sufficient to carry out their daily activities and their social incorporation.

Keywords: autistic spectrum disorder, NREM sleep, REM sleep, epilepsy, pharmacotherapy

INTRODUCTION

Sleep is a fundamental process for the whole of living beings. During this process, numerous brain functions are carried out, such as memory consolidation, tissues reparation, synthesis of substances, etc. Both quantity and quality of sleep are essential to survival, since chronic lack of sleep, or its poor quality, increases the risk of health disorders including high blood pressure, cardiovascular disease, diabetes, depression, and obesity.

SLEEP ORGANIZATION

Sleep is not a unitary process, since it is integrated by two different phases: rapid eye movement (REM) sleep and NREM sleep which has three different stages characterized by presenting different behavioral and electrophysiological characteristics (Figure 1). These types of sleep are cyclically organized, and the sleep cycles are repeated through all stages of NREM and REM sleep several times during the night. The stage 1 NREM sleep is a transitional state from passive wakefulness to sleep. During this period, considered as lighter sleep, brain activity slow down and its amplitude increases. Vertex spikes are displayed during this sleep stage while heartbeat, breathing, and eye movements slow down, and the postural muscles slightly relax displaying occasional twitches.

Stage 2 NREM sleep is another period of relatively light sleep before enter in deeper sleep. During this period heartbeat and breathing show an additional slowing, and postural muscles relax even further. The body temperature drops and eye movements stop or are minimal. Brain wave activity shows an additional slowing and sleep spindles and K-complex, characteristics of this sleep stage, are present. Stage 3 NREM sleep is considered as the period of deep sleep. During this stage, brain activity is characterized by presenting high voltage slow waves while heart beat and breathing decrease to their lowest levels across the sleep cycle. Postural muscles become hypotonic. The REM sleep stage is usually present about 90 minutes after falling asleep and it is characterized by presenting rapid eye movements and muscular atonia. The displayed mixed frequency brain wave activity is similar to that in wakefulness. Breathing becomes faster and irregular, and heart rate and blood pressure increase to near waking levels. Most of dreaming occurs during this sleep stage, although some percentage can also occur in NREM sleep. Atonia muscular is present, therefore arm and leg muscles become temporarily paralyzed, which inhibits acting out the dreams.

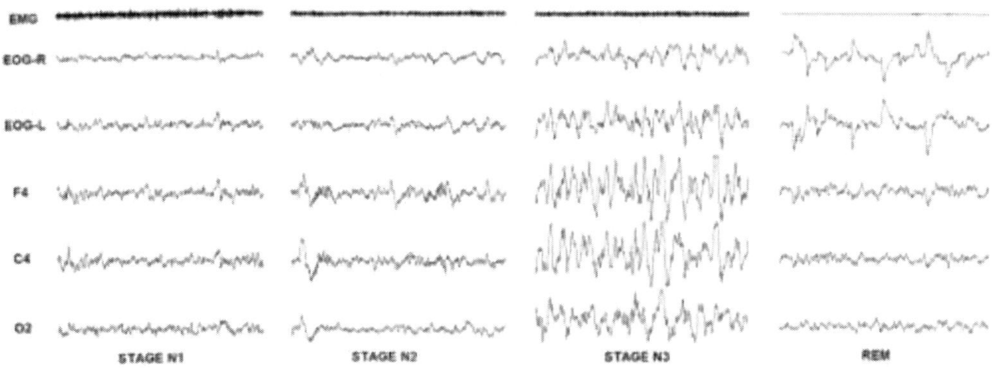

Figure 1. Stages of sleep. EMG, electromyogram; EOG, Electro-oculogram from the right (R) and the left (L) eye; F4, C4 and O2, EEG recordings.

Cerebral Regulation of Sleep

Several structures within the brain are involved with sleep regulation as follows: The hypothalamus, contains groups of neurons that act as control centers regulating the sleep-wake cycle. Among these control centers, the suprachiasmatic nucleus receives information about light exposure directly from the eyes and controls the behavioral rhythm, therefore it is important for the circadian distribution of sleep. Groups of neurons placed at the brain stem level, interact with the hypothalamus to control the transitions wake-sleep and vice versa. Sleep-promoting GABAergic neurons within the hypothalamus and the brain stem act to reduce the activity of arousal centers placed in the hypothalamus and the brain stem. The brain stem, in particular the pons and medulla, also participates in REM sleep regulation; it sends signals to relax postural muscles inducing the muscular atonia characteristic of the REM sleep stage (Figure 2, 3).

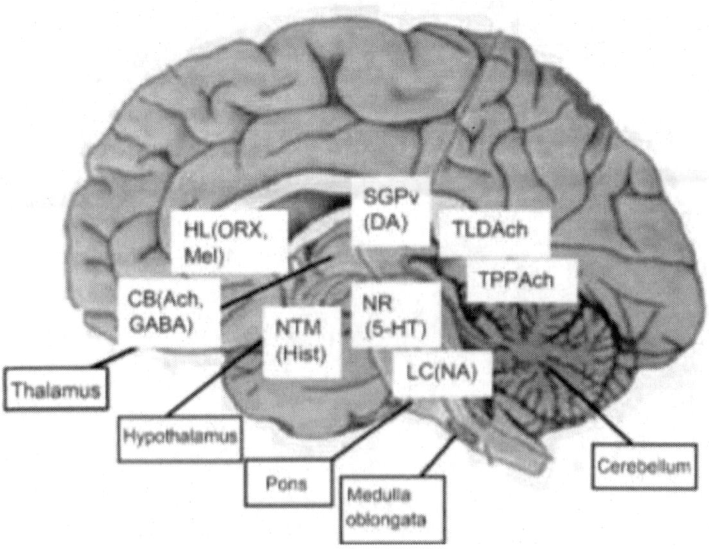

Figure 2. The ascending activating reticular system consists of noradrenergic neurons (NA) of the locus coeruleus (LC), cholinergic neurons (Ach) in the pedunculo-pontine tegmental nucleus (TPP) and in the laterodorsal tegmental nucleus (TLD), serotonergic neurons (5-HT) in the nucleus of the dorsal raphe (NR), dopaminergic neurons (DA) of the ventral periaqueductal gray matter (SGPv) and of histaminergic neurons (Hist) of the tuberomamillary nucleus (NTM). This system produces two cortical alert pathways: the dorsal route through the thalamus and the ventral route through the hypothalamus and basal forebrain (CB). The ventral route receives contribution from the lateral hypothalamus (HL) of orexin (ORX) and melanin (Mel), as well as GABAergic and cholinergic neurons of the basal brain (CB).

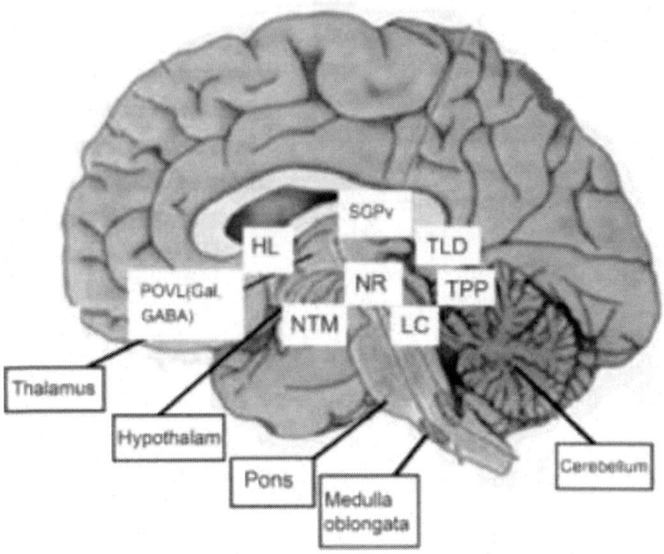

Figure 3. Induction of sleep. Scheme of projection of the ventrolateral preoptic nucleus (POVL) that inhibits the main components of the ascending activating reticular system to induce sleep. GABAergic and galaninergic neurons of the POVL nucleus remain active during sleep.

The thalamus participates as a relay for the entering information from the sensory receptors towards the cerebral cortex. During most stages of sleep, the thalamus becomes quiet, filtering the information coming from the environmental world. But during REM sleep, the thalamus is reactivated, receiving the cortex several sensations that facilitates the presence of dreams. The pineal gland also participates in the sleep regulation, receiving signals from the suprachiasmatic nucleus and increasing the release of the hormone melatonin, that facilitates the sleep onset after the light is off. Blind individuals who are not able to coordinate their natural wake-sleep cycle using natural light can stabilize their sleep patterns by taking melatonin at the same time each day before going to bed. The basal forebrain, also promotes sleep and wakefulness; the release of adenosine (considered as hipnotoxin) from cells in the basal forebrain and probably other brain regions regulates the sleep drive; in contrast, groups of neurons of the midbrain act as an arousal system. On the other hand, the amygdala, becomes activated during REM sleep, being probably related to emotional dreams including the nightmares.

Circadian and Homeostatic Regulation of Sleep

Two internal biological mechanisms act together to regulate wake-sleep oscillations throughout the 24-hr period as follows: Circadian rhythms regulate a wide variety of functions from daily fluctuations in wakefulness to body temperature, metabolism, and the release of different kinds of hormones. These rhythms control the timing of sleep and cause sleepiness at the beginning of the night and tendency to wake in the morning. Activity of the suprachiasmatic nucleus, considered as a biological clock, oscillates on a roughly 24-hour day and it controls most circadian rhythms which are synchronized with environmental cues, particularly with light and temperature. Sleep-wake homeostasis regulates the need for sleep and the homeostatic sleep drive induces the body to sleep after a certain time and also regulates sleep intensity. This sleep drive gets stronger when time on wakefulness increases and causes compensatory increment of sleep after its deprivation. The amount of this increment depends on the deprivation time. Sleep is a very sensitive biological function, therefore there are multiple factors that modify the sleep-wake organization such as health conditions, medications, mood states, sleep environmental stressors, etc. Both quantity and quality of sleep are critical for brain maturation and synaptic connectivity, therefore sleep disturbances affecting both quality and/or quantity characteristics may originate detrimental effects on mental abilities in children, affecting cognition (Gozal 1998), learning, memory (Maquet 2001) and attention, as well as mood regulation (Goodlin-Jones et al. 2008; Mazurek and Sohl 2016). A high prevalence of sleep disturbances in children has been observed at worldwide, affecting in particular to children with neurodevelopmental disorders (Goodlin-Jones et al. 2008; Robinson-Shelton and Malow 2016). Both quantity and quality of sleep are critical for brain maturation and synaptic connectivity, therefore sleep disturbances affecting both quality and/or quantity characteristics may originate detrimental effects on mental abilities in children, affecting cognition (Mazurek and Sohl 2016). Autism spectrum disorder (ASD) is a heterogeneous, behaviorally defined neurodevelopmental syndrome that occurs in 1 of 150 children (Amaral, Schumann, and Nordahl 2008). Individuals suffering from autism have deficits in social interaction and verbal and nonverbal communication, as well as in restricted or stereotyped patterns of behavior (APA, DSM-IV 2000). These patients might also have co-morbid disorders such as intellectual impairment, epilepsy and anxiety. Epilepsy is associated with pathology of multiple brain regions where the cerebral cortex, amygdale, cerebellum and hippocampal formation are included. These brain regions have also been implicated in autism (Amaral, Schumann, and Nordahl 2008). Postmortem and magnetic resonance imaging studies have highlighted several brain regions including the frontal lobes, amygdale and cerebellum as pathological in autism. However, there is no unambiguous and consistent pathology that supports with precision this neurodevelopmental disorder. Additionally, recent studies emphasize that disturbances in the time course of brain development is fundamental in the ASD (Wegiel et al. 2010). It has been suggested that the heterogeneity of both the core and co-morbid features is related to the heterogeneous pattern of neuropathology observed in ASD patients. In this context, defined phenotypes in larger samples of children and well-characterized brain damage are indispensable for clarification of the neuroanatomical alterations involved at different impairment levels of the ASD. The neurological and social deficits observed in autistic patients contribute to varying levels of impairment. Recent advancements in autism research have improved the understanding of the disorder; however numerous questions about the pathophysiology of ASD remain without precise answers. The expression of ASD is extremely heterogeneous, owing to the complex interactions between genes, the central nervous system, and behavior throughout development. Therefore, some researchers refer to 'the autisms' or ASD rather than a single autism phenotype (Geschwind and Levitt 2007). Within the ASD patients, variability in the social domain varies from a near absence of interest in interacting with others, to more slight difficulties managing complex social interactions, where an understanding of other people's goals and intentions and other cues of social context are implicated. Likewise, stereotyped and repetitive behaviors vary from simple motor stereotypes, to much more sophisticated rituals, accompanied by emotional disruption when these rituals are interrupted. Some patients with ASD lack of basic speech abilities, while others can have language deficits that are only limited to pragmatic language. While a majority of individuals with ASD exhibit some level of intellectual impairment, intelligence quotients vary from the severe and profoundly impaired to above average.

SLEEP AND EPILEPSY IN ASD PATIENTS

On the other hand, sleep disturbances have commonly been observed in children with intellectual disabilities (Espie and Tweedie 1991; Quine 1991; Stores 1992). Numerous studies have evidenced that sleep problems in mental retarded children, are frequently related to the specific type of syndrome or to its etiological characteristics (Richdale 1999). In pediatric patients with ASD, findings from questionnaire studies have revealed the existence of several sleep problems as observed in other groups of pediatric patients, who have intellectual disabilities, (Schreck and Mulick 2000; Wiggs and Stores 2004). Parents of ASD children report a high prevalence of sleep problems, such as difficulty falling asleep, fragmented sleep by frequent awakenings and reduced total nocturnal sleep time (Gail, Sears, and Allard 2004). These sleep disturbances have been considered to have a predictive role in the presence of autistic behavior during the day (Schreck, Mulick, and Smith 2004). While parental sleep concerns have been well documented in children with ASD by means of sleep questionnaires, few studies have related these concerns to polysomnographic (PSG) recordings, technique that provides objective evidence of sleep disturbances in these patients. Although sleep disturbances are not part of the diagnostic criteria for ASD, these disturbances are commonly present in autistic children (Hering et al. 1999; Richdale 1999; Schreck et al. 2004; Malow et al. 2006). Additional research is needed to clarify the amount and quality of the sleep problems experienced by these patients compared to healthy children (Durand 2002). It has been described a variety of several daytime behaviors, related to sleep disturbances for autistic patients. Some authors have suggested that the intensity of developmental disturbances (Glascoe, MacLean, and Stone 1991) and general ASD symptoms (Malow and McGrew 2006) are increased in relation to sleeping problems. Studies pointing out the relationships among some of the key diagnostic factors in ASD (e.g., communication difficulties, social problems, stereotypic behavior and difficulties breaking routines) have indicated that intensified symptoms in these areas are related to disturbed sleep (Schreck, Mulick, and Smith 2004). Other general daytime behavior troubles have been reported as related to sleep problems for people with an ASD, such as overactivity, psychopathology, and disruptive behavior (O´Brien 2009). On the other hand, about a third of people with autism also suffer from epilepsy. Therefore long-duration electroencephalograms that include an adequate amount of delta sleep should be carried out in order to detect epileptiform activity. (Tuchman and Rapin 2002) described that the incidence of epilepsy in children with autism ranged from 5% to 40% while (Hughes and Melyn 2005) reported that the electroencephalograms of 75% of the children with autism showed abnormal patterns of electrical activity in the brain and 46% had seizures (Figure 4).

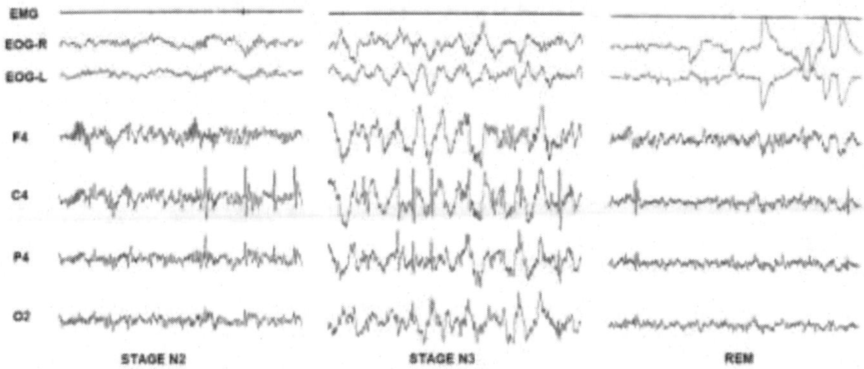

Figure 4. EEG spikes of a child with ASD during the sleep stages.

The relationship between epilepsy and ASD is poorly understood. It is not known whether epileptic activity in these patients is a secondary phenomenon or whether, in some situations, it is responsible for some symptoms that occur in people with ASD. This is, in part, because epileptic syndromes are often associated with cognitive, language and behavioral dysfunction, all of which are traits of this neurodevelopmental disorder. On the other hand, it has been hypothesized that both epilepsy (Tuchman, Moshé and Rapin 2009; Tuchman, Alessandri, and Cuccaro 2010) and autism may result from severe alterations in cortical-subcortical systems connectivity. Alterations in neocortical minicolumns observed in autism support the idea of the presence of cerebral underconnectivity (Tuchman and Cuccaro 2011). According to Tuchman, Moshé and Rapin 2009, alterations in subcortical systems such as basal ganglia-substantia nigra connectivity that may decrease the seizure threshold, contribute to cognitive impairments and to the motor stereotypes commonly found in autism. However, there are also hormonal influences on cortical and subcortical systems connectivity in the developing brain, such as testosterone that have been implicated both in autism and epilepsy (Velisková et al. 2004; Knickmeyer et al. 2005). These findings have shifted the center of attention from only anatomical alterations to morphological and functional disturbances within a developmental process. Findings from different authors suggest that molecular abnormalities of neurons as well as network abnormalities would be responsible of the epileptogenesis and the cognitive and socio-communicative disturbances usually present in autistic patients (Rubenstein and Merzenich 2003). Particularly, alteration in cortical minicolumns with selective sparsely of gabaergic interneurons is likely responsible for the increased seizure susceptibility (Casanova et al. 2002; Casanova, Buxhoeveden, and Brown 2002).

It has been described that epileptic seizures are present in up to one-third of ASD individuals (McDermott et al. 2005). On the same way, autistic characteristics are commonly observed in severe forms of epilepsy associated with mental retardation (Dravet 2002). It has been suggested that the association between epileptic and autistic phenotypes is because of these diseases may share common predisposing genes. Some of the known epilepsy-predisposing genes are related to voltage-gated or ligand-gated ion channels (Noebels 2003). Defects in synaptic proteins implicated in neurotransmitter release and synaptic vesicle flow are also involved in this mechanism (Rosahl et al. 1995; Janz et al. 1999). These findings provide the knowledge about the interaction between epilepsy and autism, and to have a better understanding of the pathophysiology of these diseases that alter brain development. They will also probably contribute to the development of new treatment strategies. The strong association between ASDs and epilepsy, suggests that common mechanisms of synaptic dysfunction may underlie both diseases. In the healthy brain, a balance of excitation and inhibition is essential for the all cerebral functions, from physiological network activity to cognitive processes. Taking into consideration these findings, monitoring children with ASDs for seizure and subclinical seizure activity is recommended. ASD is considered as a multi-genetic disorder showing a high probability to being heritable, particularly to males (Parikshak et al. 2013; Werling, Parikshak, and Geschwind 2016; Yang et al. 2016). Considerable heterogeneity has been described related to the degree of mental abilities, social and communication impairments. According to Christensen et al. (2016) cognitive abilities varies with 31% of individuals with ASD having IQ scores in the range of intellectual disability (IQ $\leq$ 70), 23% in the borderline range (IQ = 71–85), and 46% in the average or above mean range of intellectual ability (IQ > 85) (Christensen et al. 2016). Advances in the comprehension of the etiology of autism over the last years has been significant, considering now to autism as a spectrum disorder displaying several phenotypes. About 3 decades ago, specialists reached consensus about the essential deficits to its definition and reached an agreement between the two principal diagnostic systems, DSM-IV and the 10th edition of the World Health Organization (WHO) International Classification of Diseases. The DSM-V consider now two conjuncts of characteristics: (A) persistent deficits in social communication and social interactions across contexts and (B) restricted, repetitive patterns of behavior and unusual responses to sensory stimuli (American Psychiatric Association (APA).

Insomnia in children with ASD occurs across all cognitive levels, and its prevalence is located about 60 and 86%, that corresponds near to three times greater than typically developing children (Schreck, Mulick, and Smith 2004).; Wiggs and Stores 2004; Richdale and Schreck 2009; Owens and Mindell 2011).

Sleep Disorders

Insomnia in ASD patients is one of the principal problems observed by parents and one of the most important motives to seek medical support (Malow et al. 2009). Moreover, insomnia and sleep interruptions (Figure 5) in children with ASD has been described to alter parents' sleep and originate familiar stress deteriorating the quality of life (Meltzer 2008).

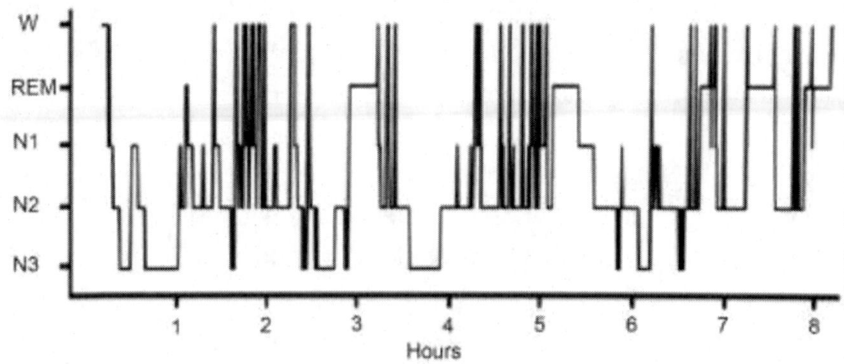

Figure 5. Hypnogram of an ASD patient. Sleep fragmentation is observed.

Recent meta-analysis carried out on sleep data in children with ASD showed some differences. Using actigraphic and polysomnographic techniques in ASD patients, it was observed sleep time decrement, longer sleep latency, and sleep efficiency decrement in comparison to control children (Elrod and Hood 2015). These differences are sometimes not observed by other researchers, due in part to methodological discrepancies such as age of patients, level of the autistic impairment, number of recording nights, etc. Sleep disturbances in ASD patients are originated by multiple causes including neurobiological, neurochemical, medical and behavioral factors, (Bourgeron 2007; Malow et al. 2014). Insomnia caused by behavioral factors is among the most frequent types of sleep disorders in children, particularly in ASD patients (Meltzer et al. 2010; Malow et al. 2014). Both external and internal stressing stimuli induce sleep disturbances in ASD patients (Bourgeron 2007, Tordjman et al. 2013). Among the factors related to intrinsic causes of insomnia in ASD patients are included (1) troubles in brain maturation, (2) sensory dysregulation, (3) abnormal melatonin release, and (4) alteration in circadian genes.

EEG AND BRAIN MATURATION

The explanation regarding an intrinsic cause of insomnia involves possible differences in brain wave organization and neurological maturation that may be detected by means polysomnography, technique considered as the best tool for objective measurement of sleep. It includes simultaneous recording of electroencephalogram (EEG), electro-oculogram (EOG), electromyogram (EMG), electrocardiogram (EKG), airflow, and oxygen saturation (AARC-APT 1995). Brain waves with a frequency of 12–15 Hz during REM sleep have been recorded both in premature and children ages between 3 and 8 months. Similarly, in a study carried out in 1969 in ASD patients ages between19 and 47 months it was observed brain wave frequencies that oscillated between 10.5 and 15 Hz during rapid eye movement (REM) sleep as compared to an almost complete absence in same age control children. These findings suggest the presence of maturational brain differences in children with ASD (Ornitz et al. 1969). Moreover, Tanguay et al. (1976) observed that eye movements do not become organized into bursts before 40-weeks of gestation (Tanguay et al. 1976). They found that the mean eye movement burst length and the ratio of eye movements within bursts to eye movements outside of bursts was significantly lesser in children with ASD than age-matched controls. These data stand out the immaturity of eye movement organization in ASD patients (Tanguay et al. 1976). Furthermore, spindle EEG activity was observed during REM sleep in ASD patients but not in their respective control, showing that sleep stages were not well differentiated (Tanguay 1976). Additionally, other studies reported a high proportion of undifferentiated sleep, excessive muscle twitching during REM sleep, high rates of periodic limb movements, and REM sleep behavior disorder when compared to control children (Diomedi et al. 1999; Thirumali et al. 2002).

Children with ASD were compared with their respective control by means two consecutive nights of sleep recordings and a sleep questionnaire (Malow et al. 2009). Patients were separated in two groups as "good sleepers" and "poor sleepers," according to their parent reports. There were no significant differences between the ASD good sleepers and controls. In contrast, poor sleepers with ASD showed prolonged sleep latency and decreased sleep efficiency on the first night of PSG and increased amounts of insomnia according to the sleep questionnaire. However, there were no significant differences between the 2nd night of sleep recording for the poor ASD sleepers and control children. Moreover, when comparing children with ASD, mean age of 4.8 years with control children (15 normal children and 13 children with developmental delay) (Buckley et al. 2010), the findings showed lower amount of total sleep time, greater SWS percentage, and significant decrement in REM sleep percentage. It is concluded that a deficiency of REM sleep may indicate an abnormality in brain maturation as well as in general neural organization in young children with ASD. Additionally, higher proportions of parasomnias have been described in children with ASD in comparison with the typically developed. All these findings suggest the participation of complex intrinsic neurobiological and neurochemical mechanisms alter the amount and architecture of sleep in children with ASD. A considerable amount of polysomnographic studies carried out in children with ASD have mainly focused on abnormalities related to REM sleep (Tanguay et al. 1976; Thirumalai et al. 2002). In this context, it has been diagnosed REM sleep behavior disorder without muscle atonia, in autistic children who had symptoms of disrupted sleep and nocturnal awakenings (Thirumalai et al. 2002). Although it was noted a decreased in the amount of REM sleep, there were not observed abnormalities in muscle tone during this sleep phase. According to numerous reports, these findings in the autistic children may reflect that they were not under medication. The absence of certain drugs, usually administered to patients with ASD, such as serotonin reuptake inhibitors may contribute to the lack of muscle atonia that normally is installed during REM sleep. As in other reports, we have observed reduced sleep time; sleep fragmentation provoked by intermittent awakenings, and reduced sleep efficiency (Ayala-Guerrero et al. 2014). However, our study evidenced a relative absence of other sleep disorders, such as sleep apnea and parasomnias, described by different authors. Disrupted sleep continuity has been reported, with total sleep time correlating negatively with social and communication abilities. Thus, it is probable that implementation of techniques for sleep continuity may improve these abilities. Similarly, ASD patients, compared to age and sex-matched controls, have showed prolonged sleep latency, increased duration of N1 and N2 stage sleep and decreased delta sleep.

Genetic Participation on Sleep Regulation

There are other explanations related to the intrinsic factors that induce insomnia in children with ASD which describe circadian gene anomalies that produce biological rhythm disturbances (Wimpory et al. 2002; Botbol et al. 2013; Tordjman et al. 2015). As described in humans and other animals, the sleep- wakefulness cycle is regulated by the suprachiasmatic nucleus that behaves as an internal circadian clock (Espie and Tweedie 1991) which regulates a circadian rhythm system that modifies clock gene expression (Espie and Tweedie 1991). Clock genes PERIOD (PER 1, 2, 3 basic framework), TIMELESS (resets clock), NPAS2, and Cryptochromes (CRY1, 2) interact and are activated by clock proteins CLOCK, ARNTL/BMAL1 (Travnickova-Bendova et al. 2002; Sasaki et al. 2009). Clock/clock-related gene anomalies have been considered as factors that participate in the etiology of principal deficits of ASD involving temporary synchrony troubles (Wimpory et al. 2002). Findings from genetic linkage studies suggest the chromosome 2q as a possible site, and region 2q37.3 has been considered to have the hPER2 gene (Wimpory et al. 2002). Additionally, hPER2 was identified as one of the possible genes for 2q37 involved in the deletion syndrome in humans that originates the phenotype of ASD, as well as the sleep problems (Le Roy et al. 2013). It has been suggested that the clock genes hPER1, hPER2, hPER3, and clock gene-gene interactions may carry out a complex role in relation to sleep and the control of oscillators associated with human social communication concomitantly with brain development (American Psychiatric Association 2013). It has been postulated (Veatch et al. 2017) that genes involved in synaptic homeostasis, including the processes of synaptic development and pruning, may be involved in both ASD and sleep-wake control, sharing regulating mechanisms (Veatch et al. 2017). It is probable that a greater understanding of the roles of genes influencing both sleep and neurological disorders could lead to new treatment strategies for ASD patients (Veatch et al. 2017).

PARTICIPATION OF MELATONIN IN SLEEP INDUCTION

It has been described that low melatonin levels are related to insomnia in ASD patients. Melatonin released from the epiphysis cerebri or pineal gland, exerts its effect on the suprachiasmatic nucleus through the receptors MTNR1A and MTNR1B, which participates in multiple functions such as sleep triggering, circadian and seasonal rhythm regulation, as well as in immunological functions (Espie and Tweedie 1991). Thus, a delayed melatonin oscillating rhythm may be related to an elongated sleep latency, while low melatonin amount has been related to insomnia in ASD patients (Veliskova et al. 2004) as described in numerous scientific publications. (Ritvo et al. 1993; Nir et al. 1995; Kulman et al. 2000; Veliskova et al. 2004; Melke et al. 2008; Owens and Mindell 2011). It has been used short-term administration of exogenous melatonin (dose 0.5 to15 mg or more frequently 3 to 5 mg), 30 min before the desired sleep time onset, to improve sleep in ASD patients (Jan, Espezel, and Appleton 1994; Ishizaki, Sugama, and Takeuchi 1999; Paavonen et al. 2003; Andersen et al. 2008; Galli-Carminati, Deriaz, and Bertschy 2009; Richdale and Schreck 2009; De Leersnyder, Zisapel, and Laudon 2011). However, some of these patients do not respond to melatonin and it has also been reported that the effect as sleep inductor decreases over time (Nir et al. 1995; Hughes and Melyn 2005). Additionally, Goldman et al. (2014) failed to find any differences in endogenous melatonin samples comparing insomniac children with ASD and normal control in maximal melatonin concentration (C(max)) and time to peak concentration (T(max)).

ENDOGENIC ACTIVATION OF INSOMNIA

It has also been described that insomnia in ASD patients results from an internalization of emotional brain activation (Hall et al. 1997; Feige et al. 2013), and recent research has focused the investigation on cognitive and physiological activation. The hypothesis of cognitive activation considers that an increment of cognitive activity while trying to fall asleep avoids the initiation of the sleep (Feige et al. 2013). While the hypothesis of physiological activation considers that there is greater activation of the sympathetic nervous system in patients with insomnia compared to good sleepers (Healy et al. 1981). In this context, increased heart rate, body temperature, heart rate variability, and higher levels of nor-adrenaline have been correlated with insomnia (Healy et al. 1981; Morin, Rodrigue and Ivers 2003).

Sympathetic Activation and Insomnia in ASD Patients

It has been observed that ASD patients have difficulties with neurophysiological functions related to attention and arousal. Additionally, there is increasing information supporting the idea that ASD is associated with arousal and autonomic nervous system dysregulation (Hutt et al. 1964). In studies that have evaluated sensory arousal and sympathetic tone using electrodermal activity in ASD patients compared to children controls, the results have been inconsistent and sometimes contradictories. However, Ming et al. (2005) investigated autonomic dysfunction using a neuroscope to analyze the brainstem activity and quantifier cardiac vagal tone, blood pressure, and heart rate evidenced that all the ASD patients displayed elevated sympathetic tone compared to control children (Ming et al. 2005).

Insomnia Induced by Excessive Sensorial Activation

Sensory hyper-reactivity may also contribute to insomnia in ASD patients, as observed in those patients aged 6–12 years that had a greater prevalence of atypical sensory behaviors and sleep disturbances (Reynolds, Lane, and Thacker 2012) than healthy children (McIntosh, Miller, and Shyu 1999; (Reynolds, Lane, and Thacker 2012). These findings suggest that children with ASD may present over-reactivity to stimuli resulting in difficulty falling or staying asleep. Similarly, Mazurek and Petroski (2015) observed that greater levels of sensory over-responsivity, measured by the sensory profile, were directly correlated with more sleep problems measured by the Children's Sleep Habits Questionnaire CSHQ (Mazurek and Petroski 2015). Findings suggest that sleep difficulties are related to problems in specific sensory areas since children with a high sensory threshold who require high levels of sensory input sleep better. There is evidence that insomnia in ASD patients may be at least partially related to poor sensory processing, and thus, assessment of sensory reactivity may be useful when addressing insomnia. Strategies to manage sensory hyper-reactivity such as decreasing environmental stimuli in the bedroom and adding calming strategies prior to bedtime may be appropriate to get a good sleep.

PRESENCE OF PSYCHIATRIC DISORDERS

Recently, there has been evidenced psychiatric disorders in ASD patients. A high prevalence of anxiety and Attention Disperse Hyperactivity Disorders, which are considered to induce high arousal disorders, have been identified in numerous studies carried out in ASD patients (Kerns et al. 2014; Kerns et al. 2015; White et al. 2015). Additionally, in a recent study, electrodermal activity was measured both in control and ASD patients. Participants carried out tasks that may originate anxiety, high attention levels, response inhibition, and social cognition processes. The ASD patients were divided into low- and high-anxiety participants, and the groups were compared analyzing the mean of the electrodermal activity level. The high-anxiety ASD patients presented significantly decreased the mean of the electrodermal activity compared to both the low-anxiety ASD and the control groups. The results showed atypical autonomic function in children with ASD, specifically with respect to sympathetic activity. The results confirm autonomic dysfunction in ASD patients and highlight the role of anxiety and autonomic features that contribute to understand the variability in the autism spectrum (Panju et al. 2015).

COGNITIVE ABILITIES AND SLEEP DISORDERS RELATED TO THE NORADRENERGIC SYSTEM

It has been suggested that the locus coeruleus-noradrenaline system could participate for both brain activation and sensory dysregulation and also for cognitive deficits observed in ASD patients. Numerous studies have evidenced a dominant role of noradrenaline in regulating the working memory and attentional processes where the prefrontal cortex participates (Brennan and Arnsten 2008). Nucleus of locus coeruleus is the origin of a divergent projection system that provides the major part of noradrenaline to the brain and is also the only source of noradrenaline in the brain cortex. This system is activated by diverse sensory and autonomic stimuli and whose activation is related to forebrain awakening (Aston–Jones and Bloom 1981). Numerous Electrophysiological studies carried out in rodents and non-human primates show that there is a particular relationship between the locus coeruleus, neuronal discharge, and prefrontal cortex cognitive abilities. This relationship allows a better understanding of the arousal dysregulation that originates the excessive and /or diminished arousal symptoms displayed by ASD patients. According to the characteristics of the noradrenaline system, ASD symptoms related to brain activation and abnormal sensory responses could be originated by locus coeruleus-noradrenaline system dysfunction. Moreover, the wake and arousal-promoting actions of noradrenergic neurotransmission could have clinical consequences under certain circumstances associated with the disorganization of sleep-waking cycle. Thus, inadequate excitatory drive on this system, which can arise from upstream cerebral areas, being included the prefrontal cortex and amygdale, could contribute to a failure of the sleep regulating mechanisms giving origin to insomnia (Berridge, Schmeichel, and España 2012).

Causes of Insomnia in ASD Patients

Behavioral factors of childhood are considered as the most common causes of sleep deficits in children both with and without ASD (Casanova et al. 2002; Casanova, Buxhoeveden, and Brown 2002; Rubenstein and Merzenich 2003). There exist insomnia models that describe predisposing, precipitating, and perpetuating factors related to the development and maintenance of insomnia (Spielman, Caruso, and Glovinsky 1987). Among the Predisposing factors it may be included genetic, physiological, and psychological causes (Spielman, Caruso, and Glovinsky 1987). While precipitating factors include physiological, environmental, or psychological stressors (Spielman, Caruso, and Glovinsky 1987) and perpetuating factors include behavioral, psychological, environmental, and physiological causes that avoid the re-establishing of normal sleep (Spielman, Caruso, and Glovinsky 1987). Precipitating factors of insomnia in ASD patients including (1) environmental and, (2) psychological stressors, as well as (3) physiological stressors. Frequently, when treating the child's insomnia, parents may paradoxically facilitate the presence of perpetuating factors as per example allowing the child to have diurnal naps.

Treatments

Various educational and behavioral treatments have been implemented to the management of ASD patients. There is a general agreement that the treatment for ASD should be individualized. Treatment for deactivating symptoms such as aggression, agitation, hyperactivity, inattention, irritability, repetitive behavior etc. the educational and behavioral interventions is a treatment method usually utilized giving positive results (Spielman, Caruso, and Glovinsky 1987). Numerous treatments have been proposed for Autism Spectrum Disorders with the most effective being combined treatments involving specialized educational programming, communication training, social skills support, and behavioral intervention. Occupational and physical therapy also may me be useful for improving comorbid difficulties of motor coordination and sensory deficits. Behavior modification has the most empirical support for a single treatment, with confirmed improvements in language, social, play, and academic skills, as well as reduction in severe behavioral problems. Pharmacological treatment has been significantly increasing from the past decades. Different types of medical management include numerous categories of drugs such as antidepressants, mood stabilizers, antiepileptic, antipsychotics, etc. (Posey and McDougle 2001; Adler et al. 2015). However, any of the tested drugs have showed to be effective in improving the social communication (Farmer, Thurm, and Grant 2013). Antidepressants have been the most commonly used drugs as well as stimulants and antipsychotics. The high prevalence of comorbidities exhibit by ASD patients is treated by means of psychotropic drugs while the antipsychotic agents have been effective in treating the repetitive behaviors only in young children with ASD, because there is not enough proofs about the efficacy and security in adolescent and adult patients (Kerbeshian, Burd, and Avery 2001; McPheeters et al. 2011). Risperidone (Risperdal) and aripiprazole (Abilify) are the only Food and Drug Administration (FDA)-approved drugs for ASD treatment, particularly for the irritability in 5–16-year-old ASD patients. No medications are currently established to treat ASD core symptoms. However, other not regulated medications are often prescribed for co-occurring behavior disorders such as sleep problems, inattention, impulsivity, hyperactivity, repetitive behaviors, anxiety, mood, agitation, aggression, and disruptive and self-injurious behaviors but may have significant side effects (Richards et al. 2012). Response rates to drugs used to treatment of comorbid problems in children with ASD are lower than for children without ASD. Complementary and alternative medicines are also commonly reported, but their effectiveness remains uncertain. Currently there are alternative options for pharmacological treatment using opiate antagonist, hormones, vitamins, etc. (Kerbeshian et al. 2001). However, the autistic symptoms do not can be controlled with pharmacologic therapy in some patients. Other alternative therapeutic treatments should be implemented in the future. Melatonin is one of most common substances recommended for treatment of sleep disorders in ASD patients (Owens, Rosen and Mindell 2003). Melatonin is a hormone synthesized in the pineal gland or epiphysis cerebri, that participates in the regulation of circadian and body temperature rhythms (Pandi-Perumal et al. 2007) and its- levels decline with advancing age therefore sleep disturbances are present in aged people. Its release is inhibited in presence of light and is regulated by a complex pathway through the suprachiasmatic nucleus. Endogenous melatonin has sedative–hypnotic effects (Owens, Rosen and Mindell 2003). It has also been described that pharmacokinetic profiles of endogenous and supplemental melatonin in children with ASD are normal in comparison to healthy children (Goldman et al. 2014). Several studies carried out in patients with ASD, have showed the efficacy of melatonin. Melatonin appears to be effective in reducing time the sleep latency onset in adults for initial insomnia. This effect appears to last from days to weeks but not long-term. Thus, melatonin is not recommended for long-term use. It is also efficient for improvements in behavior as well as a reduction in stereotypical and compulsive phenomena (Malow et al. 2012). It increases the total sleep duration and reduced sleep onset time. Most studies published to the present have not reported any serious safety concerns (Giannotti et al. 2006; Tuchman, Alessandri, and Cuccaro 2010). Generally reported adverse effects include diurnal sleepiness, headache, dizziness, and hypothermia. It has also been reported suppression of the hypothalamic pituitary axis with long term use (Luboshitzky et al. 1995). Melatonin receptor agonists act selectively at MT1 and MT2 melatonin receptors and have demonstrated usefulness in children with autism spectrum disorder (Stigler, Posey,. and McDougle 2006). It has been showed that Antipsychotics are efficient for treatment of behavioral disorders, in patients with ASD. These drugs are usually prescribed for mood and behavioral comorbidities but secondarily improve sleep disturbances. Some Typical antipsychotics, such as haloperidol, fluphenazine, and thioridazine, are associated with higher incidence of daytime somnolence. Newer, second-generation antipsychotics, as olanzapine, risperidone, and quetiapine, have a lower propensity for extrapyramidal side effects and have less sedating effects. Slow wave sleep is increased by olanzapine, ziprasidone, and Risperidone, whereas rapid eye movement (REM) sleep is suppressed by ziprasidone and Risperidone (Krystal, Goforth, and Roth 2008). Of the atypical antipsychotic drugs, risperidone and olanzapine have been prescribed for sleep disturbances in children (Meltzer et al. 2007).

Antidepressants

There are scarce data regarding use and efficacy of sedating antidepressant drugs, selective serotonin reuptake inhibitors and tricyclic antidepressants, for treatment of sleep disturbances in children with autism spectrum disorder, in particular if insomnia is associated with comorbid psychiatric disorders. The antidepressants promote sleep by antagonizing the effect of wake-promoting neurotransmitters, such as dopamine, histamine, noradrenaline, serotonin and acetylcholine. However, most of these antidepressants suppress REM sleep and induce lasting daytime sedation.

Anticonvulsants

There are scarce data in relation to the utilization of anticonvulsant drugs in the treatment of insomnia in ASD patients. Most of the obtained data have reported improvement in some behavioral topics such as irritability and aggression, but there has also been observed adverse effects in these studies ranging from insomnia to sedation which is usually dose- related (Hellings et al. 2005). The antiepileptic drugs that have usually been used include, lamotrigine, gabapentin, valproic acid and carbamazepine.

Gabapentin, in particular, has also demonstrated efficacy in adults with restless legs syndrome (Aurora et al. 2012). On the other hand, several drugs used usually for Alzheimer treatment have been also applied for treating autism. As per example donepezil, which is effective in improving behavioral and attention disturbances in ASD patients. Moreover, donepezil induces increments in the REM sleep amount both in healthy and demented adult patients (Meltzer et al. 2010; dos Santos Moraes et al. 2006). These findings are important because REM sleep is fundamental for brain plasticity. In children with autism, REM sleep abnormalities, such as immature organization, decreased quantity, and abnormal twitches, have been described (Thirumalai, Shubin, and Robinson 2002). It has been demonstrated an increase-related dose in REM sleep percentage and decrease in REM latency with use of donepezil (Bucley et al. 2011). Antihistamines drugs are one of the most usually reported non-prescript substances for treating sleep disorders (Owens, Rosen and Mindell 2003). Despite widespread use of antihistamines, the clinical trials in patients with autism spectrum disorder have been limited. Sedative and Hypnotics Benzodiazepines (BZDs) are frequently prescribed in adults with insomnia. However, they are prescribed less frequently in the pediatric population because of their side-effects profile includes sedation, headaches, dizziness, cognitive impairment, insomnia rebound, and physical and behavioral dependence. One of the few benzodiazepines studied in sleep disorders in children with Autism is clonazepam which has showed to be effective in treating parasomnias, fragmented sleep, periodic limb movement disorder, and REM sleep behavior disorder in children with neurodevelopmental disorders (Hollway and Aman 2011). Its half-life ranges between 18 to 39 h, with time to peak level of 1–2 h (Griffin et al. 2013). The nonbenzodiazepines, zaleplon, zolpidem, and eszopiclone, have similar pharmacologic effects to BZDs, but are not chemically related to. These drugs have a relatively short life. In contrast to BZD, they do not cause significant residual daytime sedation, cognitive, or memory impairment (Olson 2008). In addition, they do not usually cause insomnia rebound with abrupt discontinuation, which is one of the worsening adverse effects of BZD. Zaleplon and zolpidem have been used in children while information related to use of eszopiclone is scarce (Pelayo and Yuen 2012). Significantly low serum ferritin levels are associated with sleep disorders, such as low sleep efficiency (7 ng/mL) (Youssef et al. 2013), periodic limb movements of sleep (27 ng/mL), sleep fragmentation (24 ng/mL). Iron plays an important role in the dopamine production. In patients with periodic limb movements of sleep, low cerebrospinal iron levels and low iron in the substantia nigra has been detected. Iron supplementation is effective in the treatment of low ferritin levels in patients with sleep disorders as showed in children with autism (Dosman et al. 2007).

Parents of ASD patients and themselves, are waiting for new treatments that allow them to carry out their dairy activities and social incorporation on an independent way, improving their life quality.

Acknowledgments

This work was supported by grant IG200619 from DGAPA-PAPIIT, UNAM, Mexico.

References

AARC-APT (American Association of Respiratory Care-Association of Polysomnography Technologists) clinical practice guideline. (1995). *Respiratory Care.* 40(12):1336-43.

Adler, B. A., Wink, L. K., Early, M., Shaffer, R., Minshawi, N., McDougle, C. J. and Erickson, C. A. (2015). Drug-refractory aggression, self-injurious behavior, and severe tantrums in autism spectrum disorders: a chart review study. *Autism* 19:102-6. doi:10.1177/ 1362361314524641.

Amaral, D. G., Schumann, C. M. and Nordahl, C. W. (2008). Neuroanatomy of autism. *Trends in Neurosciences* 31(3):137-45. doi:10.1016/j.tins.2007.12.005.

American Psychiatric Association (AAP). (2013). *Diagnostic and statistical manual of mental disorders. 5th.* Arlington: American Psychiatric Publishing.

American Psychiatric Association. (2000). *Diagnostic and Statistical Manual of Mental Disorders DSM-IV-Tr.* (Fourth edn), (Text revision). Washigton D.C.

Andersen, I. M., Kaczmarska, J., McGrew, S. G. and Malow, B. A. (2008). Melatonin for insomnia in children with autism spectrum disorders. *Journal of Child Neurology* 23(5):482-5. doi:10.1177/088307380 7309783.

Aston-Jones, G. and Bloom, F. E. (1981). Activity of norepinephrine-containing locus coeruleus neurons in behaving rats anticipates fluctuations in the sleep-waking cycle. *The Journal of Neuroscience* 1(8):876-86.

Aurora, R. N., Kristo, D. A., Bista, S. R., Rowley, J. A., Zak, R. S., Casey, K. R., Lamm, C. I., Tracy, S. L. and Rosenberg, R. S. (2012). The treatment of restless legs syndrome and periodic limb movement disorder in adults—An update for 2012: Practice parameters with an evidence-based systematic review and meta-analyses. *Sleep* 35(8):1039-62. doi:10.5665/sleep.1988.

Ayala-Guerrero, F., Mexicano, G. and Huicochea-Arredondo, S. (2014). Sleep characteristics in children with autism spectrum disorder. *Journal of Neurological Disorders & Stroke* 2(3):1070.

Berridge, C. W., Schmeichel, B. E. and España, R. A. (2012). Noradrenergic modulation of wakefulness/arousal. *Sleep Medicine Reviews* 16(2):187-97. doi:10.1016/j.smrv.2011.12.003.

Botbol, M., Cabon, P., Kermarrec, S. and Tordjman, S. (2013). Biological and psychological rhythms: An integrative approach to rhythm disturbances in autistic disorder. *Journal of Physiology* Paris 107(4):298-309. doi:10.1016/j.jphysparis.2013.03.009.

Bourgeron, T. (2007). The possible interplay of synaptic and clock genes in autism spectrum disorders. *Cold Spring Harbor Symposia on Quantitative Biology* 72:645–54. doi:10.1101/sqb.2007.72.020.

Brennan, A. R. and Arnsten, A. F. (2008). Neuronal mechanisms underlying attention deficit hyperactivity disorder: The influence of arousal on prefrontal cortical function. *Annals of the New York Academy of Sciences* 1129:236-45. doi:10.1196/annals.1417.007.

Buckley, A. W., Rodríguez, A. J., Jennison, K., Buckley, J., Thurm, A., Sato, S. and Swedo, S. (2010). Rapid eye movement sleep percentage in children with autism compared with children with developmental delay and typical development. *Archives of Pediatrics and Adolescent Medicine* 164(11):1032-7. doi:10.1001/archpediatrics.2010.202.

Buckley, A. W., Sassower, K., Rodríguez, A. J., Jennison, K., Wingert, K., Buckley, J., Thurm, A., Sato, S. and Swedo, S. (2011). An open label trial of donepezil for enhancement of rapid eye movement sleep in young children with autism spectrum disorders. *Journal of Child and Adolescent Psychopharmacology* 21(4):353-7. doi:10.1089/cap.2010. 0121.

Casanova, M. F., Buxhoeveden, D. P. and Brown, C. (2002). Clinical and macroscopic correlates of minicolumnar pathology in autism. *Journal of Child Neurology* 17(9):692-695. doi:10.1177/088307380201700908.

Casanova, M. F., Buxhoeveden, D. P., Switala, A. E. and Roy, E. (2002). Minicolumnar pathology in autism. *Neurology* 58(3):428-432. doi:10.1212/WNL.58.3.428.

Christensen, D. L., Baio, J., Van Naarden Braun, K., Bilder, D., Charles, J., Constantino, J. N., Daniels, J., Durkin, M. S., Fitzgerald, R. T., Kurzius-Spencer, M., Lee, L., Pettygrove, S., Robinson, C., Schulz, E., Wells, C., Wingate, M. S., Zahorodny, W. and Yeargin-Allsopp, M. (2016). Prevalence and characteristics of autism spectrum disorder among children aged 8 years. Autism and developmental disabilities monitoring Network, 11 sites, United States, 2012. *MMWR Surveillance Summaries* 65(13):1-23. doi:10.15585/mmwr.ss6513a1.

De Leersnyder, H., Zisapel, N. and Laudon, M. (2011). Prolonged-release melatonin for children with neurodevelopmental disorders. *Pediatric Neurology* 45(1):23-6. doi:10.1016/j.pediatrneurol.2011.02.001.

Diomedi, M., Curatolo, P., Scalise, A., Placidia, F., Caretto, F. and Gigli, G. L. (1999). Sleep abnormalities in mentally retarded autistic subjects: Down's syndrome with mental retardation and normal subjects. *Brain and Development* 21(8):548-53. doi:10.1016/S0387-7604(99)00077-7.

dos Santos Moraes, W. A., Poyares, D. R., Guilleminault, C., Ramos, L. R., Bertolucci, P. H. F. and Tufik, S. (2006). The effect of donepezil on sleep and REM sleep EEG in patients with Alzheimer disease: a double-blind placebo-controlled study. *Sleep* 29(2):199-205. doi:10.1093/sleep/29.2.199.

Dosman, C. F., Brian, J. A., Drmic, I. E., Senthilselvan, A., Harford, M. M., Smith, R. W., Sharieff, W., Zlotkin, S. H., Moldofsky, H. and Roberts, S. W. (2007). Children with autism: Effect of iron supplementation on sleep and ferritin. *Pediatric Neurology* 36(3):152-8. doi:10.1016/j. pediatrneurol.2006.11.004.

Dravet, C. (2002). The behavioral disorders in epilepsy. *Revue Neurologique* 158(5 Pt 2):4S33-8.

Durand, V. M. (2002). Treating sleep terrors in children with autism. Journal of *Positive Behavior Interventions* 4(2):66-72. doi:10.1177/ 109830070200400201.

Elrod, M. G. and Hood, B. S. (2015). Sleep differences among children with autism spectrum disorders and typically developing peers: A meta-analysis. *Journal of Developmental and Behavioral pediatrics* 36(3):166-77. doi:10.1097/ DBP.0000000000000140.

Espie, C. A. and Tweedie, F. M. (1991). Sleep patterns and sleep problems amongst people with mental handicap. *Journal of Mental Deficiency Research* 35:25-36. doi:10.1111/j.1365-2788.1991.tb01028.x.

Farmer, C., Thurm, A. and Grant, P. (2013). Pharmacotherapy for the core symptoms in autistic disorder: current status of the research. *Drugs* 73:303-14. doi:10.1007/s40265-013-0021-7.

Feige, B., Baglioni, C., Spiegelhalder, K., Hirscher, V., Nissen, C. and Riemann, D. (2013). The microstructure of sleep in primary insomnia: an overview and extension. *International Journal of Psychophysiology* 89(2):171-80. doi:10.1016/j.ijpsycho.2013.04.002.

Gail, W. P., Sears, L. L. and Allard, A. (2004). Sleep problems in children with autism. *Journal of Sleep Research* 13:265-68. doi:10.1111/j. 1365-2869.2004.00405.x.

Galli-Carminati, G., Deriaz, N. and Bertschy, G. (2009). Melatonin in treatment of chronic sleep disorders in adults with autism: A retrospective study. *Swiss Medical Weekly* 139(19-20):293-6. doi:10.4414/smw.2009.12342.

Geschwind, D. H. and Levitt, P. (2007). Autism spectrum disorders: developmental disconnection syndromes. *Current Opinion in Neurobiology* 17:103-11. doi: 10.1016/j.conb.2007.01.009.

Giannotti, F., Cortesi, F., Cerquiglini, A. and Bernabei, P. (2006). An open-label study of controlled-release melatonin in treatment of sleep disorders in children with autism. *Journal of Autism and Developmental Disorders* 36(6):741-52. doi:10.1007/s10803-006-0116-z.

Glascoe, F. P., MacLean, W. E. and Stone, W. L. (1991). The importance of parents' concerns about their child's behavior. *Clinical Pediatrics* 30(1):8-11. doi:10.1177/000992289103000101.

Goldman, S. E., Adkins, K. W., Calcutt, M. W., Carter, M. D., Goodpaster, R. L., Wang, L., Shi, Y., Burgess, H. J., Hachey, D. L. and Malow, B. A. (2014). Melatonin in children with autism spectrum disorders: Endogenous and pharmacokinetic profiles in relation to sleep. *Journal of Autism and Developmental Disorders* 44(10):2525-35. doi:10.1007/s10803-014-2123-9.

Goodlin-Jones, B. L., Tang, K., Liu, J. and Anders, T. F. (2008). Sleep patterns in preschool-age children with autism, developmental delay, and typical development. *Journal of the American Academy of Child and Adolescent Psychiatry* 47(8):930-8. doi: 10.1097/CHI.ObO13e3181 799f7c.

Gozal, D. (1998). Sleep-disordered breathing and school performance in children. *Pediatrics* 102(3):616-620. doi:10.1542/peds.102.3.616.

Griffin, C. E. 3rd, Kaye, A. M., Bueno, F. R. and Kaye, A. D. (2013). Benzodiazepine pharmacology and central nervous system-mediated effects. *The Ochsner Journal* 13(2):214-23.

Hall, M., Buysse, D. J., Dew, M. A., Prigerson, H. G., Kupfer, D. J. and Reynolds, C. F. 3rd. (1997). Intrusive thoughts and avoidance behaviors are associated with sleep disturbances in bereavement-related depression. *Depression and Anxiety* 6(3):106-12. doi:10.1002/(SICI)1520-6394(1997)6:3<106::AID-DA3>3.0.CO;2-B.

Healey, E. S., Kales, A., Monroe, L. J., Bixler, E. O., Chamberlin, K. and Soldatos, C. R. (1981). Onset of insomnia: Role of life-stress events. *Psychosomatic Medicine* 43(5):439-51. doi:10.1097/00006842-198110000-00007.

Hellings, J. A., Weckbaugh, M., Nickel, E. J., Cain, S. E., Zarcone, J. R., Reese, R. M., Hall, S., Ermer, D. J., Tsai, L. Y., Schroeder, S. R. and Cook, E. H. (2005). A double-blind, placebo-controlled study of valproate for aggression in youth with pervasive developmental disorders. *Journal of Child and Adolescent Psychopharmacology* 15(4):682-92. doi:10.1089/cap.2005.15.682.

Hering, E., Epstein, R., Elroy, S., Lancu, D. R. and Zelnik, N. (1999). Sleep patterns in autistic children. *Journal of Autism and Developmental Disorders* 29(2):143-47. doi:10.1023/A:1023092627223.

Hollway, J. A. and Aman, M. G. (2011). Pharmacological treatment of sleep disturbance in developmental disabilities: a review of the literature. *Research in Developmental Disabilities*, 32(3):939-962. doi:10.1016/j.ridd.2010.12.035.

Hughes, J. R. and Melyn, M. (2005). EEG and seizures in autistic children and adolescents: further findings with therapeutic implications. *Clinical EEG and Neuroscience* 36(1):15-20. doi:10.1177/155005940 503600105.

Hutt, C., Hutt, S. J., Lee, D. and Ounsted, C. (1964). Arousal and childhood autism. *Nature* 204:908-9. doi: 10.1038/204908a0.

Ishizaki, A., Sugama, M. and Takeuchi, N. (1999). Usefulness of melatonin for developmental sleep and emotional/behavior disorders–studies of melatonin trial on 50 patients with developmental disorders. *No To Hattatsu* 31(5):428-37.

Jan, J. E., Espezel, H. and Appleton, R. E. (1994). The treatment of sleep disorders with melatonin. *Developmental Medicine & Child Neurology* 36(2):97-107. doi:10.1111/j.1469-8749.1994.tb11818.x.

Janz, R., Goda, Y., Geppert, M., Missler, M. and Südhof, T. C. (1999). SV2A and SV2B function as redundant Ca2+ regulators in neurotransmitter release. *Neuron* 24(4):1003-1016. doi:10.1016/S0896-6273(00) 81046-6.

Kerbeshian, J., Burd, L. and Avery, K. (2001). Pharmacotherapy of autism: A review and clinical approach. *Journal of Developmental and Physical Disabilities* 13(3):199-228. doi:10.1023/A:1016686802786.

Kerns, C. M., Kendall, P. C., Berry, L., Souders, M. C., Franklin, M. E., Schultz, R. T., Miller, J. and Herrington, J. (2014). Traditional and atypical presentations of anxiety in youth with autism spectrum disorder. *Journal of Autism and Developmental Disorders* 44(11):2851-61. doi:10.1007/s10803-014-2141-7.

Kerns, C. M., Maddox, B. B., Kendall, P. C., Rump, K., Berry, L., Schultz, R. T., Souders M. C., Bennett, A., Herrington, J. and Miller, J. (2015). Brief measures of anxiety in non-treatment-seeking youth with autism spectrum disorder. *Autism* 19(8):969-79. doi:10.1177/136236131 4558465.

Knickmeyer, R., Baron-Cohen, S., Raggatt, P. and Taylor, K. (2005). Foetal testosterone, social relationships, and restricted interests in children. *Journal of Child Psychology and Psychiatry* 46(2):198-210. doi:10.1111/j.1469-7610.2004.00349.x.

Krystal, A. D., Goforth, H. W. and Roth, T. (2008). Effects of antipsychotic medications on sleep in schizophrenia. *International Clinical Psychopharmacology* 23(3):150-60. doi:10.1097/YIC.0b013 e3282f39703.

Kulman, G., Lissoni, P., Rovelli, F., Roselli, M. G., Brivio, F. and Sequeri, P. (2000). Evidence of pineal endocrine hypofunction in autistic children. *NeuroEndocrinology Letters* 21(1):31-4.

LeRoy, C., Landais, E., Briault, S., David, A., Tassy, O., Gruchy, N., Delobel, B., Grégoire, M. J., Leheup, B., Taine, L., Lacombe, D., Delrue, M. A., Toutain, A., Paubel, A., Mugneret, F., Thauvin-Robinet, C., Arpin, S., Le Caignec, C., Jonveaux, P., Beri, M., Leporrier, N., Motte, J., Fiquet, C., Brichet, O., Mozelle-Nivoix, M., Sabouraud, P., Golovkine, N., Bednarek, N., Gaillard, D. and Doco-Fenzy, M. (2013). The 2q37-deletion syndrome: An update of the clinical spectrum including overweight, brachydactyly and behavioural features in 14 new patients. *European Journal of Human Genetics* 21(6):602-12. doi:10.1038/ejhg.2012.230.

Luboshitzky, R., Lavi, S., Thuma, I. and Lavie, P. (1995). Increased nocturnal melatonin secretion in male patients with hypogonadotropic hypogonadism and delayed puberty. *The Journal of Clinical Endocrinology and Metabolism* 80(7):2144-8. doi:10.1210/jcem.80.7. 7608268.

Malow, B. A, Marzec, M. L., McGrew, S. G., Wang, L., Henderson, L. M. and Stone, W. L. (2006). Characterizing sleep in children with autism spectrum disorders: a multidimensional approach. *Sleep* 29(12):1563-71. doi:10.1093/sleep/29.12.1563.

Malow, B. A., Adkins, K. W., Reynolds, A., Weiss, S. K., Loh, A., Fawkes, D., Katz, T., Goldman, S. E., Madduri, N., Hundley, R. and Clemons, T. (2014). Parent-based sleep education for children with autism spectrum disorders. *Journal of Autism and Developmental Disorders* 44(1):216-28. doi:10.1007/s10803-013-1866-z.

Malow, B. A. and McGrew, S. G. (2006). Sleep and autism spectrum disorders. In *Autism: a neurological disorder of early brain development* edited by R. Tuchman, I. Rapin 189-201. London: MacKeith Press with the International Child Neurology Association.

Malow, B. A., Crowe, C., Henderson, L., McGrew, S. G., Wang, L., Song, Y. and Stone, W. L. (2009). A sleep habits questionnaire for children with autism spectrum disorders. *Journal of Child Neurology* 24(1):19-24. doi:10.1177/0883073808321044.

Malow, B. A., Marzec, M. L., McGrew, S. G., Wang, L., Henderson, L. M. and Stone, W. L. (2006). Characterizing sleep in children with autism spectrum disorders: A multidimensional approach. *Sleep* 29(12):1563-71. doi: 10.1093/sleep/29.12.1563.

Malow, B., Adkins, K. W., McGrew, S. G., Wang, L., Goldman, S. E., Fawkes, D. and Burnette, C. (2012). Melatonin for sleep in children with autism: A controlled trial examining dose, tolerability, and outcomes. *Journal of Autism and Developmental Disorders* 42(8):1729-37. doi:s10803-011-1418-3.

Maquet, P. (2001). The role of sleep in learning and memory. *Science* 294(5544):1048-1052. doi:10.1126/science.1062856.

Mazurek, M. O. and Petroski, G. F. (2015). Sleep problems in children with autism spectrum disorder: examining the contributions of sensory over-responsivity and anxiety. *Sleep Medicine* 16(2):270-9. doi:10. 1016/j.sleep.2014.11.006.

Mazurek, M. O. and Sohl, K. (2016). Sleep and behavioral problems in children with autism spectrum disorder. *Journal of Autism and Developmental Disorders* 46:1906-15. doi:10.1007/ s10803-016-2723-7.

McDermott, S., Moran, R., Platt, T., Wood, H., Isaac, T. and Dasari, S. (2005). Prevalence of epilepsy in adults with mental retardation and related disabilities in primary care. *American Journal on Mental Retardation* 110(1):48-56. doi: 10.1352/0895-8017(2005)110<48:POEIAW>2.0. CO;2.

McIntosh, D. N., Miller, L. J. and Shyu, V. (1999). Development and validation of the short sensory profile. In *Sensory profile examiner´s manual* edited by W. Dunn, 59-73. San Antonio, TX: Psychological Corporation.

McPheeters, M. L., Warren, Z., Sathe, N., Bruzek, J. L., Krishnaswami, S., Jerome, R. N. and Veenstra-VanderWeele, J. (2011). A systematic review of medical treatments for children with autism spectrum disorders. *Pediatrics* 127(5), e1312-e1321. doi:10.1542/peds.2011-0427.

Melke, J., Goubran Botros, H., Chaste, P., Betancur, C., Nygren, G., Anckarsater, H., Rastam, M., Ståhlberg, O., Gillberg, I. C., Delorme, R., Chabane, N., Mouren-Simeoni, M. C., Fauchereau, F., Durand, C. M., Chevalier, F., Drouot, X., Collet, C., Launay, J. M., Leboyer, M., Gillberg, C. and Bourgeron, T. (2008). Abnormal melatonin synthesis in autism spectrum disorders. *Molecular Psychiatry* 13(1):90-8. doi:10.1038/sj.mp.4002016.

Meltzer, L. J. (2008). Brief report: Sleep in parents of children with autism spectrum disorders. *Journal of Pediatric Psychology* 33(4):380-6. doi:10.1093/jpepsy/jsn005.

Meltzer, L. J., Johnson, C., Crosette, J., Ramos, M. and Mindell, J. A. (2010). Prevalence of diagnosed sleep disorders in pediatric primary care practices. *Pediatrics* 125(6):e1410-8. doi:10.1542/peds. 2009-2725.

Meltzer, L. J., Mindell, J. A., Owens, J. A. and Byars, K. C. (2007). Use of sleep medications in hospitalized pediatric patients. *Pediatrics* 119(6):1047-55. doi:10.1542/peds.2006-2773.

Ming, X., Julu, P. O., Brimacombe, M., Connor, S. and Daniels, M. L. (2005). Reduced cardiac parasympathetic activity in children with autism. *Brain and Development* 27(7):509-16. doi:10.1016/j.braindev.2005.01. 003.

Morin, C. M., Rodrigue, S. and Ivers, H. (2003). Role of stress, arousal, and coping skills in primary insomnia. *Psychosomatic Medicine* 65(2):259-67. doi:10.1097/01.PSY.0000030391.09558.A3.

Nir, I., Meir, D., Zilber, N., Knobler, H., Hadjez, J. and Lerner, Y. (1995). Brief report: Circadian melatonin, thyroidstimulating hormone, prolactin, and cortisol levels in serum of young adults with autism. *Journal of Autism and Developmental Disorders* 25(6):641-54. doi:10.1007/BF02178193.

Noebels, J. L. (2003). Exploring new gene discoveries in idiopathic generalized epilepsy. *Epilepsia* 44:16-21.

O'Brien, L. M. (2009). The neurocognitive effects of sleep disruption in children and adolescents. *Child and Adolescent Psychiatric Clinics of North America* 18(4):813-23. doi:10.1016/j.chc.2009.04.008.

Olson, L. G. (2008). Hypnotic hazards: Adverse effects of zolpidem and other z-drugs. *Australian Prescriber* 31(6):146-9. doi:10.18773/ austprescr.2008.084.

Ornitz, E. M., Ritvo, E. R., Brown, M. B., La Franchi, S., Parmelee, T. and Walter, R. D. (1969). The EEG and rapid eye movements during REM sleep in normal and autistic children. *Electroencephalography and Clinical Neurophysiology* 26(2):167-75. doi:10.1016/0013-4694(69) 90207-7.

Owens, J. A. and Mindell, J. A. (2011). Pediatric insomnia. *Pediatric Clinics* 58(3):555-569. doi:10.1016/j.pcl.2011.03.011.

Owens, J. A., Rosen, C. L. and Mindell, J. A. (2003). Medication use in the treatment of pediatric insomnia: results of a survey of community-based pediatricians. *Pediatrics* 111(5):e628-e635. doi:10.1542/peds. 111.5.e628.

Paavonen, E. J., Nieminen-von, W. T., Vanhala, R., Aronen, E. T. and von Wendt, L. (2003). Effectiveness of melatonin in the treatment of sleep disturbances in children with Asperger disorder. *Journal of Child and Adolescent Psychopharmacology* 13(1):83-95. doi:10.1089/104454 603321666225.

Pandi-Perumal, S. R., Srinivasan, V., Spence, D. W. and Cardinali, D. P. (2007). Role of the melatonin system in the control of sleep. *CNS drugs* 21(12):995-1018. doi:10.2165/00023210-200721120-00004.

Panju, S., Brian, J., Dupuis, A., Anagnostou, E. and Kushki, A. (2015). Atypical sympathetic arousal in children with autism spectrum disorder and its association with anxiety symptomatology. *Molecular Autism* 6:64. doi:10.1186/s13229-015-0057-5.

Parikshak, N. N., Luo, R., Zhang, A., Won, H., Lowe, J. K., Chandran, V., Horvath, S. and Geschwind, D. H. (2013). Integrative functional genomic analyses implicate specific molecular pathways and circuits in autism. *Cell* 155(5):1008-1021. doi:10.1016/j.cell.2013.10.031.

Pelayo, R. and Yuen, K. (2012). Pediatric sleep pharmacology. *Child and Adolescent Psychiatric Clinics of North America* 21(4):861-83. doi:10.1016/j.chc.2012.08.001.

Posey, D. J. and McDougle, C. J. (2001). Pharmacotherapeutic management of autism. *Expert opinion on Pharmacotherapy* 2:587-600. doi:10.1517/14656566.2.4.587.

Quine, L. (1991). Sleep problems in children with mental handicap. *Journal of Mental Deficiency Research* 35:269-90. doi:10.1111/j.1365-2788.1991.tb00402.x.

Reynolds, S. Lane, S. J. and Thacker, L. (2012). Sensory processing, physiological stress, and sleep behaviours in children with and without autism spectrum disorders. *OTJR: Occupation, Participation and Health* 32:246-57. doi:10.3928/15394492-20110513-02.

Richards, C., Oliver, C., Nelson, L. and Moss, J. (2012). Self-injurious behaviour in individuals with autism spectrum disorder and intellectual disability. *Journal of Intellectual Disability Research* 56(5):476-489. doi:10.1111/j.1365-2788. 2012.01537.x.

Richdale, A. L. (1999). Sleep problems in autism: prevalence, cause, and intervention. *Developmental Medicine and Child Neurology* 41(1):60-66. doi:10.1017/ S0012162299000122.

Richdale, A. L. and Schreck, K. A. (2009). Sleep problems in autism spectrum disorders: Prevalence, nature, & possible biopsychosocial aetiologies. *Sleep Medicine Reviews* 13(6):403-11. doi:10.1016/j. smrv.2009.02.003.

Ritvo, E. R. R. R., Yuliwer, A., Brothers, A., Freeman, B. J. and Plotkin, S. (1993). Elevated daytime melatonin concentration in autism. *European Child & Adolescent Psychiatry* 2:75-8. doi:10.1007/ BF02098862.

Robinson-Shelton, A. and Malow, B. A. (2016). Sleep Disturbances in Neurodevelopmental Disorders. *Current Psychiatry Reports* 18(1):6. doi:10.1007/s11920-015-0638-1.

Rosahl, T. W., Spillane, D., Missler, M., Herz, J., Selig, D. K., Wolff, J. R., Hammer, R. E., Malenka, R. C. and Südhof, T. C. (1995). Essential functions of synapsins I and II in synaptic vesicle regulation. *Nature* 375(6531):488-9. doi:10.1038/375488a0.

Rubenstein, J. L. R. and Merzenich, M. M. (2003). Model of autism: increased ratio of excitation/inhibition in key neural systems. *Genes, Brain and Behavior* 2(5):255-267. doi:10.1034/j.1601-183X.2003. 00037.x.

Sasaki, M., Yoshitane, H., Du, N. H., Okano, T. and Fukada, Y. (2009). Preferential inhibition of BMAL2-CLOCK activity by PER2 reemphasizes its negative role and a positive role of BMAL2 in the circadian transcription. *The Journal of Biological Chemistry* 284(37):25149-59. doi:10.1074/jbc.M109.040758.

Schreck, K. A. and Mulick, J. A. (2000). Parental report of sleep problems in children with autism. *Journal of Autism and Developmental Disorders* 30(2):127-35. doi: 10.1023/A:1005407622050.

Schreck, K. A., Mulick, J. A. and Smith, A. F. (2004). Sleep problems as possible predictors of intensified symptoms of autism. *Research in Developmental Disabilities* 25(1):57-66. doi: 10.1016/j.ridd.2003. 04.007.

Spielman, A. J., Caruso, L. S. and Glovinsky, P. B. (1987). A behavioral perspective on insomnia treatment. *Psychiatric Clinics of North America* 10(4):541-53. doi:10.1016/S0193-953X(18)30532-X.

Stigler, K. A., Posey, D. J. and McDougle, C. J. (2006). Ramelteon for insomnia in two youths with autistic disorder. *Journal of Child and Adolescent Psychopharmacology* 16(5):631-6. doi:10.1089/cap.2006. 16.631.

Stores, G. (1992). Annotation: sleep studies in children with a mental handicap. *Journal of Child Psychology and Psychiatry* 33(8):1303-1317. doi:10.1111/j.1469-7610.1992.tb00951.x.

Tanguay, P. E., Ornitz, E. M., Forsythe, A. B. and Ritvo, E. R. (1976). Rapid eye movement (REM) activity in normal and autistic children during REM sleep. *Journal of Autism and Childhood Schizophrenia* 6(3):275-88. doi:10.1007/BF01543468.

Thirumalai, S. S., Shubin, R. A. and Robinson, R. (2002). Rapid eye movement sleep behavior disorder in children with autism. *Journal of Child Neurology* 17(3):173-8. doi:10.1177/088307380201700304.

Tordjman, S., Davlantis, K. S., Georgieff, N., Geoffray, M. M., Speranza, M., Anderson, G. M., Xavier, J., Botbol, M., Oriol, C., Bellissant, E., Vernay-Leconte, J., Fougerou, C., Hespel, A., Tavenard, A., Cohen, D., Kermarrec, S., Coulon, N., Bonnot, O. and

Dawson, G. (2015). Autism as a disorder of biological and behavioral rhythms: Toward new therapeutic perspectives. *Frontiers in Pediatrics* 3:1. doi:10.3389/fped.2015.00001.

Tordjman, S., Najjar, I., Bellissant, E., Anderson, G. M., Barburoth, M., Cohen, D., Jaafari, N., Schischmanoff, O., Fagard, R., Lagdas, E., Kermarrec, S., Ribardiere, S., Botbol, M., Forgerou, C., Bronsard, G. and Vernay-Leconte, J. (2013). Advances in the research of melatonin in autism spectrum disorders: Literature review and new perspectives. *International Journal of Molecular Sciences* 14(10):20508-42. doi:10.3390/ijms141020508.

Travnickova-Bendova, Z., Cermakian, N., Reppert, S. M. and Sassone-Corsi, P. (2002). Bimodal regulation of mPeriod promoters by CREB-dependent signaling and CLOCK/BMAL1 activity. *Proceedings of the National Academy of Sciences of the United States of America* 99(11):7728-33. doi:10.1073/pnas.102075599.

Tuchman, R. and Cuccaro, M. (2011). Epilepsy and autism: neurodevelopmental perspective. *Current neurology and neuroscience reports* 11(4):428-434. doi:10.1007/s11910-011-0195-x.

Tuchman, R. and Rapin, I. (2002). Epilepsy in autism. *The Lancet Neurology* 1(6):352-358. doi: 10.1016/s1474-4422(02)00160-6.

Tuchman, R., Alessandri, M. and Cuccaro, M. (2010). Autism spectrum disorders and epilepsy: moving towards a comprehensive approach to treatment. *Brain and Development* 32(9):719-730. doi:10.1016/j. braindev.2010.05.007.

Tuchman, R., Moshé, S. L. and Rapin, I. (2009). Convulsing toward the pathophysiology of autism. *Brain and Development* 31(2):95-103. doi:10.1016/j.braindev.2008.09.009.

Veatch, O. J., Keenan, B. T., Gehrman, P. R., Malow, B. A. and Pack, A. I. (2017). Pleiotropic genetic effects influencing sleep and neurological disorders" *The Lancet Neurology* 16(2):158-70. doi:10.1016/S1474-4422(16)30339-8.

Velíšková, J., Claudio, O. I., Galanopoulou, A. S., Lado, F. A., Ravizza, T., Velíšek, L. and Moshé, S. L. (2004). Seizures in the developing brain. *Epilepsia* 45:6-12. doi:10.1111/j.0013-9580.2004.458002.x.

Wegiel, J., Kuchna, I., Nowicki, K., Imaki, H., Wegiel, J., Marchi, E., Ma, S. Y., Chauhan, V., Bobrowicz, T. W., de Leon, M., Louis, L. A., Cohen, I. L., London, E., Brown, W. T. and Wisniewski, T. (2010). The neuropathology of autism: defects of neurogenesis and neuronal migration, and dysplastic changes. *Acta Neuropathologica* 119(6):755-70. doi:10.1007/s00401-010-0655-4.

Werling, D. M., Parikshak, N. N. and Geschwind, D. H. (2016). Gene expression in human brain implicates sexually dimorphic pathways in autism spectrum disorders. *Nature communications* 7:10717. doi:10.1038/ncomms10717.

White, S. W., Lerner, M. D., McLeod, B. D., Wood, J. J., Ginsburg, G. S., Kerns, C., Ollendick, T., Kendall, P. C., Piacentini, J., Walkup, J. and Compton, S. (2015).

Anxiety in youth with and without autism spectrum disorder: Examination of factorial equivalence. *Behavior Therapy* 46(1):40-53. doi:10.1016/ j.beth.2014.05.005.

Wiggs, L. and Stores, G. (2004). Sleep patterns and sleep disorders in children with autistic spectrum disorders: insights using parent report and actigraphy. *Developmental Medicine and Child Neurology* 46(6):372-380. doi:10.1017/S0012162204000611.

Wimpory, D., Nicholas, B. and Nash, S. (2002). Social timing, clock genes and autism: A new hypothesis. *Journal of Intellectual Disability Research* 46(Pt 4):352-8. doi:10.1046/j.1365-2788.2002.00423.x.

World Health Organization (WHO). (2016). *International Classification of Diseases (ICD-10-CM).*

Yang, Z., Matsumoto, A., Nakayama, K., Jimbo E. F., Kojima, K., Nagata, K. Iwamoto, S. and Yamagata, T. (2016). Circadian-relevant genes are highly polymorphic in autism spectrum disorder patients. *Brain and Development* 38(1):91-9. doi:10.1016/j.braindev.2015.04.006.

Youssef, J., Singh, K., Huntington, N., Becker, R. and Kothare, S. V. (2013). Relationship of serum ferritin levels to sleep fragmentation and periodic limb movements of sleep on polysomnography in autism spectrum disorders. *Pediatric Neurology* 49(4):274-8. doi:10.1016/j. pediatrneurol.2013.06.012.

In: Sleep, Cognition and Emotion
Editors: I. Ramírez Salado et al.

ISBN: 978-1-53617-561-5

Chapter 11

EFFECT OF TRANSCRANIAL MAGNETIC STIMULATION ON SLEEP AND BEHAVIORAL DISORDERS IN PATIENTS WITH AUTISM

Fructuoso Ayala-Guerrero**, PhD
and Horacio Balam Alvarez-García
Laboratorio de Neurociencias, Facultad de Psicología,
Universidad Nacional Autónoma de México,
Ciudad de México, México

ABSTRACT

The Transcraneal Magnetic Stimulation (TMS) is a recently implemented non-invasive therapeutic procedure that consists of brain stimulation through the induction of a magnetic field that causes minimal electrical discharges in determined cerebral áreas. The brain stimulation may be applied under three different modalities: 1) a single pulse, 2) paired pulses and 3) trains of pulses. Under this last modality, pulses may be applied unilaterally or bilaterally at a high (>1 Hz) or low (<1 Hz) frequency on determined cerebral áreas. This stimulation technique has had a wide variety of applications for treating depressive disorders in adults, with positive results in pharmaco-resistent patients. It is being currently utilized in children that present neuropsychiatric disorders such as Autism, Attention Deficit Hyperactivity Disorder (ADHD), affective disorders, schizophrenia, etc. with variable results. It is considered that the positive effects exerted by the TMS are related to the brain plasticity.

The secondary effects of the application of the repetitive Transcranial Magnetic Stimulation have been considered as not important for the health of individuals; however, in some patients it exists the possibility of inducing seizures, although their probability is minimal (< 1%). Autism treatment with Transcranial Magnetic Stimulation is still under

* Corresponding Author's Email: fayala@unam.mx.

investigation, therefore it can not regularly be used as an appropriated therapeutic treatment. Since there are various etiopathogenic hypotheses about autism, this technique may have a wide variety of applications depending on the specifically of the type of autism spectrum disorder. Furthermore, this technique is emerging as a possible therapeutic tool for the regulation of sleep patterns in patients with Autism Spectrum Disorders due to the functional relationship existing between the groups of neurons that participate in the regulation of circadian cycles, including sleep, and the control of behavioral and social cognition.

Keywords: autism spectrum disorder, sleep, transcranial magnetic stimulation, neuropsychiatric disorders, brain plasticity

INTRODUCTION

The Transcranial Magnetic Stimulation (TMS) is a relatively new non-invasive therapeutic procedure that consists of brain stimulation through the induction of a magnetic field that triggers minimal electrical discharges in determined cerebral areas. A reference signal of its effect is a motor response produced by the stimulation of the primary motor cortex, manifested per example by finger movements. It has become a technical tool that expands the range of possibilities for study and research in neurosciences in general, and in the treatment of various neuropsychiatric diseases in particular. This technique may also be utilized for emotional states modulation. It allows the safe and painless stimulation of the nervous system (cerebral cortex, spinal cord, afferent-efferent pathways, peripheral nerves, etc.). In addition to regulating cerebral activity under controlled conditions, it may modulate the brain plasticity. Numerous changes have been described at different levels from the interaction of the TMS with neurons as the following processes: electrophysiological (EEG, membrane potentials), biochemical-molecular (signaling, neurotransmitters, genes, etc.), and functional cellular plasticity (growth, differentiation, etc.). Furthermore, the TMS has effects on behavioral modulation, emotional states, memory processes, myelination and neurophysiological activities (Pascual and Tormos 2008; Verdugo and Drucker 2014).

HISTORICAL DEVELOPMENT OF THE TMS

Hans Christian Ørsted (August, 1777- March, 1851) demonstrated a relationship between magnetic and electrical processes, and he coined, in 1821, the concept of electromagnetism. In the biennial 1830-1832 Michael Faraday (September, 1791-August, 1867) described the interaction between a magnetic and an electrical field that was integrated and consolidated by James Clerk Maxwell (June, 1831- November, 1879).

In the middle of the past century, (Kolin, Brill, and Broberg 1959) showed that a fluctuating magnetic field could stimulate a peripheral muscle in experiments performed in a chirurgical preparation of a frog. In 1980, Patrick Merton (Patrick Anthony Merton (1920–2000) and his co-worker Bert Morton, were the first neuroscientists to use this technique to stimulate the motor cerebral cortex, and they recorded motor action potentials as responses to the stimulation. Five years after, (Barker, Jalinous R. and Freeston 1985) built a stimulating device capable of depolarizing cortical neurons and evoking contralateral hemicorporal movements, by activating corticospinal pathways. Using this method, they analyzed the functionality of the central motor pathways activities (Pascual and Tormos 2008). Subsequently, these authors applied this type of stimulation in patients that suffered from multiple sclerosis and they observed a slowing down of the motor pathway activity, and also described the advantages of this technique. After this historical development, there has been an exponential increase in the applications of the TMS under experimental and clinical conditions (Barker, Jalinous R. and Freeston 1985). The technique of Transcranial Magnetic Stimulation is based on the principle of electromagnetic induction, where every electric field produces a magnetic area perpendicular to it, and vice versa (Figure 1).

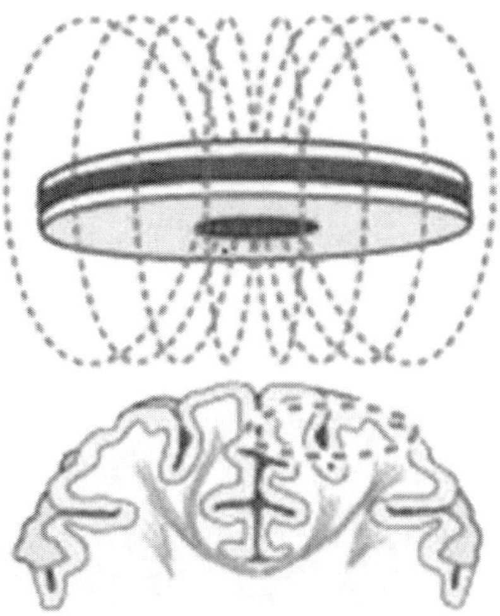

Figure 1. Technique of transcranial magnetic stimulation.

In the TMS, pulses of electrical current generated in a capacitor are applied, and they flow across a coil of copper conductive wire, electrically isolated, covered with a plastic wrap. The coil is placed on the patient's head where it generates a magnetic field, distributed perpendicularly. The frequency of pulses of this magnetic field determines the induction of a secondary electric current in a nearby conductive material (Pascual and Tormos 2008; Kobayashi and Pascual-Leone 2003). When an electrical pulse is driven by the stimulation coil, a magnetic field is generated passing through the scalp and cranial bone without attenuation. The most used stimulation coils are of circular type (Figure 2a) and another with the shape of the number 8 (Figure 2b). The circular coil produces a wider electric field, allowing simultaneous stimulation of the 2 cerebral hemispheres, while the 8-shaped coil induces a more localized stimulating field (Rossi et al. 2009; Emara et al. 2010).

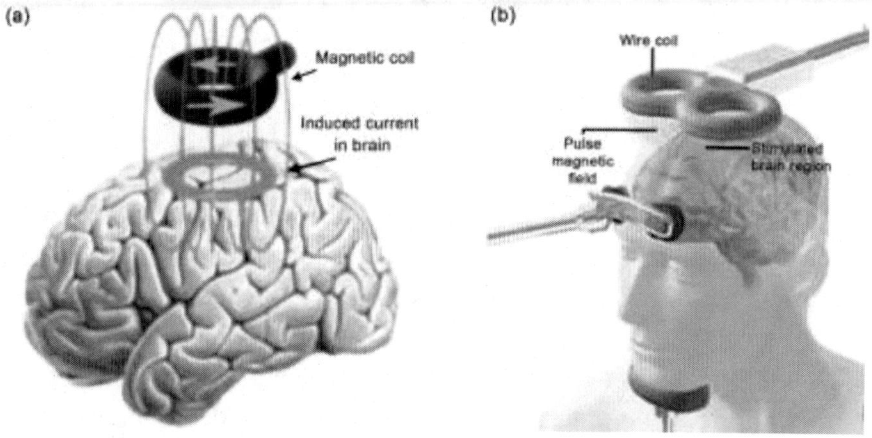

Figure 2. Distribution of the electric fields induced by a circular stimulation coil (a) and an 8-shaped coil (b).

These specific magnetic pulses originate a selective depolarization of neurons of the cerebral cortex, situated between 1.5 and 2 cm below the cranial vault. The electric current may inhibit or stimulate the neurons, modulating their function. These effects depend on the shape, size, type and orientation of the coil, as well as the intensity of the magnetic field and the frequency and duration of the magnetic pulses (Bayón 2011; Kakuda et al. 2011; Corti, Patten, and Triggs 2012). Three modalities of stimulation may be applied: 1) a single pulse, 2) paired pulses and 3) a series of pulses (Figure 3).

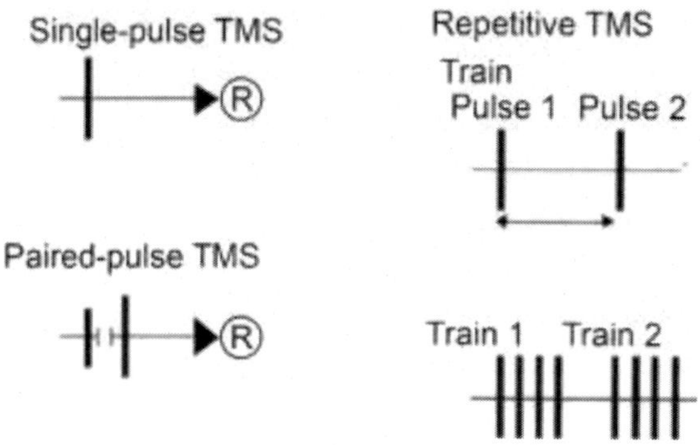

Figure 3. Types of pulses.

A single stimulus applied over a determined brain region, depolarizes cortical neurons, and causes a motor evoked potential in a muscular area of the contralateral hemicorporal region. On the other hand, the TMS with paired pulses, generates 2 stimuli of identical or different intensity (separated by a range of several milliseconds), which are applied over one or different cortical areas. This method allows one to analyze intra and inter cortical excitability, as well as the integrity of interhemispheric connectivity. Repetitive transcranial magnetic stimulation (rTMS) generates a pulse series of low (≤ 1 Hz, with a range of 0.5-1 Hz) or high frequency (≥ 5 Hz, with a range of 5-20 Hz), during very short times (milliseconds), inducing changes in corticospinal excitability, and on the regulatory effects of the neuronal activity that remain beyond the time covered by the rTMS session itself. This stimulation method, is the most used TMS modality for therapeutic purposes. This technique may unilaterally be applied as follows: high frequency (> 1 Hz) stimulation is usually used in the left lateral prefrontal cortex and low frequency (<1 Hz) in the right lateral prefrontal cortex. But bilateral stimulation may be also used.

Effects Triggered by Repetitive Transcranial Magnetic Stimulation

The cellular and molecular mechanisms, through which rTMS produces its therapeutic effects, are beginning to become apparent. One of these mechanisms is related to its ability as a modulator to activate certain genes of early expression, which are involved in the initial response to brain damage, inducing neuroplasticity processes. These genes regulate the expression of several growth factors, such as the brain-derived neurotrophic factor involved in neuroplasticity. In addition, rTMS interferes with apoptosis and induces mitochondrial energy production by favoring an intraneuronal oxidative balance within the brain tissue. These actions modify the regulation and activity of determined transcription factors associated with apoptosis (Wassermann and Zimmermann 2012; Chervyakov et al. 2015; Ljubisavljevic et al. 2015; Hwang et al. 2015; Tang, Thickbroom, and Rodger 2015).

Role of Sensorimotor Integration

Sensorimotor integration consists of continuous brain processing of afferent information, which reaches multiple associative areas of the cerebral cortex and causes motor activity. After a brain injury, remote cortical areas to the injured region can be rearranged to regain lost functions. The rTMS applied to the ipsolateral or contralateral area to a brain injury can generate varied motor patterns according to its stimulation frequency (Wassermann and Zimmermann 2012; Chervyakov et al. 2015; Ljubisavljevic et al. 2015; Hwang et al. 2015; Tang, Thickbroom, and Rodger 2015).

USE OF TRANSCRANIAL MAGNETIC STIMULATION

Applications in Neurophysiology

The physiological effects of TMS are related to the fact that a single stimulus of threshold intensity, produces neuronal depolarization followed by an action potential that generates an excitatory post-synaptic response, followed by an inhibitory post-synaptic potential. Thus, the local effect of the TMS consists of inhibiting the normal neural activity, by means of increasing the time of the refractory period and regulating the normal discharge pattern. The TMS in neurophysiological applications may be used for the exploration of cortical motor areas and for the conduction of the corticospinal tracts (Liew et al. 2014; Cunningham et al. 2015; Raguer 2008, Rossini et al. 2015).

Applications in Neural Networks

Neuropsychological, neurophysiological and neuroimaging studies in both animals and humans have shown that cognitive and behavioral processes are the result of interactions between brain regions linked to networks functional neural trees. The impact exerted by the use of TMS in a specific brain area, affects the cortical and subcortical regions of both brain hemispheres, and due to brain interconnectivity, this effect can reach deep areas of the brain i.e., the activation or inhibition of a specific area can produce distant effects (Di Pino et al. 2014; Malcolm, Vaughn and Greene 2015; Cassidy et al. 2015; Demirtas-Tatlidede et al. 2015; Page et al. 2015; Blesneag, Popa and Stan 2015; Klomjai et al. 2015).

Therapeutic Applications

Nowadays it is known that the rTMS, has a favorable impact on neuroplasticity, and induces neuroprotective effects that help, at least temporarily, people suffering from different neuropsychiatric disorders. Therefore, the possible use of the rTMS as an adjunctive therapy of the following disorders has been considered by different researchers (Claflin, Krishnan and Khot 2015; Klein et al. 2015; Jin et al. 2015; Rubio 2014; Lee et al. 2015; Van den Noort et al. 2015; Roland et al. 2016; Lipton et al. 2010).

Psychiatric Processes

Autism, schizophrenia, attention deficit hyperactivity disorder, mood disorders (major depression refractory to pharmacotherapy, postpartum depression, dysthymia, mania, bipolar disorder, etc.), anxiety disorders (obsessive-compulsive disorder, disorder post-traumatic stress, etc.), drug abuse disorders, etc.

Neurological Diseases

Alzheimer's disease, Parkinson's disease, Huntington's disease, stroke, pharmaco-refractory epilepsy, multiple sclerosis, essential tremor, head trauma, migraine with aura, trigeminal neuralgia, etc.

Pain Disorders

Neuropathic pain, visceral pain, chronic pain in phantom limb syndrome, fibromyalgia, atypical facial pain, tinnitus, etc.

Repetitive Transcranial Magnetic Stimulation Applied in Patients with Stroke

Stroke incidence is higher in adult men, and is the main cause of their death and also it is the first cause of disability in adults and the second leading cause of dementia (Fuentes et al. 2012). After a stroke, a neuroplasticity process is generally established. This process depends upon genetic factors, patient age, precocity of the neurorehabilitation, development of complications, nature and extent of that brain injury. The objective of the Neurorehabilitation is to recover the maximum possible lost capabilities, (Figueroa, Villamayor, and Antelo 2015; Raffin and Siebn 2014; Auriat et al. 2015; Hwang et al. 2015; Klomjai et al. 2015). TMS in stroke victims can be used as a regenerative therapy technique. As for its therapeutic effects, TMS can be conducted for improving neuroplasticity abilities, and to decrease the symptoms and the severity level associated with the stroke (Cunningham et al. 2015; Blesneag, Popa, and Stan 2015; Klomjai et al. 2015; Claflin, Krishnan, and Khot 2015).

Neuroplasticity

There are numerous reports showing significant effects of TMSr on neuroplasticity abilities, which have a dual function either to facilitate or inhibit synaptic connections among neurons as well as in neurotransmitter releasing. Most of the neurophysiological disorders generated by stroke events are not originated by the injury itself, but by the hyperactivity developed in the intact hemisphere, which indirectly inhibits the injured one. The low-frequency EMTr (-1 Hz) applied to the healthy hemisphere normalizes the diffuse cortical activation of the primary and secondary motor areas of both brain hemispheres. Thus, the injured cortical area that had been inhibited, increases its excitability as well as its motor recovery. In contrast, the high-frequency EMTr (-5 Hz) increases cortical excitability and can be applied to induce a neuronal activation of the brain cortex of the injured hemisphere. That is, the TMSr favors the reparation of neuroplasticity mechanisms, rearranging brain connections, leading to greater efficiency of the interneuronal networks of the affected brain area (Demirtas-Tatlidede et al. 2015; Blesneag, Popa, and Stan 2015; Klomjai et al. 2015; Claflin, Krishnan, and Khot 2015).

Motor Recovery

The EMTr favors the neuroplasticity mechanisms for motor recovery by activation of glutamatergic receptors of NMDA type and inhibition of the GABAergic system (Medina, Pascual and Túnez 2014; Simonetta-Moreau 2014; Demirtas-Tatlidede et al. 2015; Page et al. 2015; Klomjai et al. 2015). Reports from studies carried out in patients with chronic hemiparesis provoked by a stroke showed that the 10 Hz TMSr induced an increase in the amplitude of motor evoked potentials in patients treated but not in those not treated with this procedure. This finding coincided with an improvement in motor capabilities as observed in finger executions (Kim et al. 2006). Additionally, it has been observed that the application of EMTr at high frequencies in the injured hemisphere, along with application of neurorehabilitation techniques, improves the motor activity in the upper limbs. The effectiveness of the EMTr, depending on the type of stroke. has a better result when it is subcortical (Ameli et al. 2009). When EMTr is applied on the contralateral side to the injury area, it seems to improve the motor activity. Thus, patients receiving inhibitory Transcraneal Magnetic Stimulation (1 Hz) in the contralateral hemisphere to the injured one, and undergoing motor rehabilitation, have shown improvements in motor deficit, as the injured area was inhibited through the transcallosal pathway (Takeuchi et al. 2005). The results of studies analyzing the effectiveness of the TMSr indicate that low-frequency TMSr, targeted on the contralateral cerebral hemisphere to the injured one, shows better effects (Lüdemann-Podubecká et al. 2015; Buetefisch 2015). The application of *theta burst* stimulation, a modality of TMS consisting of repetitive magnetic discharges at high frequency and low intensity bursts (3 pulses of 50 Hz repeated every 200 ms), to the healthy motor cortex of patients with stroke, increases the muscular evoked potentials and decreases the cortical excitability of the healthy hemisphere. Furthermore, it facilitates the neuroplasticity mechanisms and determines a greater functional recovery. There are 2 stimulation modalities, one intermittent with excitatory effect on the cortex (2 separate applications 10 s) and another continuous one that causes cortical inhibition. The intermittent modality applied in the area injured was effective for the motor regeneration in comparison to the continuous one applied in the contralateral area to the injury. The positive effects of the continuous modality disappeared after application (Huang et al. 2007; Ackerley et al. 2010). For their part, studies carried out using Transcranial Magnetic Stimulation of *theta burst* modality, applied to the contralateral region to the injured one, on 2 different areas of the cortex (motor and somatosensory) concluded that, although it improved motor function, it was more effective when applied in the somatosensory area (Meehan et al. 2011).

Perceptual and Cognitive Difficulties

Unilateral spatial negligence is a perceptual disorder with a poor prognosis of functional recovery in stroke patients (Lim, Kang, and Paik 2010; Kim, Jung, and Shin 2015). The rTMS has been effective when applied in these type of post-natal disorders. It was described that, the hemispatial negligence produced by a stroke which affected the right hemisphere, improved after the application of a single 1 Hz session on the left parietal area, prior to a 10-day behavioral therapy. Regarding the effect of rTMS exerted on the cognitive functions in this type of patient, a prospective clinical trial revealed that the application of this type of stimulation on the prefrontal cortex did not obtain significant results on the executive and cognitive functions in studied patients, but it did show the improvement of the mood of the patients after the application of high frequency repetitive stimuli (10 Hz for 10 days) on the left dorsolateral prefrontal cortex areas 9 and 46 of Brodmann, (Kim et al. 2010).

Depression

Depression has been the most studied mood disorder using TMS. It is also the most prevalent emotional disturbance in stroke patients. Although most studies agree that it is a safe and well tolerated method when applied daily and for several weeks on the left prefrontal area, other studies suggest that the antidepressant effect depends on the area where the coil is placed. In patients who have had a stroke, the application of high-frequency rTMS (10 Hz) in 10 sessions for 2 weeks, in the left dorsolateral prefrontal area, resulted in a significant improvement in mood disorders (Kim et al. 2010).

SIDE EFFECTS

The secondary effects of the application of rTMS (repetitive Transcranial Magnetic Stimulation) usually are not important; however, in some patients there exists the possibility of inducing seizures, although their probability is minimal (< 1%). Although, the rTMS is a safe technique, some patients may experience adverse effects after application, which are considered mild and transient. One of these adverse effects is the presence of headaches that are mitigated by taking conventional analgesics. On the other hand, the risk of presenting epileptic seizures during rTMS is very low and it has not been shown to increase the risk of developing epileptic seizures after the end of the stimulation session in controlled epileptic patients (Nitsche 2016; Kumar et al. 2016; Agosta et al. 2016).

CONTRAINDICATIONS

The main contraindications for application of TMS have been observed in pregnant women and children under 2 years old as well as in patients presenting uncontrolled epilepsy and those with electronic body devices (Leon-Sarmiento, Granadillo and Bayona 2013). The rTMS has proven to be a technique safe and effective to treat some of the alterations that appear after presenting a stroke. Similarly, the rTMS has been especially valuable in helping to promote brain regeneration by the neuroplasticity mechanisms. The excitatory and inhibitory electromagnetic pulses applied in the ipsilateral or contralateral cerebral hemisphere to the lesion, as well as at the transcallosal level to modulate cerebral interhemispheric communication, gives the possibility of optimizing the cerebral functions, as well as achieving a recovery of the damaged brain area in a shorter time. Different studies carried out using rTMS, have corroborated the improvement of motor disorders, as well as of the depression and perceptual and cognitive difficulties originated in stroked patients (Nitsche 2016; Kumar et al. 2016; Agosta et al. 2016). On the other hand, considering the wide variety of applications that has had this type of stimulation in treating depressive disorders in adults with positive results in pharmaco-resistant patients, it is now being utilized in children that present neuropsychiatric disorders such as autism, ADHD, affective disorders, schizophrenia, etc. with variable results. In this context, several application models based on etiopathogenic hypothesis have been implemented, especially those referring to a possible alteration of brain plasticity in patients with autism spectrum disorders.

Effect of Transcranial Magnetic Stimulation on Sleep and Behavioral Disorders in Patients with Autism

In recent years, great interest has been generated in the use of rMTS to treat autism spectrum disorders. The main objectives of research in this field are related to the improvement of disruptive behavior. The most frequently brain area stimulated by this technique has been the CPFDL due to its multiple pathways of connection with subcortical areas that participate in higher brain functions, some of which do not function properly in patients with ASD (Oberman, Rotenberg and Pascual-Leone 2015). This area began to be explored under the hypothesis of improving the functional efficacy of long-range connections, since it was observed through different neuroimaging studies, the existence of a high level of connectivity in the CPFDL and a low connectivity level with other brain structures such as the orbitofrontal lateral cortex, fusiform gyrus, ventral striatum and the amygdala (Pascual-Leone and Tormos-Muñoz 2008). This could partly explain the high performance of some patients with ASD in cognitive activities, but with alterations in social interaction (Martínez-Morga et al. 2018). Based on the above, a series of investigations have been carried out to explore the efficacy of TMS in the improvement of the symptoms of patients with ASD, such as irritability, stereotyped behaviors, and hyperactivity. In most of the reported studies, a low frequency stimulation has been applied, in a range of 6 to 20 sessions using an eight-shaped coil placed on the CPFDL area (Barahona-Corréa et al. 2018). Different types of behavioral evaluations have been applying in various studies for the analysis of disruptive patterns. Among such evaluations are the revised scale of repetitive behaviors (RBS-R), the aberrant behavior checklist (ABC), specifically the stereotypic subscale or the autistic spectrum questionnaire (AQ). These studies have been complemented with neuroimaging techniques such as tomography by Positron emission (PET), functional magnetic resonance imaging (RMF) and electrophysiological recordings such as electroence-phalogram (Oberman et al. 2016). Most ASD patients show several types of sensory dysfunction with hypersensitivity (Crane, Goddard, and Pring 2009), which may be originated by the inhibitory dysfunction on sensory inputs. In addition, this impaired inhibition may explain in part the difficulties in social communication observed in these patients (Althaus et al. 1994). For example, some children with ASD tend to activly seek interactions with others, but in an unusual way (Scheeren, Koot, and Begeer 2012), possibly owing to both a neurophysiological excitation/inhibition imbalance and to the presence of clinical symptoms such as sensory deficits and/or problems in social communication (Selimbeyoglu et al. 2017). It is considered that the processes of the excitation/inhibition balance are continuously controlled by the interaction of GABA-ergic and glutamatergic mechanisms. The GABA receptor-mediated dysfunction has been proposed as a pathological hypothesis in ASD, which considers that the clinical symptoms including a deterioration in the quality of sensory information, may be caused by the failure to suppress irrelevant information (Hussman 2001). As described by other authors, inhibition, involving the sensorimotor gating through GABA-ergic neurotransmission, (Inui et al. 2018) is impaired in adults with ASD (Perry et al. 2007). Magnetic resonance spectroscopy studies have shown reduced GABA levels in several brain regions including the anterior cingulate cortex, left auditory cortex, and left motor cortex of ASD patients in comparison to control individuals (Cochran et al. 2015; Schür et al. 2016). Furthermore, the decrement in GABA levels in the anterior cingulate cortex in ASD patients have been correlated with greater severity of clinical symptoms, including social cognition and motor stereotypies (Harris et al. 2016). In postmortem studies, it was observed that the GABA(A) receptors were reduced in the parietal, cerebellar, and frontal regions in ASD patients in comparison with healthy controls (Dickinson, Jones, and Milne 2016). All of these findings indicate that the dysfunction of GABA receptors in the neural circuits probably are associated with the pathophysiology of ASD (Chen et al. 2014). On the other hand, the transcranial magnetic stimulation is a useful tool that facilitates the exploration of the dynamic functional properties of neural circuits in a non-invasive way. By means of this technique, it is possible to differentially examine the function of the neural circuits, that involve various neurotransmitters related with cortical functions in both healthy individuals, and in neurological and psychiatric patients (Kobayashi and Pascual-Leone 2003). Thus, the transcranial magnetic stimulation may be an important technical tool to assess the neurophysiological mechanisms involved in the excitation/inhibition balance in ASD patients. Autism treatment with Transcranial Magnetic Stimulation is still under investigation, therefore it cannot regularly be used as an appropriate therapeutic treatment. Since there are various etiopathogenic hypotheses about autism, this technique may have a wide variety of applications depending on the specifically of the type of autism spectrum disorder. The neurochemical studies carried out by magnetic resonance spectroscopy and postmortem brain analysis, suggest the existence of an alteration of the mechanisms involved in the excitation/inhibition balance, in neural circuits at various brain areas in patients with ASD (Dickinson, Jones, and Milne 2016; Uzunova, Pallanti, and Hollander 2016). However, according to studies carried out by means of magnetic resonance spectroscopy, it is not yet clear whether the excitation/inhibition balance participates to excite or inhibit the involved neural circuits in ASD.

SLEEP IN PATIENTS WITH ASD

The transcranial magnetic stimulation technique is emerging as a possible therapeutic tool for the regulation of sleep patterns in patients with ASD, due to the functional relationship existing between the neurons that participate in the regulation of circadian cycles, including sleep, and the control of behavioral and social cognition. This type of approach will facilitate the development of multiple investigations within the field of sleep medicine in neuropsychiatric patients (Lin et al. 2019). It is important to mention that sleep disorders are health problems of high prevalence in patients with ASD who present fragmented sleep, long REM sleep latency and low efficiency (Ayala-Guerrero, Mexicano, and Huicochea-Arredondo 2014) (Figure 4).

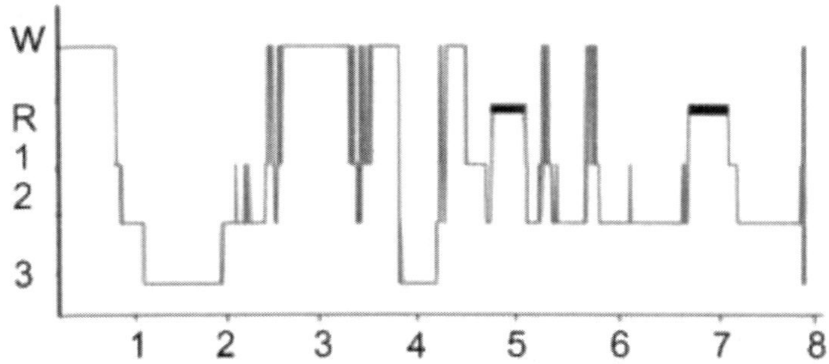

Figure 4. Hypnogram of an autistic patient. W, wakefulness; R, REM sleep; 1, 2, 3, N1, N2 and N3 NREM sleep phases.

Among the most common cormobid sleep alterations are obstructive sleep apnea syndrome (OSAS), insomnia, periodic limb movement and Bruxism (Moore et al. 2017). The comorbidity of these disorders impacts cognitive functioning, social performance and emotional states that are important functional abilities for the daily activities of the patients with ASD (Gail-Williams, Sears, and Allard 2004). At a molecular level, these sleep disorders have been associated with specific alterations in the Per1 and NAPS2 genes related to the circadian cycle and hormonal secretion in particular of melatonin releasing (Glickman 2010). Usually, the study of sleep disorders in patients with ASD, has been carried out through polysomnographic records, as well as with applications of behavioral questionnaires about sleep habits that allow evidencing the level of severity of sleep disturbances. Among the reported findings, it is an important highlight that there is concordance with the deficiencies in cognitive abilities presented by healthy individuals, when they have been deprived of sleep (Pierce 2011; Lambert et al. 2013). The changes in sleep characteristics observed in patients with ASD, also involve the brain activity microstructure, since it has reported alterations in the frequency and topographic distribution of mu waves (García-Vite et al. 2018) (Figure 5).

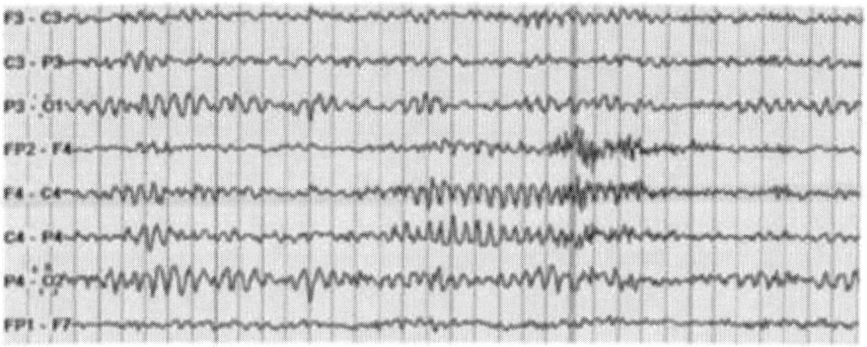

Figure 5. Trains of mu waves are observed in several derivations.

These waves have been considered as an important electrophysiological sign related to the social communication, whose alteration might explain, in part, the presence of this kind of problems observed in ASD patients. In addition, in these patients there is a reduction in the fronto-cortical connectivity during rapid eye movement sleep (Lazar et al. 2010). Furthermore, they present fragmented sleep due to the presence of multiple nocturnal awakenings, compromising the various functions in which sleep participates, such as secretion of hormones, degradation of noxious substances, memory consolidation, learning, etc. By the moment, it is not knowing the most adequate duration of treatment using the transcranial magnetic stimulation, as well as the exact moment of intervention and the most suitable standardized protocol. Probably in the coming years, with better designed studies, with larger samples of patients and longer follow-up times, a higher level of scientific evidence will be obtained. Future treatments of ASD and other neuropsychiatric diseases by TMS means, must be implemented in a multidisciplinary and interdisciplinary way, along with other physical and neurocognitive rehabilitation techniques. In the future it will be necessary to establish standardized protocols that facilitate the comparison of findings obtained by different groups of researchers. Although, with the utilization of TMS, several positive changes have been obtained at the neurophysiological level, and these changes have not been sufficiently correlated with behavioral improvements of autistic symptoms. Furthermore, it is missing longitudinal investigations to determine the duration of the reached improvements.

ACKNOWLEDGMENT

This work was supported by grant IG200619 from DGAPA-PAPIIT, UNAM, Mexico.

References

Ackerley, S. J., Stinear, C. M., Barber, P. A., Byblow, W. D. 2010. "Combining theta burst stimulation with training after subcortical stroke." *Stroke* 41(7):1568-72. doi:10.1161/STROKEAHA.110.
583278.

Agosta, S., Galante, E., Ferraro, F, Pascual-Leone, A., Oster J., Battelli, L. 2016. "Report of a delayed seizure after low frequency repetitive transcranial magnetic stimulation in a chronic stroke patient." *Clinical Neurophysiology* 127(2):1736-1737. doi:10.1016/j.clinph.2014.11.029.

Althaus, M., Renk, M., Minderaa, R. B., De Vries, H. D. 1994. "Initiatory and reactive behaviour in young children with a pervasive developmental disorder: two independent components in communicative behavior for distinguishing subgroups." *European Child & Adolescent Psychiatry* 3(4):242-254. doi: 10.1007/BF01978113.

Ameli, M., Grefkes, C., Kemper, F., Riegg, F. P., Rehme, A. K., Karbe, H., Fink, G. R., Nowak, D. A. 2009. "Differential effects of high-frequency repetitive transcranial magnetic stimulation over ipsilesional primary motor cortex in cortical and subcortical middle cerebral artery stroke." *Annals of Neurology: Official Journal of the American Neurological Association and the Child Neurology Society* 66(3):298-309. doi:10.1002/ana.21725.

Auriat, A. M., Neva, J. L., Peters, S., Ferris, J. K., Boyd, L. A. 2015. "A review of transcranial magnetic stimulation and multimodal neuroimaging to characterize post-stroke neuroplasticity." *Frontiers in Neurology* 6:226. doi:10.3389/fneur.2015.00226.

Ayala-Guerrero, F., Mexicano, G., and Huicochea-Arredondo, S. 2014. "Sleep characteristics in patients with autism spectrum disorders." *Journal of Neurological Disorders & Stroke*, 2(3):1-6.

Barahona-Corréa, J. B., Velosa, A., Chainho, A., Lopes, R., Oliveira-Maia, A. J. 2018. "Repetitive Transcranial Magnetic Stimulation for Treatment of autism spectrum disorder: a systematic review and meta-analysis." *Frontiers in Integrative Neuroscience* 12:27. doi: 10.3389/fnint.2018.00027.

Barker, A. T., Jalinous R., and Freeston I. L. 1985. "Non-invasive magnetic stimulation of the human motor cortex." *The Lancet* 1(8437):1106-1107.

Bayón, M. 2011. "Transcranial magnetic stimulation in stroke rehabilitation." *Rehabilitation* 45(3):261-267. doi:10.1016/j.rh.2011.03.014.

Blesneag, A. V., Popa, L., Stan, A. D. 2015. "Non-invasive brain stimulation in early rehabilitation after stroke." *Journal of Medicine and Life* 8:52-56.

Buetefisch, C. M. 2015. "Role of the contralesional hemisphere in post-stroke recovery of upper extremity motor function." *Frontiers in Neurology,* 6:214. doi: 10.3389/fneur.2015.00214.

Cassidy, J. M., Chu, H., Anderson, D. C., Krach, L. E., Snow, L., Kimberley, T. J., Carey, J. R. 2015. "A comparison of primed low-frequency repetitive transcranial magnetic stimulation treatments in chronic stroke." *Brain Stimulation* 8(6):1074-1084. doi: 10.1016/j.brs.2015.06.007.

Chen, C. H., Huang, C. C., Cheng, M. C., Chiu, Y. N., Tsai, W. C., Wu, Y. Y., Liu, S. K., Gau, S. S. F. 2014. "Genetic analysis of GABRB3 as a candidate gene of autism spectrum disorders." *Molecular Autism* 5(1):36. doi: 10.1186/2040-2392-5-36.

Chervyakov, A. V., Chernyavsky, A. Y., Sinitsyn, D. O., Piradov, M. A. 2015. "Possible mechanisms underlying the therapeutic effects of transcranial magnetic stimulation. *Frontiers in Human Neuroscience* 9:303. doi:10.3389/fnhum.2015.00303.

Clanflin, E. S., Krishnan, C., Khot, S. P. 2015. "Emerging treatments for motor rehabilitation after stroke." *The Neurohospitalist* 5(2):77-88. doi: 10.1177/1941874414561023.

Cochran, D. M., Sikoglu, E. M., Hodge, S. M., Edden, R. A., Foley, A., Kennedy, D. N., Moore, C. M., Frazier, J. A. 2015. "Relationship among glutamine, γ-aminobutyric acid, and social cognition in autism spectrum disorders." *Journal of Child and Adolescent Psychophar-macology* 25(4):314-322. doi:10.1089/cap.2014.0112.

Corti, M., Patten, C., and Triggs, W. 2012. "Repetitive transcranial magnetic stimulation of motor cortex after stroke: a focused review." *American Journal of Physical Medicine & Rehabilitation* 91(3):254-270. doi:10.1097/PHM.0b013e318228bf0c.

Crane, L., Goddard, L., Pring, L. 2009. "Sensory processing in adults with autism spectrum disorders." *Autism* 13(3):215-228. doi: 10.1177/1362361309103794.

Cunningham, D. A., Machado, A., Janini, D., Varnerin, N., Bonnett, C., Yue, G., Jones, S., Lowe, M., Beall, E., Sakaie, K., Plow, E. B. 2015. "Assessment of inter-hemispheric imbalance using imaging and noninvasive brain stimulation in patients with chronic stroke." *Archives of Physical Medicine and Rehabilitation* 96(4):S94-S103. doi: 10.1016/j.apmr.2014.07.419.

Demirtas-Tatlidede, A., Alonso-Alonso, M., Shetty, R. P., Ronen, I., Pascual-Leone, A., Fregni, F. 2015. "Long-term effects of contralesional rTMS in severe stroke: safety, cortical excitability, and relationship with transcallosal motor fibers." *NeuroRehabilitation* 36(1):51-59. doi: 10.3233/NRE-141191.

Di Pino, G., Pellegrino, G., Assenza, G., Capone, F., Ferreri, F., Formica, D., Ranieri, F., Tombini, M., Ziemann, U., Rothwell, J. C., Di Lazzaro, V. 2014. "Modulation of brain plasticity in stroke: a novel model for neurorehabilitation." *Nature Reviews Neurology* 10(10):597-608. doi: 10.1038/nrneurol.2014.162.

Dickinson, A., Jones, M., Milne, E. 2016. "Measuring neural excitation and inhibition in autism: different approaches, different findings and different interpretations." *Brain Research* 1648:277–289. doi: 10.1016/j.brainres.2016.07.011.

Emara, T. H., Moustafa, R. R., Elnahas, N. M., Elganzoury, A. M., Abdo, T. A., Mohamed, S. A., Eletribi, M. A. 2010. "Repetitive transcranial magnetic stimulation at 1Hz and 5Hz produces sustained improvement in motor function and disability after ischaemic stroke." *European Journal of Neurology* 17(9):1203-1209. doi: 10.1111/j.1468-1331.2010.03000.x.

Figueroa, J., Villamayor, B., Antelo, A., 2015. "Rehabilitation of cerebral stroke: evaluation, prognosis and treatment." In Functional re-education after a stroke edited by J. A. Castillo Sánchez, I. Jiménez Martín, 89-104. Barcelona, Spain: S.L.U Elsevier.

Fuentes, B., Gállego, J., Gil-Nuñez, A., Morales, A., Purroy, F., Roquer, J., Segura, T., Tejada, J., Lago, A., Díez-Tejedor, E., Alonso de Leciñana, M., Álvarez-Sabin, J., Arenillas, J., Calleja, S., Casado, I., Castellanos, M., Castillo, J., Dávalos, A., Díaz-Otero, F., Egido, J.A., López-Fernández, J.C., Freijo, M., García Pastor, A., Gilo, F., Irimia, P., Maestre, J., Masjuan, J., Martí-Fábregas, J., Martínez-Sánchez, P., Martínez-Vila, E., Molina, C., Nombela, F., Ribó, M., Rodríguez-Yañez, M., Rubio, F., Serena, J., Simal, P., Vivancos J. 2012. "Guidelines for the preventive treatment of ischaemic stroke and TIA (I). Update on risk factors and life style." *Neurología (English Edition)* 27(9):560-574. doi: 10.1016/j.nrleng.2011.06.004.

Gail-Williams, P., Sears, L. L., and Allard, A. 2004. "Sleep problems in children with autism." *Journal of Sleep Research* 13(3):265-268. doi:10.1111/j.1365-2869.2004.00405.x.

Garcia-Vite, T. K., Ayala-Guerrero, F., Mateos Salgado, E. L., Cebreros Paniagua, R. 2018. "Characterization of the Mu rhythm during the sleep of children with autism spectrum disorder level 1." *Mental health* 41(3):109-116. doi:10.17711/SM.0185-3325.2018.017.

Glickman, G. 2010. "Circadian rhythms and sleep in children with autism." *Neuroscience and Biobehavioral Reviews* 34(5):755–768. doi: 10.1016/j.neubiorev.2009.11.017.

Harris, A. D., Singer, H. S., Horska, A., Kline, T., Ryan, M., Edden, R. A. E., Mahone, E. M. 2016. "GABA and glutamate in children with primary complex motor stereotypies: an 1H-MRS study at 7T." *American Journal of Neuroradiology* 37(3):552-557. doi: 10.3174/ajnr.A4547.

Huang, Y. Z., Rothwell, J. C., Edwards, M. J., Chen, R. S. 2007. "Effect of physiological activity on an NMDA-dependent form of cortical plasticity in humans." *Cerebral Cortex* 18(3):563-570. doi: 10.1093/cercor/bhm087.

Hussman, J. P. 2001. "Suppressed GABAergic inhibition as a common factor in suspected etiologies of autism." *Journal of Autism and Developmental Disorders* 31(2):247-248.

Hwang, J. M., Kim, Y. H., Yoon, K. J., Uhm, K. E., Chang, W. H. 2015. "Different responses to facilitatory rTMS according to BDNF genotype." *Clinical Neurophysiology* 126(7):1348-1353. doi:10.1016/j.clinph.2014.09.028.

Inui, K., Takeuchi, N., Sugiyama, S., Motomura, E., Nishihara, M. 2018. "GABA-ergic mechanisms involved in the prepulse inhibition of auditory evoked cortical responses in humans." *PloS ONE* 13(1):e0190481.

Jin, Y., Xing, G., Li, G., Wang, A., Feng, S., Tang, Q., Liao, X., Guo, Z., McClure, M. A., Mu, Q. 2015. "High frequency repetitive transcranial magnetic stimulation therapy for chronic neuropathic pain: A meta-analysis." *Pain Physician* 18(6):1029-1046.

Kakuda, W., Abo, M., Momosaki, R., Morooka, A. 2011. "Therapeutic application of 6-Hz-primed low-frequency rTMS combined with intensive speech therapy for post-stroke aphasia." *Brain Injury* 25(12):1242-1248. doi:10.3109/02699052.2011.608212.

Kim, B. R., Kim, D. Y., Chun, M. H., Yi, J. H., Kwon, J. S. 2010. "Effect of repetitive transcranial magnetic stimulation on cognition and mood in stroke patients: a double-blind, sham-controlled trial." *American Journal of Physical Medicine & Rehabilitation* 89(5):362-368. doi:10.1097/PHM.0b013e3181d8a5b1.

Kim, Y. H., You, S. H., Ko, M. H., Park, J. W., Lee, K. H., Jang, S. H., Yoo, W. K., Hallett, M. 2006. "Repetitive transcranial magnetic stimulation–induced corticomotor excitability and associated motor skill acquisition in chronic stroke." *Stroke* 37(6):1471-1476. doi:10.1161/01.STR.0000221233.55497.51.

Kim, Y. K., Jung, J. H., Shin, S. H. 2015. "A comparison of the effects of repetitive transcranial magnetic stimulation (rTMS) by number of stimulation sessions on hemispatial neglect in chronic stroke patients." *Experimental Brain Research* 233(1):283-289. doi:10.1007/s00221-014-4112-9.

Klein, M. M., Treister, R., Raij, T., Pascual-Leone, A., Park, L., Nurmikko, T., Lenz, F., Lefaucheur J. P., Lang, M., Hallett, M., Fox, M., Cudkowicz, M., Costello, A., Carr, D., Ayache, S., Oaklander, A. 2015. "Transcranial magnetic stimulation of the brain: guidelines for pain treatment research." *Pain* 156(9):1601-1614. doi:10.1097/j.pain.0000000000000210.

Klomjai, W., Lackmy-Vallee, A., Roche, N., Pradat-Diehl, P., Marchand-Pauvert, V., Katz, R. 2015. "Repetitive transcranial magnetic stimulation and transcranial direct current stimulation in motor rehabilitation after stroke: an update." *Annals of Physical and Rehabilitation Medicine* 58(4):220-224. doi:10.1016/j.rehab.2015.05.006.

Kolin, A., Brill, N. Q., and Broberg, P. J. 1959. "Stimulation of irritable tissues by means of an alternating magnetic field." *Proceedings of the Society for Experimental Biology and Medicine 102*(1):251-253. doi:10.3181/00379727-102-25209.

Kobayashi, M. and Pascual-Leone, A. 2003. "Transcranial magnetic stimulation in neurology." *The Lancet Neurology* 2(3):145-156. doi:10.1016/S1474-4422(03)00321-1.

Kumar, N., Padma, M. S., Verma, R., Sharma, H., Modak, T. 2016. "Can low-frequency repetitive transcranial magnetic stimulation precipitate a late-onset seizure in a stroke patient?" *Clinical Neurophysiology: Official Journal of the International Federation of Clinical Neurophysiology* 127(2):1734-1736. doi:10.1016/j.clinph.2015.06.033.

Lambert, A., Tessier, S., Chevrier, É., Scherzer, P., Mottron, L., Godbout, R. 2013. "Sleep in children with high functioning autism: Polysomnography, questionnaires and diaries in a non-complaining sample." *Sleep Medicine* 14:e137-e138. doi:10.1016/j.sleep.2013.11.310.

Lázár, A. S., Lázár, Z. I., Bíró, A., Győri, M., Tárnok, Z., Prekop, C., Keszei, A., Stefanik, K., Gádoros, J., Halász, P., Bódizs, R. 2010. "Reduced fronto-cortical brain connectivity during NREM sleep in Asperger syndrome: An EEG spectral and phase coherence study." *Clinical Neurophysiology* 121(11):1844-1854. doi:10.1016/j.clinph.2010.03.054.

Lee, J. H., Byun, J. H, Choe, Y. R., Lim, S. K., Lee, K. Y, Choi, I. S. 2015. "Successful treatment of phantom limb pain by 1 Hz repetitive transcranial magnetic stimulation over affected supplementary motor complex: a case report." *Annals of Rehabilitation Medicine* 39(4):630-633. doi:10.5535/arm.2015.39.4.630.

Leon-Sarmiento, F. E., Granadillo E., and Bayona E. A. 2013. "Present and future of transcranial magnetic stimulation." *Clinical research* 54(1):74-89.

Liew, S. L., Santarnecchi, E., Buch, E. R., Cohen, L. G. 2014. "Non-invasive brain stimulation in neurorehabilitation: local and distant effects for motor recovery." *Frontiers in Human Neuroscience* 8:378. doi:10.3389/fnhum.2014.00378.

Lim, J. Y., Kang, E. K., Paik, N. J. 2010. "Repetitive transcranial magnetic stimulation for hemispatial neglect in patients after stroke: an open-label pilot study." *Journal of Rehabilitation Medicine* 42(5):447-452. doi: 10.2340/16501977-0553.

Lin, J., Liu, X., Li, H., Yu, L., Shen, M., Lou, Y., Xie, S., Chen, J., Zhang, R., Yuan, T. F. 2019. "Chronic repetitive transcranial magnetic stimulation (rTMS) on sleeping quality and mood status in drug dependent male inpatients during abstinence." *Sleep Medicine* 58:7-12. doi:10.1016/j.sleep.2019.01.052.

Lipton, R. B., Dodick, D. W., Silberstein, S. D., Saper, J. R., Aurora, S. K., Pearlman, S. H., Fischell, R. E., Ruppel, P.L., Goadsby, P. J. 2010. "Single-pulse transcranial magnetic stimulation for acute treatment of migraine with aura: a randomized, double-blind, parallel-group, sham-controlled trial." *The Lancet Neurology* 9(4):373-380. doi:10.1016/S1474-4422(10)70054-5.

Ljubisavljevic, M. R., Javid, A., Oommen, J., Parekh, K., Nagelkerke, N., Shehab, S., Adrian, T. E. 2015. "The effects of different repetitive transcranial magnetic stimulation (rTMS) protocols on cortical gene expression in a rat model of cerebral ischemic-reperfusion injury." *PLoS ONE* 10(10):e0139892. doi:10.1371/journal.pone.0139892.

Lüdemann-Podubecká, J., Bösl, K., Theilig, S., Wiederer, R., Nowak D.A. 2015. "The effectiveness of 1 Hz rTMS over the primary motor area of the unaffected hemisphere to improve hand function after stroke depends on hemispheric dominance." *Brain Stimulation* 8(4):823-830. doi:10.1016/j.brs.2015.02.004.

Malcolm, M. P., Vaughn, H. N., and Greene, D. 2015. "Inhibitory and Excitatory Motor Cortex Dysfunction Persists in the Chronic Post-Stroke Recovery Phase. *Journal of Clinical Neurophysiology: Official Publication of The American Electroencephalographic Society* 32(3):251. doi:10.1097/WNP.0000000000000143.

Martinez-Morga, M., Quesada-Rico, M., Bueno, C., Martinez, S. 2018. "Neurobiological basis of Autism Spectrum Disorder and Attention Deficit / Hyperactivity Disorders: Neural Differentiation and Synaptogenesis." *Journal of Neurology* 66(1):97-101.

Medina, F. J., Pascual, and A., Túnez, I. 2014. "Mechanisms of action in transcranial magnetic stimulation." In *Transcranial magnetic stimulation and neuromodulation. Present and future in neuroscience*, edited by I. Túnez Fiñana, A. Pascual Leone, 21-30, Barcelona, Spain: Elsevier.

Meehan, S. K., Dao, E., Linsdell, M. A., Boyd, L. A. 2011. "Continuous theta burst stimulation over the contralesional sensory and motor cortex enhances motor learning post-stroke." *Neuroscience Letters* 500(1):26-30. doi:10.1016/j.neulet.2011.05.237.

Moore, M., Evans, V., Hanvey, G., Johnson, C. 2017. "Assessment of sleep in children with autism spectrum disorder." *Children* 4(8):72. doi:10.3390/children4080072.

Nitsche, M. A. 2016. "Co-incidence or causality? Seizures after slow rTMS in stroke patients." *Clinical Neurophysiology: Official Journal of the International Federation of Clinical Neurophysiology* 127(2):1020-1021. doi:10.1016/j.clinph.2015.07.019.

Oberman, L. M., Enticott, P. G., Casanova, M. F., Rotenberg, A., Pascual-Leone, A., McCracken, J. T., TMS in ASD Consensus Group. 2016. "Transcranial magnetic stimulation in autism spectrum disorder: challenges, promise, and roadmap for future research." *Autism Research* 9(2):184-203. doi:10.1002/aur.1567.

Oberman, L. M., Rotenberg, A., and Pascual-Leone, A. 2015. "Use of transcranial magnetic stimulation in autism spectrum disorders." *Journal of Autism and Developmental Disorders* 45(2):524-536. doi:10.1007/s10803-013-1960-2.

Page, S. J., Cunningham, D. A., Plow, E., Blazak, B. 2015. "It takes two: noninvasive brain stimulation combined with neurorehabilitation." *Archives of Physical Medicine and Rehabilitation* 96(4):S89-S93. doi:10.1016/j.apmr.2014.09.019.

Pascual-Leone, A. and Tormos-Muñoz, J. M. 2008. "Transcranial magnetic stimulation: fundamentals and potential of modulating specific neural networks". *Journal of Neurology* 46(1):S3-10. doi:10.33588/rn.46S01.2008081.

Perry, W., Minassian, A., Lopez, B., Maron, L., Lincoln, A. 2007. "Sensorimotor gating deficits in adults with autism." *Biological psychiatry* 61(4):482-486. doi:10.1016/j.biopsych.2005.09.025.

Pierce, K. 2011. "Early functional brain development in autism and the promise of sleep fMRI." *Brain Research* 1380:162-174. doi:10.1016/j.brainres.2010.09.028.

Raffin, E. and Siebner, H. R. 2014. "Transcranial brain stimulation to promote functional recovery after stroke." *Current Opinion in Neurology* 27(1):54-60. doi:10.1097/WCO.0000000000000059.

Raguer Sanz, N. 2008. "Transcranial and peripheral magnetic stimulation" In *Manual of Clinical Electromyography* edited by E. Gutierrez-Rivas, M.D. Jimenez Hernandez, J. Pardo Fernandez, M. Romero-Acebal 1991-2003. Ergon: Majadahonda.

Roland, L. T., Peelle, J. E., Kallogjeri, D., Nicklaus, J., Piccirillo, J. F. 2016. "The effect of noninvasive brain stimulation on neural connectivity in Tinnitus: a randomized trial." *The Laryngoscope* 126(5):1201-1206. doi:10.1002/lary.25650.

Rossi, S., Hallett, M., Rossini, P. M., Pascual-Leone, A., Safety of TMS Consensus Group. 2009. "Safety, ethical considerations, and application guidelines for the use of transcranial magnetic stimulation in clinical practice and research. *Clinical Neurophysiology* 120(12):2008-2039. doi: 10.1016/j.clinph.2009.08.016.

Rossini, P. M., Burke, D., Chen, R., Cohen, L. G., Daskalakis, Z., Di Iorio, R., Lazzaro, V. D., Ferreri, F., Fitzgerald, P. B., George, M.S., Hallett, M., Lefaucheur, J. P., Langguth, B., Matsumoto, H., Miniussi, C., Nitsche, M. A., Pascual-Leone, A., Paulus, W., Rossi, S. Rothwell, J. C., Siebner, H. R., Ugawa, Y., Walsh, V., Ziemann, U. 2015. "Non-invasive electrical and magnetic stimulation of the brain, spinal cord, roots and peripheral nerves: basic principles and procedures for routine clinical and research application. An updated report from an IFCN Committee." *Clinical Neurophysiology* 126(6):1071-1107. doi: 10.1016/j.clinph.2015.02.001.

Rubio, B. 2014. "New perspectives of transcranial magnetic stimulation in psychiatric disorders of childhood and adolescence." In *Transcranial magnetic stimulation and neuromodulation. Present and future in neuroscience* edited by I. Túñez Fiñana and A. Pascual Leone, 135-159. Barcelona, Spain: ElsevierSpain.

Scheeren, A. M., Koot, H. M., and Begeer, S. 2012. "Social interaction style of children and adolescents with high-functioning autism spectrum disorder." *Journal of Autism and Developmental Disorders* 42(10):2046-2055. doi: 10.1007/s10803-012-1451-x.

Schür, R. R., Draisma, L. W., Wijnen, J. P., Boks, M. P., Koevoets, M. G., Joëls, M., Klomp, D. W., Kahn, R. S., Vinkers, C. H. 2016. "Brain GABA levels across psychiatric disorders: A systematic literature review and meta-analysis of 1H-MRS studies." *Human Brain Mapping* 37(9):3337-3352. doi: 10.1002/hbm.23244.

Selimbeyoglu, A., Kim, C. K., Inoue, M., Lee, S. Y., Hong, A. S., Kauvar, I., Ramakrishnan, C., Fenno, L. E., Davidson, T. J., Wright, M., Deisseroth, K. 2017. "Modulation of prefrontal cortex excitation/inhibition balance rescues social behavior in CNTNAP2-deficient mice." *Science Translational Medicine* 9(401):eaah6733. doi: 10.1126/scitranslmed.aah6733.

Simonetta-Moreau, M. 2014. "Non-invasive brain stimulation (NIBS) and motor recovery after stroke." *Annals of Physical and Rehabilitation Medicine* 57(8):530-542. doi: 10.1016/j.rehab.2014.08.003.

Takeuchi, N., Chuma, T., Matsuo, Y., Watanabe, I., Ikoma, K. 2005. "Repetitive transcranial magnetic stimulation of contralesional primary motor cortex improves hand function after stroke." *Stroke* 36(12):2681-2686. doi: 10.1161/01.STR.0000189658.51972.34.

Tang, A., Thickbroom G., and Rodger, J. 2015. "Repetitive transcranial magnetic stimulation of the brain: Mechanisms from animal and experimental models." *The Neuroscientist* 23(1):82-94. doi: 10.1177/1073858415618897.

Uzunova, G., Pallanti, S., and Hollander, E. 2016. "Excitatory/inhibitory imbalance in autism spectrum disorders: implications for interventions and therapeutics." *The World Journal of Biological Psychiatry* 17(3):174-186. doi:10.3109/15622975.2015.1085597.

Van den Noort, M., Bosch, P., Yeo, S., Lim, S. 2015. "Transcranial magnetic stimulation for Parkinson's disease." *Movement Disorders* 30(14):1973. doi: 10.1002/mds.26439.

Verdugo-Diaz, L. and Drucker-Colin, R. 2014. "Magnetic fields: uses in biology and medicine". In *Transcranial magnetic stimulation and neuromodulation. Present and future in neuroscience* edited by I. Túñez Fiñana and A. Pascual Leone, 1-19. Barcelona, Spain: Elsevier Spain.

Wassermann, E. M., and Zimmermann, T. 2012. "Transcranial magnetic brain stimulation: therapeutic promises and scientific gaps." *Pharmacology & Therapeutics* 133(1):98-107. doi:/10.1016/j.pharmthera.2011.09.003.

In: Sleep, Cognition and Emotion
Editors: I. Ramírez Salado et al.

ISBN: 978-1-53617-561-5

Chapter 12

DEVELOPMENTAL MODE AND REM SLEEP IN MAMMALS SHOW CORRELATED EVOLUTION

Pilar Chiappa*

Departamento de Etología, Instituto Nacional de Psiquiatría Ramón de la Fuente Muñiz, Tlalpan, Ciudad de México, Mexico

ABSTRACT

In this chapter, I present a new hypothesis about the relationship between the developmental mode and the amount of REM sleep, which states that REM sleep is one of the features that characterize neotenous-altricial cognitive paedomorphosis. I test this new hypothesis within the framework of phylogenetic methods. I obtained data from a published resource about relative REM sleep, relative age at weaning, and relative youth period. A phylogenetic generalized least squares analysis sustained partially the predicted explanatory power of relative age at weaning age and relative youth period for relative REM sleep. The hypothesis about the relationship between the mode of development and the relative amount of REM sleep that I presented in this chapter cannot be rejected because other operationalizations remain possible. Therefore, it could still be useful in explaining some of the failures of two previous hypotheses. The notion of neotenous-altricial cognitive paedomorphosis or juvenile-like adult cognition has the potential to sustain the assumption of a constant function of REM sleep during lifespan.

Keywords: sleep, developmental mode, evolution, mammals, comparative method, PGLS (phylogenetic general least squares)

* Corresponding Author's Email: pilar.chiappa@gmail.com.

INTRODUCTION

Sleep is a common behavioral state in many animal species (Campbell and Tobler 1984; Rattenborg et al. 2017). In marsupial and placental mammals EEG techniques are useful to identify two specific states during sleep (Allison and Van Twyver 1970): non-REM (NREM) sleep and rapid eye movement (REM) sleep (Capellini et al. 2008a). Great variation in the duration of both sleep states, as well as in the duration of total sleep time (TST) have been observed among mammalian species (Allison and Van Twyver 1970; Zepelin and Rechtschaffen 1974; Allison and Cicchetti 1976; Campbell and Tobler 1984; Capellini et al. 2008a; Lesku et al. 2006; Lesku et al. 2008; Rattenborg et al. 2017). Measurements of the duration of REM sleep, non-REM sleep, and TST are known as sleep quotas (Acerbi, McNamara and Nunn 2008).

From an evolutionary perspective, the variation in sleep quotas among mammals could indicate the participation of different selective processes among species. In addition to the variation between species, there are three data that indicate that sleep has a function linked to evolutive processes: 1) there is no convincing case of a species that does not sleep, 2) sleep loss occurs with a compensatory rebound, and 3) sleep loss has negative consequences (Cirelli and Tononi 2008). Several comparative attempts have been made to identify factors, both intrinsic and extrinsic, that could had promoted the differential evolution of relative duration of sleep states among mammals and shed light on the adaptive functions of sleep (Zepelin and Rechtschaffen 1974; Berger 1975; Allison and Cicchetti 1976; Campbell and Tobler 1984; Elgar et al. 1988; Elgar et al. 1990; Berger 1990; Siegel 1995; Rechtschaffen 1998; Ruth et al. 2006; Lesku et al. 2006; Lesku et al. 2008; Capellini et al. 2008a; Capellini et al. 2008b; Capellini et al. 2009; Preston et al. 2009; Nunn et al. 2010; Samson and Nunn 2015; Barton and Capellini 2016). However, no clear adaptive function of sleep has been identified (Rattenborg et al. 2017).

Earlier comparative studies of sleep comprehend several attempts to associate the developmental mode with REM sleep. In this line of research, two major hypotheses have been tested so far. The first posits that REM sleep provides artificial stimulation for cortical development and predicts that REM sleep will last longer in animal species that have a less developed brain at birth (Roffwarg, Muzio and Dement 1966). Sustained by the characteristic low level of the muscular tone of REM sleep (Rattenborg et al. 2017), the second hypothesis assumes that this stage is very dangerous for the physical integrity of an animal, and it postulates that risk exposure of the sleeping place might mediate the relationship between REM sleep and developmental mode. Therefore, it leads to the expectation that REM sleep quotas and developmental modes are related to the security of sleeping sites (Allison and Cicchetti 1976).

Lesku et al. (2006) tested these hypotheses with the first comparative study that accounted for the phylogenetic relationships among species. They approached the degree of development of the brain at birth with the residuals for the gestational period relative to body mass in 54 mammal species. They found that adults of species that are born relatively precocial have significantly lesser REM sleep duration and percentage than adults of species that are born relatively altricial. The authors pointed out that REM sleep might be important for early brain development, however, they recognized that it is not clear why, under this hypothesis, altricial species would engage in more REM sleep as adults than precocial species (see also Siegel 2004). They found no evidence to support the second hypothesis. Subsequently, Lesku et al. (2008) replicated this finding in 83 mammalian species.

Similarly, Capellini et al. (2008a) found a negative association between REM sleep quotas and the residuals of gestation length relative to body mass in 40 mammal species. They also found no evidence of an association between REM sleep quotas and relative brain mass at birth, which is a closer proxy for neonatal brain maturation. As Capellini et al. (2008a) pointed out, the lack of association between REM sleep quotas and neonatal brain undermines the idea of a functional role of REM sleep in neonatal brain maturation. However, they did not suggest any explanation for their first result.

The recurrent finding of an evolutionary association between developmental mode and adult REM sleep quotas remain without explanation (Siegel 2004, Lesku et al. 2006, Lesku et al. 2008). Furthermore, the length of gestation represents an indirect measure of the extension of altricial life. It has been used in comparative studies of sleep because it is thought to influence the developmental states of neonates (Pagel and Harvey 1990).

In the understanding that the study of the processes that were involved in the evolution of sleep can shed light on some medical problems related to various sleep disorders (Nunn, Samson and Krystal 2016), comparative studies are still required in the specialized literature (for instance, Nunn, Samson and Krystal 2016; Rattenborg et al. 2017).

In this chapter, I present a new hypothesis about the relationship between the mode of development and the duration of REM sleep relative to total sleep, and I test this new hypothesis through a direct measure of altriciality within the framework of phylogenetic methods (Harvey and Pagel 1991).

REM Sleep, Altricial Development and Neotenous-Altricial Cognitive Paedomorphosis

The human species has many evolutionary particularities in REM sleep, cognitive capabilities, and developmental mode. Regarding the particularities of human REM sleep, recently it has been assessed that human sleep is highly derived (namely, that the characteristics of human sleep originated in the human lineage and therefore are novel). Human sleep is shorter, deeper, and exhibits a higher proportion of REM than expected (Samson and Nunn 2015).

As regards to human developmental mode, a remarkable feature is that babies are highly altricial, so much so that raising one of them usually involves more than one adult (Blaffer Hrdy 1999). This neonatal condition is linked to the evolution of development through its delay compared to the ancestral pattern (Gould 1977). This form of developmental evolution is known as neoteny (Gould 1977). Human neotenous development is responsible of human paedomorphosis, which is the presence of infantile traits in adults (Gould 1977). Albeit neotenous human paedomorphosis has been hypothesized to comprehend some human cognitive features (Gould 1977), for decades the most striking aspect of human neotenous paedomorphosis has been morphological, which account for features such as a rounded head, a broad forehead and a shortage of hair among many others (Bolk 1926). Only recently, the neurological aspects of neotenous human paedomorphosis have been highlighted, accounting for features such as the rate and period of cerebral growth relative to lifespan, the increased-synaptic plasticity and the timing of cortical development in some regions of the brain (Bufill, Agustí and Blesa 2011). Neotenous paedomorphosis in the human brain provides an extended window for neural plasticity, and therefore a longer period for learning (Shi and Su 2019). Humans can afford a slow development in part by the extended period of caregiving provided by adults (Tottenham 2019). Human neotenous paedomorphosis and human altricial developmental mode are two sides of the same coin.

The human species is not the only one that has evolved in this way: 1) there are others live beings with extraordinary proportions of REM sleep, for instance, the North American opossum, *Virginia opossum*, has 45% of the total sleep devoted to REM (Lesku et al. 2006); 2) many other species have extended periods of altricial life in comparison to close phylogenetically relatives, as the South American Hystricomorph Rodent, *Octodon degus*, whose breastfeeding time represents 0.70 percent of lifespan in comparison with the Volcano Mouse, *Neotomodon alstoni*, who occupies 3.75 percent of his life before being weaned (data from Table 1), and finally, 3) the evolution of cognitive capabilities has occurred in many branches of the tree of life; for instance, both chimpanzees, *Pan troglodytes*, and capuchins, *Sapajus apella*, have evolved a risk-prone patient strategy that is dissimilar from more closely related species within the great ape clade and New World monkey clade, respectively (De Petrillo and Rosati 2019).

A new hypothesis arises from the considerations set so far, which states that REM sleep is one of the features that characterize neotenous-altricial cognitive paedomorphosis. Therefore, in a comparative framework, it might be expected to find a positive association between the degree of altriciality in offspring, the duration of youth period and the proportion of REM sleep relative to total sleep in adults.

My aim in the present chapter was to explore such a possibility by means of a comparative analysis of the amount of REM sleep relative to total sleep and two developmental variables addressing the degree of altriciality (age at weaning relative to lifespan) and the degree of neotenous development (extension of juvenile period relative to lifespan), respectively, in adults from different species.

METHODS

Data Selection

Sleep Data

I obtained the sleep data from a published resource (McNamara et al. 2008). This is an impressive database constructed expressly for comparative studies of sleep. Besides the sleep quotas for animal species, the database contains relevant data of each entry, as laboratory conditions, duration of the recording period and age of subjects. Because comparative studies of sleep must apply criteria to homogenize data to the fullest (Capellini 2008a; McNamara et al. 2008; Barton and Capellini 2016), I selected the entries corresponding to adult land animals with a period of adaptation to recording conditions and a 24 hours period of EEG recording. I also selected the more recent studies with highest number of subjects.

Developmental Mode Data

Species place in the precocial-altricial spectrum has been determined through different features (i.e., locomotion, thermoregulation, sensation, and alimentation) (Starck and Ricklefs 1998). However, this method of classification could reduce inter specific variation. Chiappa, Singh and Pellicer (2018) used the period between birth and weaning relative to lifespan as a direct of the altricial period. Following them, the analysis includes the age at weaning (i.e., duration of lactation). Life history traits are known to covary systematically across species (Stearns 1983). Therefore, in addition to the age at weaning in days, the analysis also includes the lifespan in days. Because other features may indicate the speed of development, I included the length of the youth period in relation to life expectancy and the age at which sexual maturity is reached in relation to life expectancy.

For species that met the criteria for selecting sleep data, I performed a semi-structured search to obtain the mean in days of the age at weaning and the mean lifespan (i.e., the time from birth to death) in days. (i.e., the time from birth to death). I used three large databases to conduct our main search: Human Ageing Genomic Resources (Tacutu et al. 2013), Animal Diversity (Myers et al. 2017), and Pantheria (Jones et al. 2009). I prioritized data on wild over captive conditions. When more than one value or one range was available, I calculated the mean. The search did not produce information on some species. To fill the gaps, I used Google Scholar to search for papers published in English. The entries we used included "Latin name of the species" "weaning"; "Latin name of the species" "lifespan"; "Latin name of the species" "sexual maturity"; and "Latin name of the species" "longevity." I focused on the more recent publications and on those containing data on several species.

Phylogenetic Data

A phylogenetic tree was generated for the studied species with information from current online versions of OneZoom Tree of Life Explorer (Rosindell and Harmon 2012), Timetree of Life (Kumar et al. 2017), and 10kTrees (Arnold, Matthews and Nunn 2010). Additional divergence dates were obtained from Benton et al. (2015) and Swanson, Oliveros and Esselstyn (2019).

Data Analysis

Species cannot be treated as independent points because they have a nested phylogenetic history (Harvey and Pagel 1991). Therefore, besides exploring the hypothesized relationship with ordinary least squares regression through the origin (OLS), I worked with the phylogenetic generalized least squares (PGLS) regression through the origin (Martins and Hansen 1997; MacLean and Nunn 2017), using several packages (i.e., ape (Paradis et al. 2004); MASS (Venables and Ripley 2002), mvtnorm (Genz 2018), caper (Orme et al. 2011), and nlme (Pinheiro et al. 2018) for the statistical software R 3.5.3 (2019-03-11) (Team R 2000). Lambda was estimated with maximum likelihood (ML) to calculate branch length transformations and optimize residual error structure (MacLean and Nunn 2017). Plot method (Orme 2013) was used for diagnostic purposes.

RESULTS

I obtained data of sleep quotas and life history traits in 26 mammal species (Table 1).

Variable Construction

Relative REM Sleep

Relative REM sleep was obtained firstly by dividing the mean of daily duration of REM sleep by the mean of daily duration of total sleep and then by using the arcsine transformation.

Relative Age at Weaning

The age at weaning relative to longevity was obtained dividing the mean age at weaning in days by the mean of maximum longevity in days. Then, the arcsine transformation was applied to the result.

Table 1. Raw data used in this study. From left to right: Species name, mean of daily total sleep, mean of daily REM sleep, reference number of sleep quotas, mean of longevity in days, reference number of longevity, mean age of female at sexual maturity in days, reference number of sexual maturity, mean of age at weaning in days, reference number of weaning

Species	Sleep Quotas			Longevity		Female Sexual Maturity		Weaning	
	Daily Hours								
	Total	REM		Days		Days		Days	
	Mean	Mean	Ref	Mean	Ref	Mean	Ref	Mean	Ref
Potorous_tridactylus	10.68	0.62	1	5292.5	2	400	2	154	2
Octodon_degus	9.02	0.82	1	5110	2	182	2	36	2
Meriones_unguiculatus	13.07	1.18	1	2299.5	2	43	2	24	2
Macaca_sylvanus	11.74	1.07	1	10621.5	2	1399	2	213	2
Mus_musculus	13.15	1.26	1	1460	2	42	2	22	2
Neotomodon_alstoni	17.30	1.81	1	960	3	65	4	36	6
Aotus_trivirgatus	16.97	1.82	1	10986.5	2	821	2	77	2
Chinchilla_lanigera	12.53	1.55	1	6278	2	240	2	54	2
Rattus_norvegicus	12.41	1.68	1	1387	2	90	2	25	2
Saimiri_sciureus	9.60	1.40	1	11023	2	1003	2	177	2
Phodopus_sungorus	14.52	2.14	1	1423.5	2	120	2	21	2
Pan_troglodytes	9.67	1.45	1	21681	2	3376	2	1111	2
Ovis_aries	3.84	0.58	1	8322	2	548	2	182	2
Spalax_ehrenbergi	12.43	1.90	1	5475	3	1095	5	70	7
Cavia_porcellus	7.73	1.30	1	4380	2	66	2	18	2
Tupaia_glis	15.84	2.69	1	4526	2	90	2	36	2
Bos_taurus	3.94	0.74	1	7300	2	548	2	205	8
Erinaceus_europaeus	17.64	3.62	1	4270.5	2	253	2	42	2
Chaetophractus_villosus	20.41	4.48	1	9198	2	273	2	56	2
Eutamias_sibiricus	12.67	2.90	1	3650	3	270	3	49	2
Mesocricetus_auratus	15.53	3.65	1	1423.5	2	48	2	20	2
Ictidomys_tridecemlineatus	13.80	3.41	1	2883.5	2	354	2	33	2
Equus_ferus	2.88	0.79	1	20805	2	914	2	274	2
Felis_catus	13.15	3.72	1	10950	2	289	2	28	9
Canis_lupus_familiaris	11.60	3.30	1	8760	2	510	2	63	3
Didelphis_marsupialis	19.39	5.64	1	1533	2	178	2	94	2

References: 1: McNamara et al. 2008; 2: Tacutu et al. 2013; 3: Myers et al. 2017; 4: Ceballos 2014; 5: Novikov and Burda 2013; 6: Luis et al. 2004; 7: Gazit et al. 1996; 8: Martínez González et al. 2008; 9: Cats Protection 2015.

Relative Youth Period

The length of the youth period was obtained from the subtraction of sexual maturity in days minus the weaning age in days. The relative length of the youth period to lifespan was obtained in the same way as the previous variable. The arcsine transformation was also applied to relative length of the youth period.

Phylogenetic Tree

The obtained phylogenetic tree in Newick Standard in computer-readable form is: (((((((Aotus_trivirgatus:2.57,Saimiri_sciureus:0.7081)'8':18.6,(Pan_troglodytes:1.6,Macaca_sylvanus:3.93)'11':24.44)'7':33.9,Tupaia_glis:19.6)'6':70,((Eutamias_sibiricus:12.6,Ictidomys_tridecemlineatus:1.96)'16':20.6,(((Octodon_degus:35.77899950,Chinchilla_lanigera:4.98)'22':31.3,Cavia_porcellus:35.77899950)'20':37.9,((((Rattus_norvegicus:7.92,Mus_musculus:6.29)'29':11.1,Meriones_unguiculatus:9.31)'27':23.8,((Mesocricetus_auratus:10.4,Phodopus_sungorus:0.816)'33':24.2,Neotomodon_alstoni:0.9)'32':28.3)'26':33,Spalax_ehrenbergi:10.4)'25':45)'19':67.3)'15':67.4)'5':75,(Erinaceus_europaeus:4.67,(((Felis_catus:1.93,Canis_lupus:1.12)'44':51.4,Equus_ferus:10)'43':70,(Bos_taurus:2.66,Ovis_aries:0.9963)'40':22)'39':72)'38':75)'4':85,Chaetophractus_villosus:7.8)'3':96,(Didelphis_marsupialis:18.8,Potorous_tridactylus:7.55)'50':87)'2':156.3;

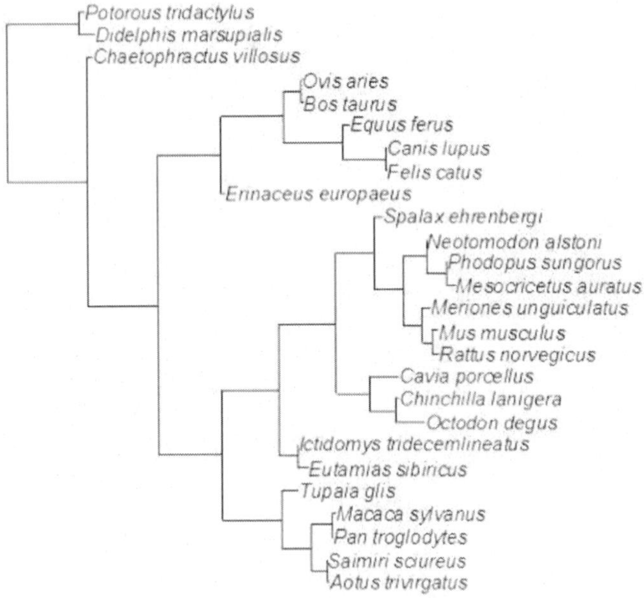

Figure 1. Phylogenetic tree of the 26 mammalian species included in the analyses. Phylogenetic tree constructed with OneZoom Tree of Life Explorer (Rosindel and Harmon 2012), Timetree of Life (Hedges et al. 2015), 10k Trees (Arnold, Matthews and Nunn 2010), Additional divergence dates were obtained from Benton et al. (2015), and from Swanson, Oliveros and Esselstyn (2019). Species names are shown on branch tips.

The phylogenetic relationships among the 26 mammals are modelled graphically in Figure 1.

Hypothesized Model (Relative REM Sleep ~ 0 + Relative Age at Weaning + Relative Youth Period)

The ordinary least square regression through the origin (OLS) for relative REM sleep predicted by relative age at weaning and relative youth period showed that both factors have explanatory power (Overall model: adjusted $R^2 = 0.855$, $F = 77.67$, df = 2 and 24, $p < 0.0001$; relative age at weaning: Estimated = 1.3901, SE = 0.5933, t = 2.343, p = 0.02776; relative youth period: Estimated = 0.9666, SE = 0.3378, t = 2.8661, p = 0.00861) (Figure 2).

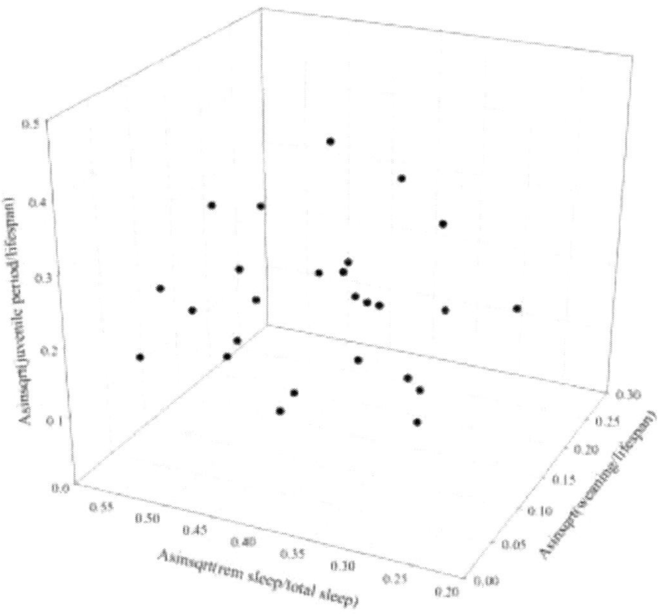

Figure 2. The figure represents the hypothesized correlated evolution between REM sleep relative to total sleep, age at weaning relative to lifespan and duration of juvenile period relative to lifespan. The hypothesized model acquired partial support in the present study (see text for details).

The phylogenetic general least squares regression through the origin (PGLS) for relative REM sleep predicted by relative age at weaning and relative youth period pointed out that the relative age at weaning was the only factor that maintained its explanatory power (Overall model: $\lambda_{ML} = 1.000$, adjusted $R^2 = 0.1762$, $F = 7.522$, df = 1 and 24, p = 0.01132; relative age at weaning: Estimated = 0.963641, SE = 0.305400, t = 3.1553, p = 0.004278; relative youth period: Estimated = -0.068392, SE = 0.198358, t = -0.3448, p = 0.733254).

The explanatory power of relative age at weaning for relative REM sleep is maintained after removing the relative youth period from the model, but the overall model loses statistical significance (Overall model: $\lambda_{ML} = 1$, adjusted R2 = 0.235, F NA=, df = 0 and 25, p = NA; relative weaning: Estimated = 0.93839, SE = 0.29122, t = 3.2223, p = 0.003518).

DISCUSSION

The analysis presented here sustained partially the predicted explanatory power of relative age at weaning and relative youth period for relative REM sleep. In fact, the ordinary least squares seemed to work, but the model losses the explanatory power of relative youth period when the phylogenetic relationships are considered.

The hypothesis about the relationship between the mode of development and the relative amount of REM sleep that I presented in this chapter cannot be rejected because other operationalizations remain possible. Therefore, it could still be useful in explaining some of the failures of two previous hypotheses (i.e., the role of REM in adults). Regarding the hypothesis that REM sleep provides artificial stimulation for brain maturation (Roffwarg, Muzio and Dement 1966), Lesku et al. (2006) noted that it does not explain the relationship between the infant condition in the precocial-altricial spectrum and the amount of REM sleep in adults; and, similarly, Capellini et al. (2008a) indicated that this hypothesis does not explain the lack of association between the duration of REM sleep in adults and the state of brain development in neonates. About the hypothesis by Allison and Cicchetti in 1976 that postulates that the gestation time (which is an approximate measure of the degree of altricial condition at birth) is reflecting another aspect of the sensitivity to of REM sleep to predatory risk, the only objection to it that I found in the specialized literature was made by Allison and Cicchetti themselves:

> Partial correlation analysis shows that the negative correlation between PS [paradoxical sleep] and gestation time is independent of the influence of other variables. Since some prey species (such as guinea pigs), have longer gestation times than would predicted from their body weights, the strong relationship between PS and gestation time may reflect another aspect of the sensitivity of PS to predatory danger. [...] This interpretation is weakened, however, by the low correlations between gestion time and predation index [...] or overall danger, and by the small partial correlations between these variables [...]. (Alison and Cicchetti 1976: 734).

Interesting, the alternative explanation they offer implicates the first hypothesis:

> Alternatively, the correlation indicates that large amounts of PS are associated with short gestation times. If the assumption is made that large amounts of PS in the adult are correlated with

> large amounts in utero [...], the correlation is consistent with the notion that PS hastens development of fetal central nervous system. (Alison and Cicchetti 1976: 734).

If their alternative explanation is accepted, the role of REM in adults remains without explanation.

In contrast, the new hypothesis assumes a constant function of REM sleep during lifespan in maintaining cognitive plasticity, and it does so based on the explicit establishment of a relationship between the degree of adult neotenous-altricial cognitive paedomorphosis and the two sides of the developmental model (i.e., the infant degree of altriciality and the extension of the juvenile period). Therefore, it could provide a good explanation for the variation of relative REM sleep in mammals. Of course, more efficient operationalizations for the variables in the model are still needed.

Gould's (1977) proposal about the evolution of human neotenous paedomorphosis by means of a process of selection targeting the juvenile cognition is consistent with a life history characterized by several features, some of which are included in various functional hypotheses of REM sleep: a prolonged altricial period, an extended juvenile period, and adult play, among many others, but more importantly, an extended window for neural plasticity, and therefore a longer period for learning (Shi and Su 2019), which is supported by sleep stages, including REM (Tamaki et al. 2019).

If the percentage of REM sleep is involved in cognitive plasticity as it has been suggested by many authors (for instance, Tamaki et al. 2019), the notion of neotenous-altricial cognitive paedomorphosis or juvenile-like adult cognition has the potential to sustain the assumption of a constant function of REM sleep during lifespan.

It has been hypothesized that learning is a good strategy to face changes in the environment, which, in turn, are expected to increase with lifespan (Dukas 1998, van Schaik, Isler and Burkart 2012). In contrast to precocial newborns, animals born as developing dependents spend more proportion of their lives time in social environments sheltered by caregivers, which offers opportunities for learning to occur safely (van Schaik, Isler and Burkart 2012). Memory reprocessing during sleep is an important component of how our memories are formed. For instance, participants in a visual texture discrimination task show improvement after post training sleep, and improvement correlated with early-night slow-wave sleep and late-night REM sleep (Stickgold 2005). This is probably due to the repetition of memory networks during sleep, strengthening the corresponding synaptic connections (Ji and Wilson 2006). During sleep, the activity of memory networks is also evident in the content of subjective experience (Wamsley and Stickgold 2019). Therefore, it is probable that the evolution of learning and other cognitive abilities that require memory concurred with the evolution of an altricial developmental mode and extended juvenile periods, as well as with the evolution of REM sleep.

Comparative studies rely on data that generally have been collected by various authors and with different methods, which affect the results (Capellini et al. 2008a). In addition, they rarely consider intraspecific variation, although it is possible that this affects the averages that are used.

Regarding the limitations of the comparative method, Nunn (2011) warned about data quality and completeness. In the case of sleep data used in this chapter, the quality is similar to that of Capellini et al. (2008a) because I followed their criteria for the inclusion in the database. However, they are not free from taxonomic sampling biases arising from the fact that some species are easier to study than others. With respect to the data on the mode of development, most of it comes from the same source (Tacutu et al. 2013). Because this source has its own criteria for inclusion, I consider them to be quite homogeneous. However, these criteria were not targeting important issues for evolutionary comparative analyses. So, the analysis presented here have been done without repairing in sources of bias, as intraspecific variation, methodological heterogeneity, and incompleteness of the taxonomic sampling. It is possible that the analysis has yielded partial results for the reasons stated. Therefore, it would be necessary to apply research strategies that can solve these problems, such as operationalize the hypothesis in other ways, expand the sample size, and restrict the analysis to certain taxonomic groups, among others. For instance, the relative length of the youth period was obtained from the subtraction of sexual maturity minus the weaning age, and then it was scaled to lifespan. Since the neoteny is a condition defined by the elongation of youth in comparison with the ancestral developmental pattern, future studies must include more precise measurements of degree of extended youth period. In addition, it would be necessary to include a cognitive variable.

ACKNOWLEDGMENTS

I am very grateful to Ana Paula Rivera García and Suneeta Singh Carbone, whose comments greatly improved the manuscript.

REFERENCES

Acerbi, A., McNamara, P., and Nunn, C. L. 2008. "To sleep or not to sleep: the ecology of sleep in artificial organisms." *BMC Ecology* 8(1):10.

Allison, T., and Cicchetti, D. V. 1976. "Sleep in mammals: ecological and constitutional correlates." *Science* 194(4266), 732-734.

Allison, Truett, and Van Twyver, Henry. 1970. "The evolution of sleep." *Natural History* 792:57-65.

Arnold C., Matthews L. J., and Nunn C. L. 2010. "The 10kTrees website: a new online resource for primate phylogeny." *Evolutionary Anthropology* 19(3):114-118.

Barton, R. A., and Capellini, I. 2016. "Sleep, evolution and brains." *Brain, Behavior and Evolution* 87(2):65-68.

Benton, M. J., Donoghue, P. C., Asher, R. J., Friedman, M., Near, T. J., and Vinther, J. 2015. "Constraints on the timescale of animal evolutionary history." *Palaeontologia Electronica* 18(1):1-106.

Berger, R. J. 1975. "Bioenergetic functions of sleep and activity rhythms and their possible relevance to aging." In: *Biology of Aging and Development.* Boston: Springer.

Berger, R. J. 1990. "Relations between sleep duration, body weight and metabolic rate in mammals." *Animal Behaviour* 40(5):989-991.

Blaffer Hrdy, S. 1999. "Mother nature: A history of mothers, infants, and natural selection." New York: Pantheon Books.

Bolk, L. (1926). "On the problem of anthropogenesis." *Proceedings of the Section of Sciences, Koninklijke Akademie van Wetenschappen te Amsterdam* 29 (3):465-475).

Bufill, E., Agustí, J., and Blesa, R. 2011. "Human neoteny revisited: the case of synaptic plasticity." *American Journal of Human Biology* 23(6):729-739.

Campbell, S. S., and Tobler, I. 1984. "Animal sleep: a review of sleep duration across phylogeny." *Neuroscience and Biobehavioral Reviews* 8(3):269-300.

Capellini, I., Barton, R. A., McNamara, P., Preston, B. T., and Nunn, C. L. 2008a. "Phylogenetic analysis of the ecology and evolution of mammalian sleep." *Evolution* 62(7):1764-1776.

Capellini, I., Nunn, C. L., McNamara, P., Preston, B. T., and Barton, R. A. 2008b. "Energetic constraints, not predation, influence the evolution of sleep patterning in mammals." *Functional Ecology* 22(5):847-853.

Capellini, I., McNamara, P., Preston, B. T., Nunn C. L., and Barton R. A. 2009. "Does sleep play a role in memory consolidation? A comparative test." *PLoS One* 4(2):e4609. doi:10.1371/journal.pone.00 04609.

Cats Protection. 2015. *Charity of England: Wales and Scotland.* Accessed August 10, 2019: https://www.cats.org.uk/uploads/documents/cat-care-leaflets-.

Ceballos, G. 2014. *Mammals of Mexico.* Baltimore, Maryland: Johns Hopkins University Press.

Chiappa, P., Singh, S., and Pellicer, F. 2018. "The degree of altriciality and performance in a cognitive task show correlated evolution." *PloS ONE* 13(10):e0205128. Accessed October 10, 2019. doi.org/10.1371/ journal.pone.0205128.

Cirelli, C., and Tononi, G. 2008. "Is sleep essential?" *PLoS Biology* 6(8):e216. Accessed October 10, 2019. doi:10.1371/journal.pbio.0060 216.

De Petrillo, F., and Rosati, A. G. 2019. "Ecological rationality: Convergent decision-making in apes and capuchins." *Behavioural Processes* 164: 201-213.

Dukas, R. 1998. "Evolutionary ecology of learning" In: Dukas, R. (Ed.). *Cognitive ecology: the evolutionary ecology of information processing and decision making*. Chicago: University of Chicago Press.

Elgar, M. A., Pagel, M. D., & Harvey, P. H. 1988. "Sleep in mammals." *Animal Behaviour* 36(5):1407-1419.

Elgar, M. A., Pagel, M D., & Harvey, P. H. 1990. "Sources of variation in mammalian sleep." *Animal Behaviour* 40(5):991-995.

Gazit, I., Shanas, U., and Terkel, J. 1996. "First successful breeding of the blind mole rat (*Spalax ehrenbergi*) in captivity." *Israel Journal of Ecology and Evolution* 42(1):3-13.

Gould, S. J. 1977. *Ontogeny and phylogeny*. Cambridge, Massachusetts: Harvard University Press.

Harvey, P. H., and Pagel, M. D. 1991. *The comparative method in evolutionary biology*. Oxford: Oxford University Press.

Ji, D., and Wilson, M. A. 2006. "Coordinated memory replay in the visual cortex and hippocampus during sleep." *Nature Neuroscience* 10:100-107.

Jones, K. E., Bielby, J., Cardillo, M., Fritz, S. A., O'Dell, J., Orme, C. D. L., Safi, K., Sechrest, W., Boakes, E. H. Carbone, C., Connolly, C., Cutts, M. J., Foster, J. K., Grenyer, R., Habib, M., Plaster, C. A., Price, S., Rigby, E. A., Rist, J., Teacher, A., Bininda-Emonds, O. R. P., Gittleman, J. L., Mace, G. M., and Purvis, A. 2009. "PanTHERIA: a species-level database of life history, ecology, and geography of extant and recently extinct mammals." *Ecology* 90(9):2648.

Kumar, S., Stecher, G., Suleski, M., and Hedges, S. B. (2017). TimeTree: a resource for timelines, timetrees, and divergence times. *Molecular Biology and Evolution* 34(7):1812-1819. doi: 10.1093/molbev/msx116. Accessed August 10, 2019. http://www.timetree.org/.

Lesku, J. A., Roth II, T. C., Amlaner, C. J., and Lima, S. L. 2006. "A phylogenetic analysis of sleep architecture in mammals: the integration of anatomy, physiology, and ecology." *The American Naturalist* 168(4):441-453.

Lesku, J. A., Roth, T. C., Rattenborg, N. C., Amlaner, C. J., and Lima, S. L. 2008. "Phylogenetics and the correlates of mammalian sleep: a reappraisal." *Sleep Medicine Reviews* 12(3):229-244.

Luis, J., Cervantes, F. A., Martínez, M., Cardenas, R., Delgado, J., and Carmona, A. 2004. "Male influence on maternal behavior and offspring of captive volcano mice (*Neotomodon alstoni*) from Mexico." *Journal of Mammalogy* 85(2):268-272.

Martínez González, J. C., Azuara Martínez, A., Hernández Meléndez, J., Parra Bracamonte, G. M., and Castillo Rodríguez, S. P. 2008. "Características pre-destete de bovinos simmental (*Bos taurus*) y sus cruces con brahman (*Bos indicus*) en el trópico mexicano." Revista Colombiana de Ciencias Pecuarias 21(3):365-371. ["Pre-weaning characteristics of simmental cattle (Bos taurus) and their crosses with brahman (Bos indicus) in the Mexican tropics." *Colombian Journal of Animal Science* 21 (3): 365-371].

McNamara, P., Capellini, I., Harris, E., Nunn, C. L., Barton, R. A., and Preston, B. 2008. "The phylogeny of sleep database: a new resource for sleep scientists." *The Open Sleep Journal* 1:11-14. doi: 10.2174/ 1874620900801010011 Accessed August 10, 2019. http://www.bu. edu/phylogeny/).

Myers, P., Espinosa, R., Parr, C. S., Jones, T., Hammond, G. S., and Dewey, T. A. 2017 *The Animal Diversity Web* [dataset]. Accessed August 10, 2019. http://animaldiversity.org.

Novikov, E. A., and Burda, G. 2013. "Ecological and evolutionary preconditions of extended longevity in subterranean rodents." *Biology Bulletin Reviews* 3(4): 325-333.

Nunn, C. L., McNamara, P., Capellini, I., Preston, P., and Barton, R. A. 2010. "Primate sleep in phylogenetic perspective." In: McNamara, P., Barton, R. A., and Nunn, C. L. (eds.). *Evolution of sleep: phylogenetic and functional perspectives.* New York: Cambridge University Press.

Nunn, C. L. 2011. *The comparative approach in evolutionary anthropology and biology*. Chicago: University of Chicago Press.

Nunn, C. L., Samson, D. R., and Krystal, A. D. 2016. "Shining evolutionary light on human sleep and sleep disorders." *Evolution, Medicine, and Public Health* (1):227-243.

Pagel, M. D., and Harvey, P. H. 1990. "Diversity in the brain sizes of newborn mammals." *BioScience* 40(2):116-122.

Preston, B. T., Capellini, I., McNamara, P., Barton, R. A., and Nunn, C. L. 2009. "Parasite resistance and the adaptive significance of sleep." *BMC Evolutionary Biology* 9(7):1-8. Accessed August 10, 2019. doi: 10.1186/1471-2148-9-7.

Rattenborg, N. C., de la Iglesia, H. O., Kempenaers, B., Lesku, J. A., Meerlo, P., and Scriba, M. F. 2017. "Sleep research goes wild: new methods and approaches to investigate the ecology, evolution and functions of sleep." *Philosophical Transactions of the Royal Society B: Biological Sciences* 372(1734): 20160251. Accessed August 10, 2019. doi:10.1098/rstb.2016.0251.

Roffwarg, H. P., Muzio, J. N., and Dement, W. C. 1966. "Ontogenetic development of the human sleep-dream cycle." *Science* 152:604-619.

Rosindell, J., and Harmon, L. J. 2012. "OneZoom: a fractal explorer for the tree of life." *PLoS Biology* 10(10):e1001406. Accessed August 10, 2019. doi:10.1371/journal.pbio.1001406.

Samson, D. R., and Nunn, C. L. 2015. "Sleep intensity and the evolution of human cognition." *Evolutionary Anthropology: Issues, News, and Reviews* 24(6):225-237.

Shi, L., and Su, B. 2019. “A transgenic monkey model for the study of human brain evolution.” *Zoological Research* 40(3):236-238.

Siegel, J. M. 1995. “Phylogeny and the function of REM sleep.” *Behavioural Brain Research* 69(1-2):29-34.

Siegel, J. M. 2004. “Sleep phylogeny: clues to the evolution and function of sleep.” In: *Sleep: circuits and functions,* edited by Luppi, P. H., 163-176. Boca Raton: CRC Press LLC.

Starck, J. M., and Ricklefs, R. E. 1998. “Patterns of development: the altricial-precocial spectrum.” In: *Avian growth and development: evolution within the altricial-precocial spectrum*, edited by Starck, J. M., and Ricklefs, R. E., 3-30. Oxford Ornithology Series, 8, Oxford: Oxford University Press.

Stearns, S. C. 1983. “The influence of size and phylogeny on patterns of covariation among life-history traits in the mammals.” *Oikos* 41(2):173-187.

Stickgold, R. 2005. “Sleep-dependent memory consolidation.” *Nature* 437:1272-1278.

Swanson, M. T., Oliveros, C. H., and Esselstyn, J. A. 2019. “A phylogenomic rodent tree reveals the repeated evolution of masseter architectures.” *Proceedings of the Royal Society B* 286(1902), 20190672. https://doi.org/10.1098/rspb.2019.0672.

Tacutu, R., Craig, T., Budovsky, A., Wuttke, D., Lehmann, G., Taranukha, D., Costa, J., Fraifeld, V. E., and de Magalhães, J. P. 2013. “Human Ageing Genomic Resources: integrated databases and tools for the biology and genetics of ageing.” *Nucleic Acids Research* 41(D1): D1027-D1033. Accessed August 10, 2019. https://doi.org/10.1093/nar/ gks1155.

Tamaki, M., Wang, Z., Barnes-Diana, T., Berard, A. V., Walsh, E., Watanabe, T. and Sasaki, Y. (2019). “Opponent neurochemical and functional processing in NREM and REM sleep in visual learning.” *bioRxiv*, 738666. doi: https://doi.org/10.1101/738666.

Tottenham, N. 2019. “Early Adversity and the Neotenous Human Brain.” *Biological Psychiatry*. Available online 29 June 2019 [October 17, 2019] https://doi.org/10.1016/j.biopsych.2019.06.018.

Van Schaik, C. P., Isler, K., and Burkart, J. M. 2012. “Explaining brain size variation: from social to cultural brain.” *Trends in Cognitive Sciences* 16(5):277-284.

Wamsley, E. J., and Stickgold, R. 2019. “Dreaming of a learning task is associated with enhanced memory consolidation: Replication in an overnight sleep study.” *Journal of Sleep Research* 28(1):e12749.

Zepelin, H. and Rechtschaffen, A. 1974. “Mammalian sleep, longevity, and energy metabolism.” *Brain Behavior and Ecology* 10(6):447-470.

EDITOR CONTACT INFORMATION

Ignacio Ramírez-Salado, PhD
Laboratorio de Cronobiología y Sueño,
Dirección de Investigaciones en Neurociencias,
Instituto Nacional de Psiquiatría Ramón de la Fuente Muñiz,
Mexico City, Mexico
Email: salado@imp.edu.mx

Ana Paula Rivera-García, PhD
Laboratorio de Cronobiología y Sueño,
Dirección de Investigaciones en Neurociencias,
Instituto Nacional de Psiquiatría Ramón de la Fuente Muñiz,
Mexico City, Mexico
Email: aprivera@imp.edu.mx

Manuel Alejandro Cruz-Aguilar, PhD
Laboratorio de Cronobiología y Sueño,
Dirección de Investigaciones en Neurociencias,
Instituto Nacional de Psiquiatría Ramón de la Fuente Muñiz,
Mexico City, Mexico
Email: cruzsleep@imp.edu.mx

INDEX

A

adaptive function of sleep, 298
adaptive functions, 298
affect, vii, 17, 19, 21, 26, 38, 45, 49, 54, 64, 73, 83, 92, 100, 102, 103, 107, 108, 109, 113, 129, 141, 145, 146, 147, 164, 203, 204, 208, 209, 211, 218, 220, 223, 225, 226, 227, 308
alpha, 70, 106, 107, 116, 119, 122, 130, 157, 179, 180, 189, 193, 197, 212
altricial, 164, 297, 299, 300, 301, 306, 307, 312
altriciality, 299, 300, 301, 307, 309
amygdala, vii, ix, 5, 13, 69, 70, 71, 72, 74, 75, 76, 77, 78, 79, 80, 81, 82, 83, 84, 85, 86, 87, 88, 89, 90, 91, 92, 93, 95, 96, 97, 147, 166, 167, 169, 172, 173, 174, 178, 179, 183, 186, 188, 189, 190, 191, 197, 199, 200, 211, 217, 220, 221, 224, 225, 241, 242, 244, 249, 285
anticonvulsants, 262
antidepressants, 56, 261, 262
anxiety, 45, 48, 49, 52, 53, 73, 74, 82, 89, 103, 116, 120, 164, 178, 211, 217, 243, 250, 259, 261, 267, 268, 269, 271, 274, 281
apnea syndrome, 47, 62, 102, 125, 132, 287
arousal, 6, 7, 11, 12, 13, 43, 91, 131, 188, 191, 207, 209, 210, 211, 224, 247, 249, 258, 259, 264, 265, 267, 270, 271
ascending reticular activating system, 4
attention deficit hyperactivity disorder, 265, 275, 281
autism, viii, x, 164, 169, 187, 188, 190, 192, 193, 195, 198, 199, 200, 245, 250, 251, 252, 253, 259, 260, 262, 263, 264, 265, 266, 267, 268, 269, 270, 271, 272, 273, 274, 275, 276, 281, 284, 285, 289, 290, 291, 293, 294, 295, 296
autism spectrum disorder(s), 169, 187, 192, 199, 260, 262, 263, 264, 265, 266, 267, 268, 269, 270, 271, 273, 274, 276, 285, 289, 290, 291, 294, 295, 296

B

basal forebrain, 3, 5, 6, 7, 8, 11, 110, 132, 184, 248, 249
basic emotions, 208, 209, 222, 223
behavioral and psychological symptoms of dementia, 102
beta, 48, 58, 71, 73, 82, 104, 116, 212
brain injury, 96, 185, 279, 281, 292
brain plasticity, ix, 80, 263, 275, 276, 285, 290
brain stem, 3, 4, 6, 7, 8, 9, 10, 11, 12, 247
brain stimulation, 99, 113, 118, 119, 120, 121, 123, 275, 276, 289, 290, 293, 294, 295, 296
brain tissue, 146, 279
brain wave activity, 247
bruxism, 287

C

cataplexy, 7
cerebral cortex, 5, 8, 71, 91, 93, 107, 112, 113, 249, 250, 276, 277, 278, 279, 291
c-Fos, 75, 76, 89, 197

cholinergic (acetylcholine), 6, 8, 9, 10, 11, 14, 15, 24, 95, 110, 112, 114, 116, 132, 133, 134, 153, 159, 166, 183, 185, 191, 192, 195, 248, 262
chronic pain, 281
cingulate cortex, 70, 91, 169, 170, 286
circadian rhythms, x, 25, 38, 102, 130, 131, 135, 143, 145, 152, 249
cognition, ix, x, 77, 91, 93, 100, 103, 104, 106, 107, 109, 114, 119, 121, 123, 163, 164, 169, 172, 174, 176, 177, 178, 188, 189, 190, 193, 195, 197, 198, 199, 200, 201, 205, 222, 224, 225, 227, 243, 244, 249, 259, 276, 286, 287, 290, 292, 297, 307, 311
cognitive empathy, 174, 175
cognitive paedomorphosis, 297, 299, 300, 307
coil, 115, 159, 277, 278, 283, 285
comparative method, 297, 308, 310
comparative studies, 298, 299, 301
complex dreams, 212, 215, 217
consciousness, 62, 93, 129, 163, 165, 166, 180, 185, 190, 194, 195, 198, 199, 200, 201, 225, 227, 230, 231, 241, 243, 244
cortex, 5, 8, 13, 19, 21, 70, 71, 73, 75, 77, 78, 79, 80, 82, 83, 84, 85, 86, 87, 90, 91, 93, 94, 95, 104, 105, 107, 108, 111, 112, 113, 114, 115, 116, 120, 123, 129, 132, 163, 166, 168, 169, 170, 171, 177, 178, 182, 183, 186, 188, 189, 191, 193, 196, 211, 221, 232, 240, 243, 249, 250, 259, 276, 277, 278, 279, 282, 283, 285, 289, 290, 291, 294, 295, 296, 310
cortical, vii, 2, 3, 4, 5, 6, 28, 42, 48, 70, 71, 75, 77, 78, 81, 82, 85, 86, 88, 91, 93, 96, 99, 100, 101, 102, 103, 104, 105, 108, 109, 110, 111, 112, 113, 114, 115, 116, 120, 122, 123, 131, 166, 178, 181, 184, 189, 218, 248, 252, 265, 277, 278, 279, 280, 282, 283, 286, 288, 289, 290, 291, 292, 293, 298, 300
cortical activity, vii, 4, 99, 100, 101, 108, 111, 115
culture, 203, 206, 207, 221, 226

D

default mode network, ix, 163, 169, 189, 197
delta, 19, 25, 29, 30, 58, 70, 79, 81, 85, 93, 95, 102, 104, 106, 107, 108, 116, 118, 130, 178, 179, 251, 256
dementias, 100, 118
depression, 22, 27, 29, 30, 33, 34, 38, 46, 48, 49, 51, 52, 53, 56, 57, 59, 61, 66, 67, 85, 87, 89, 113, 114, 120, 123, 129, 139, 190, 215, 221, 222, 223, 227, 246, 267, 281, 283, 284
developmental mode, viii, 297, 298, 299, 300, 301, 307
diabetes, 108, 117, 129, 151, 246
dream character, 163, 168, 170, 171, 172, 173, 175, 176, 177, 181, 182, 188, 240
dream content(s), ix, 179, 180, 188, 192, 196, 204, 213, 216, 219, 221, 226, 230, 243
dream enactment behaviors, 181
dream mentation, 216
dream recall, 179, 180, 184, 188, 189, 193, 198, 200, 243
dreaming, ix, 10, 42, 62, 64, 91, 92, 93, 163, 165, 167, 168, 169, 170, 171, 173, 177, 178, 179, 180, 181, 182, 183, 184, 187, 188, 190, 191, 193, 194, 195, 197, 198, 199, 200, 201, 214, 216, 217, 218, 219, 223, 224, 226, 227, 230, 232, 239, 240, 242, 243, 244, 247, 312
dreams, vii, ix, 2, 29, 77, 103, 130, 138, 163, 164, 165, 166, 167, 168, 169, 170, 171, 173, 175, 176, 177, 179, 180, 181, 182, 183, 184, 185, 186, 187, 188, 189, 193, 194, 195, 196, 197, 198, 200, 201, 203, 204, 212, 213, 214, 215, 216, 217, 218, 219, 220, 221,222, 223, 224, 225, 226, 227, 229, 230, 231, 232, 233, 235, 236, 237, 239, 240, 241, 242, 243, 244, 247, 249

E

EEG power, 28, 73
ejaculation, 232
electroencephalogram (EEG), ix, 2, 3, 4, 9, 10, 12, 13, 20, 27, 28, 29, 31, 41, 44, 47, 48, 49, 56, 57, 58, 60, 71, 73, 77, 79, 81, 87, 92, 96, 103, 104, 106, 107, 114, 115, 116, 117, 119, 120, 122, 129, 130, 132, 147, 165, 166, 172, 178, 179, 180, 182, 186, 187, 188, 190, 192, 212, 215, 247, 252, 254, 266, 267, 270, 276, 293, 298, 301
emotion, ix, 69, 70, 77, 89, 91, 92, 93, 163, 167, 174, 177, 189, 190, 195, 196, 198, 203, 204, 205, 206, 207, 208, 209, 210, 211, 212, 216, 217, 218, 219, 220, 221, 222, 223, 224, 225, 226, 227, 242, 244
emotional categories, 209
emotional expressions, 74, 216, 221
emotional memory(ies), 61, 70, 71, 73, 81, 83, 84, 86, 87, 89, 90, 93, 179, 194

emotional processing, 69, 71, 72, 73, 74, 76, 77, 79, 95, 217, 218, 222
emotional reactivity, 94, 210, 217
empathy, 174, 175, 177, 189, 194, 201, 212
encephalitis lethargica, 2
endocrine disruptors, 125
entorhinal cortex, 70, 79, 80
environment, 2, 41, 42, 49, 82, 105, 126, 127, 132, 133, 136, 137, 144, 150, 151, 153, 156, 158, 203, 204, 206, 208, 215, 219, 307
environmental pollution, vii, 132
epilepsy, 81, 164, 186, 223, 246, 250, 251, 252, 266, 269, 270, 273, 281, 284
erotic content, 239
erotic dreams, 230, 236, 237, 239, 240, 241, 242
evoked potentials, 79, 104, 111, 282
evolution, viii, ix, 101, 208, 215, 224, 231, 297, 298, 299, 300, 305, 307, 309, 310, 311, 312

F

facial expressions, 74, 174, 214, 225, 226, 240, 241
facial feedback hypothesis, 217
fast oscillations, 42, 70
feelings, 52, 55, 164, 168, 170, 171, 175, 176, 185, 195, 208, 209, 219, 221, 225, 226, 243
female sex hormones, ix, 33, 34, 36, 42, 43, 44
fragmented sleep, 102, 131, 245, 251, 263, 287, 288
Freudian model of dreams, 231

G

gamma, 9, 43, 71, 79, 82, 87, 93, 113, 138, 142, 178, 180, 201
gamma-aminobutyric acid (GABA), 9, 10, 24, 25, 43, 113, 131, 138, 143, 286, 291, 292, 295
gender, viii, 26, 61, 62, 229, 230, 233, 235
gestation, 255, 299, 306
glutamate (Glu), 9, 10, 112, 151, 185, 196, 243, 291

H

heterosexual, 229, 233, 236, 237, 239
hippocampus, 5, 22, 58, 70, 71, 75, 77, 78, 79, 80, 82, 83, 84, 85, 86, 87, 88, 89, 90, 96, 105, 110, 115, 147, 163, 166, 167, 178, 179, 186, 187, 189, 191, 200, 202, 217, 310
homosexual, 229, 233, 236, 237
hormones, ix, 17, 18, 23, 25, 27, 29, 31, 33, 34, 35, 37, 39, 42, 43, 44, 47, 48, 55, 59, 61, 65, 67, 130, 142, 143, 164, 189, 232, 242, 249, 261, 288
hypnagogic hallucinations, 213
hypothalamus, 3, 5, 6, 8, 11, 13, 19, 20, 21, 29, 30, 35, 40, 49, 71, 77, 78, 130, 146, 147, 149, 167, 189, 192, 211, 247, 248

I

Immediate Early Genes (IEGs), 71, 75
insomnia, 3, 5, 11, 17, 23, 26, 32, 34, 38, 39, 45, 47, 48, 49, 53, 56, 57, 63, 64, 66, 67, 72, 102, 116, 120, 140, 141, 253, 254, 255, 256, 257, 258, 260, 261, 262, 264, 266, 267, 270, 271, 272, 287

L

latent content, 230
lifespan, 43, 44, 118, 297, 300, 301, 304, 305, 307, 308
limbic system, ix, 70, 166, 167, 184, 186, 221, 232
locus coeruleus, 5, 10, 11, 12, 185, 189, 248, 259, 264

M

magnetic field, 115, 275, 276, 277, 278, 292
magnetoencephalography (MEG), 79, 90, 96, 178
major depression disorder (MDD), 212, 217
mammals, viii, 1, 19, 25, 126, 131, 164, 171, 172, 173, 205, 297, 298, 305, 307, 308, 309, 310, 311, 312
manifest content, 214, 230
medial septum, 70
melatonin, ix, 25, 26, 27, 28, 29, 30, 31, 32, 116, 131, 141, 143, 148, 150, 157, 232, 249, 254, 257, 261, 264, 265, 266, 267, 268, 269, 270, 271, 272, 273, 287
memory, v, ix, 1, 24, 28, 58, 61, 66, 69, 70, 71, 73, 75, 76, 78, 80, 81, 82, 83, 84, 85, 86, 87, 88, 89, 90, 91, 92, 93, 94, 95, 96, 97, 103, 105, 106, 107, 108, 110, 111, 115, 116, 118, 121, 123, 143, 144, 164, 166, 170, 177, 178, 179, 180, 185, 186, 188, 189,190, 191, 192, 193, 194, 199, 201, 202, 205, 212, 214, 217, 218, 220, 222, 223, 231, 235, 239, 246, 249, 259, 263, 269, 276, 288, 307, 309, 310, 312

memory consolidation, 69, 71, 75, 76, 83, 84, 88, 90, 94, 96, 103, 107, 166, 179, 180, 246, 288, 309, 312
menopausal transition, 33, 34, 35, 37, 38, 39, 40, 43, 45, 46, 47, 48, 49, 51, 52, 53, 54, 56, 58, 59, 60, 61, 62, 63, 64, 66, 67
mental health, 58, 164, 222, 223, 245
mirror neuron systems, 172
mode of development, x, 297, 299, 306, 308
moods, 52, 59, 209
motor recovery, 282, 293, 296
muscle activity, 129, 216, 220, 224, 226
muscle twitching, 255
muscular atony (atonia), 2, 7, 9, 42, 182, 186, 189, 212, 216, 247, 255

N

narcolepsy, 7, 11, 13, 17
narcolepsy-cataplexy, 7
negative emotions, 211, 230, 241
neonatal brain, 299
neotenous, 297, 299, 300, 301, 307, 312
neoteny, 300, 308, 309
neural control, 1, 14
neural networks, 10, 280, 294
neurodegenerative diseases (ND), ix, 99, 100, 101, 102, 103, 104, 105, 106, 107, 108, 110, 112, 113, 114, 116, 122
neuroendocrinology, vii, 17, 18, 40, 59, 64, 65, 155
neurological diseases, 281
neuropathic pain, 292
neurophysiology, 56, 91, 92, 121, 122, 158, 159, 188, 192, 199, 243, 270, 280, 289, 291, 293, 294, 295
neuroplasticity, 104, 107, 108, 109, 111, 112, 113, 114, 115, 118, 123, 279, 280, 281, 282, 284, 289
neuropsychiatric disorders, 69, 275, 276, 280, 284
nightmares, 63, 179, 189, 191, 214, 249
nitric oxide (NO), 9, 10, 131, 146, 149, 150, 151, 158, 159
noradrenaline, 5, 12, 116, 193, 196, 243, 259, 262
NREM sleep, 42, 44, 107, 108, 115, 130, 172, 174, 175, 178, 179, 180, 183, 204, 212, 216, 218, 220, 225, 245, 246, 287, 293

O

obesity, 23, 108, 246
oneiric experiences, viii, 229
oneiric sexual experiences, 230
orexins, 6, 7, 11, 14
oscillations, vii, 2, 42, 56, 69, 70, 71, 72, 73, 78, 79, 80, 81, 82, 83, 84, 85, 86, 87, 90, 92, 95, 104, 106, 114, 115, 121, 165, 178, 179, 180, 183, 190, 197, 245, 249
oxytocin, 155, 159, 232

P

paradoxical sleep, 2, 14, 28, 93, 94, 95, 130, 190, 192, 193, 197, 243, 306
pedunculopontine tegmental nucleus (PPT), 8
perifornical area (PFA), 6
periodic limb movement, 34, 46, 47, 53, 55, 62, 102, 255, 263, 264, 274, 287
pesticides, 125, 127, 128, 133, 136, 137, 142, 143, 150, 151, 152, 153, 155, 156
pharmacotherapy, 246, 266, 268, 271, 281
phase difference, 85
phase synchronization, 85
phasic activity, 8, 191
phylogenetic general least squares (PGLS), 297, 302, 305
plasticity, vii, ix, 69, 70, 71, 72, 73, 75, 76, 79, 80, 81, 82, 83, 84, 85, 86, 87, 88, 89, 92, 94, 96, 99, 109, 110, 111, 112, 113, 114, 116, 122, 123, 263, 275, 276, 285, 290, 291, 300, 307, 309
polybrominated diphenyl ethers (PBDEs), 125, 127, 146, 147
polychlorinated biphenyls (PCBs), 125, 127, 128, 140, 141, 145, 147, 150, 154, 157, 159, 160, 161
polysomnographic, 42, 47, 108, 179, 216, 251, 254, 255, 287
polysomnography (PSG), 41, 42, 44, 45, 47, 48, 49, 50, 52, 53, 54, 56, 66, 67, 79, 103, 251, 254, 255, 264, 274, 293
ponto-geniculo-occipital waves (PGOs), 7, 13, 79, 166, 191, 192
positive emotions, 211, 230, 241
positron emission tomography (PET), 76, 91, 93, 191, 197, 285
post-traumatic stress disorder (PTSD), 73, 81, 88, 91, 96, 97, 179, 217, 223

precocial species, 164, 299, 301, 306, 307, 312
prefrontal cortex, 13, 73, 75, 77, 78, 82, 83, 85, 86, 104, 114, 163, 168, 169, 171, 178, 188, 211, 221, 240, 243, 259, 279, 283, 295
primary motor cortex, 111, 276, 289, 296
prolactin, 18, 30, 31, 232, 270
psychiatric disorders, 111, 168, 259, 262, 295
P-waves, 79, 83

R

rapid eye movement sleep (REM), 2, 10, 12, 29, 41, 44, 56, 64, 71, 88, 102, 107, 115, 125, 129, 130, 164, 166, 186, 187, 212, 215, 225, 229, 230, 231, 243, 246, 247, 255, 262, 265, 270, 288, 298
rapid eye movements (REMs), 2, 8, 42, 215, 217
REM behavior disorder, 182
REM-off neurons, 9
REM-on neurons, 9
repetitive transcranial magnetic stimulation (rTMS), ix, 99, 101, 111, 112, 113, 114, 115, 116, 117, 118, 120, 121, 123, 279, 280, 281, 283, 284, 289, 290, 291, 292, 293, 294
resilience, 82, 88, 92, 96
respiratory diseases, 125, 132
reticular formation, 3, 4, 9, 10, 12, 13, 14

S

seizures, 133, 138, 143, 154, 251, 252, 267, 273, 275, 284, 294
serotonin, 5, 13, 14, 20, 25, 43, 114, 116, 143, 256, 262
sexual act, 29, 229, 237
sexual characters, 230
sexual orientation, viii, x, 229, 230, 233, 235, 237
sexual partner, 230, 233, 235, 237, 242
sharp wave-ripples, 71, 84
simulation theory, 163, 180, 181, 189, 201
sleep architecture, 24, 42, 44, 45, 47, 49, 51, 52, 96, 157, 178, 310
sleep deprivation, 19, 30, 31, 48, 69, 71, 72, 73, 74, 75, 76, 87, 88, 90, 92, 93, 94, 95, 116, 131, 164, 169, 186, 187, 195, 196, 197, 202
sleep disorders, ix, 10, 17, 22, 26, 33, 34, 39, 46, 47, 49, 52, 53, 56, 57, 65, 99, 100, 101, 102, 108, 110, 114, 117, 118, 125, 126, 132, 133, 134, 138, 141, 191, 245, 253, 254, 256, 259, 261, 263, 266, 268, 270, 274, 287, 299, 311
sleep disturbances, viii, 26, 33, 34, 39, 46, 47, 49, 52, 55, 56, 62, 72, 73, 89, 109, 140, 141, 245, 249, 251, 254, 258, 261, 262, 267, 271, 272, 287
sleep efficiency, 26, 42, 45, 50, 52, 53, 55, 102, 108, 132, 254, 255, 263
sleep mentation, 165, 168, 172, 174, 175, 178, 179, 187, 193
sleep quotas, 298, 299, 301, 302, 303
sleep spindles, 42, 44, 71, 80, 86, 91, 96, 107, 115, 178, 186, 187, 246
sleep-wake cycle (sleep-waking cycle), vii, 1, 4, 5, 10, 12, 13, 17, 18, 22, 23, 25, 26, 27, 33, 34, 59, 91, 102, 108, 116, 131, 153, 164, 191, 195, 243, 247, 264
slow wave sleep (SWS), 4, 5, 6, 7, 9, 10, 17, 18, 19, 20, 23, 24, 42, 45, 48, 71, 72, 79, 81, 86, 102, 107, 115, 116, 130, 164, 165, 166, 173, 175, 186, 187, 255, 307
snoring, 54, 125, 132, 147
social cognition, x, 163, 164, 169, 172, 174, 176, 177, 178, 188, 189, 190, 193, 195, 199, 259, 276, 286, 287, 290
spike, 81, 85
spinal cord, 4, 6, 9, 19, 276, 295
spindle EEG, 255
stress, 5, 6, 13, 21, 22, 26, 29, 30, 31, 44, 52, 53, 54, 59, 61, 66, 73, 74, 81, 82, 90, 92, 95, 96, 97, 129, 139, 147, 156, 164, 179, 189, 196, 217, 254, 267, 270, 271, 281
stroke, 264, 281, 282, 283, 284, 289, 290, 291, 292, 293, 294, 295, 296
synaptic plasticity, 70, 71, 72, 73, 76, 79, 82, 83, 85, 86, 87, 89, 92, 94, 112, 113, 116, 122, 123, 300, 309

T

temperaments, 209
thalamus, 3, 5, 8, 10, 70, 71, 78, 91, 95, 96, 105, 107, 116, 178, 199, 211, 248, 249
theory of mind, 164, 167, 168, 169, 171, 189, 240
theta, 59, 61, 66, 70, 73, 79, 81, 82, 83, 85, 86, 88, 90, 91, 93, 94, 104, 106, 108, 112, 116, 118, 119, 130, 178, 179, 180, 189, 191, 194, 197, 201, 212, 217, 224, 282, 289, 294
theta rhythm, 66, 70, 85, 90, 178, 189, 201, 212

threat simulation theory, 189, 201
tonic activity, 4
total sleep time (TST), 42, 44, 45, 48, 51, 52, 53, 82, 108, 187, 255, 298
transcranial direct current stimulation (tDCS), ix, 99, 101, 111, 113, 114, 115, 116, 123
transcranial magnetic stimulation (TMS), viii, 111, 112, 113, 114, 115, 116, 275, 276, 277, 278, 279, 280, 281, 283, 284, 285, 286, 287, 288
tuberomammillary nucleus (TMN), 7, 8

V

valence, 73, 74, 76, 80, 207, 209, 210, 211
ventromedial prefrontal cortex, 75, 169

W

wakefulness, 1, 2, 3, 4, 5, 6, 7, 8, 9, 10, 11, 13, 14, 15, 24, 27, 28, 41, 48, 56, 70, 71, 72, 77, 79, 80, 82, 85, 92, 96, 103, 105, 108, 110, 120, 122, 130, 131, 149, 153, 163, 164, 166, 173, 174, 177, 178, 179, 180, 181, 182, 183, 184, 189, 192, 195, 203, 204, 212, 214, 215, 217, 218, 219, 231, 232, 246, 247, 249, 256, 264, 287
wave-ripples, 71, 84, 86
weaning, 297, 301, 303, 304, 305, 306, 308, 311
working memory, 188, 259